Living with the Genome

Ethical and Social Aspects of Human Genetics

Edited by

Angus Clarke

and

Flo Ticehurst

palgrave macmillan

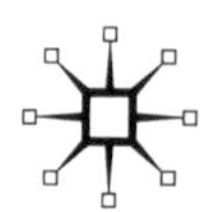

First published 2006 by
PALGRAVE MACMILLAN
Houndmills, Basingstoke, Hampshire RG21 6XS and
175 Fifth Avenue, New York, N.Y. 10010
Companies and representatives throughout the world

PALGRAVE MACMILLAN is the global academic imprint of the Palgrave
Macmillan division of St. Martin's Press, LLC and of Palgrave Macmillan Ltd.
Macmillan® is a registered trademark in the United States, United Kingdom
and other countries. Palgrave is a registered trademark in the European
Union and other countries.

ISBN-13: 978–1–4039–3620–2 hardback
ISBN-10: 1–4039–3620–X hardback
ISBN-13: 978–1–4039–3621–9 paperback
ISBN-10: 1–4039–3621–8 paperback

This book is printed on paper suitable for recycling and made from fully
managed and sustained forest sources.

A catalogue record for this book is available from the British Library.

Library of Congress Cataloging-in-Publication Data
Living with the genome : ethical and social aspects of human
genetics / edited by Angus Clarke and Flo Ticehurst.
p. cm.
Includes bibliographical references (p.).
ISBN 1–4039–3620–X (cloth) — ISBN 1–4039–3621–8 (pbk.)
1. Medical genetics—Moral and ethical aspects. 2. Medical
genetics—Social aspects. 3. Human Genome Project—Moral and
ethical aspects. 4. Human Genome Project—Social aspects.
I. Clarke, Angus, 1954– II. Ticehurst, Flo.
RB155.L58 2006
616′.042—dc22 2006045716

10 9 8 7 6 5 4 3 2 1
15 14 13 12 11 10 09 08 07 06

Printed and bound in Great Britain by
Antony Rowe Ltd, Chippenham and Eastbourne

Contents

Acknowledgements ix

List of Contributors x

Glossary xiii

Introduction 1
Angus Clarke

Part 1 The Human Genome Project: Genetic Research and Commercialization

Introduction 21
Angus Clarke and Flo Ticehurst

1 Celera Genomics: The Race for the Human Genome Sequence 27
Michael A. Fortun

2 Human Genome Project as a Social Enterprise 33
Philip L. Bereano

3 Commercialization of Human Genetic Research 39
Timothy Caulfield

4 Patenting of Genes: A Personal View 46
Jeremy Rifkin

5 Human Genome Diversity Project (HGDP): Impact on Indigenous Communities 49
Jonathan Marks

6 deCODE and Iceland: A Critique 56
Einar Árnason and Frank Wells

7 Informed Consent in Human Genetic Research 64
Philip R. Reilly

8 Gene Therapy: Expectations and Results 70
Paul A. Martin

Part 2 Genetic Disease: Implications for Individuals, Families and Populations

Introduction 77
Angus Clarke and Flo Ticehurst

9 Genetic Counseling: Psychological Issues 81
Barbara B. Biesecker

10 Genetic Testing of Children 90
Annie Procter

11 Genetic Risk 96
Lindsay Prior

12 Genetic Susceptibility 102
Carol Emslie and Kate Hunt

13 Thrifty Gene Hypothesis: Challenges 108
Robyn McDermott

14 Carrier Screening for Inherited Hemoglobin Disorders in Cyprus and the United Kingdom 114
Bernadette Modell

15 Carrier Screening of Adolescents in Montreal 122
Charles R. Scriver and John J. Mitchell

Part 3 Disability, Genetics and Eugenics

Introduction 131
Angus Clarke and Flo Ticehurst

16 Nazi Movement and Eugenics 135
Paul Weindling

17 Eugenics: Contemporary Echoes 140
Troy Duster

18 China: The Maternal and Infant Health Care Law 147
Sun-Wei Guo

19 Disability, Human Rights and Contemporary Genetics 157
Tom Shakespeare

20 Mentally Handicapped in Britain: Sexuality and Procreation 161
Jane McCarthy, Valerie Sinason and Sheila Hollins

Part 4 Genetics and Society: Information, Interpretation and Representation

Introduction 167
Angus Clarke and Flo Ticehurst

21 Gene as a Cultural Icon 171
Dorothy Nelkin

22 Heredity: Lay Understanding 177
Martin P. M. Richards

23 Inheritance and Society 183
Paul Atkinson, Aditya Bharadwaj and Katie Featherstone

24 Privacy and Genetic Information 192
Veronica English and Ann Sommerville

25 Insurance and Genetic Information 198
Tony McGleenan

26 'Race', IQ and Genes 203
Jon Beckwith and Joseph S. Alper

27 Criminal Responsibility and Genetics 210
Jennifer Bostock and Gwen Adshead

Part 5 Genetic Explanations: Understanding Origins and Outcomes

Introduction 217
Angus Clarke and Flo Ticehurst

28 Genetics as Explanation: Limits to the Human Genome Project 221
Irun R. Cohen and Henri Atlan

29 Darwin and the Idea of Natural Selection 232
Gabriel Dover

30 Sociobiology, Evolutionary Psychology, and Genetics 238
Patrick Bateson

31 Creationism 246
Michael Ruse

32 Genetics, Reductionism and Autopoiesis 251
Steven P. R. Rose

33 Personal Identity: Genetics and Determinism 255
Ruth Chadwick

34 Racism, Ethnicity, Biology and Society 259
Hannah Bradby

Part 6 Reproduction, Cloning and the Future

Introduction 267
Angus Clarke and Flo Ticehurst

35 *In Vitro* Fertilization: Regulation 271
Derek Morgan and Mary Ford

36 Reproductive Choice 279
John Harris

37 Feminist Perspectives on Human Genetics and Reproductive Technologies 286
Donna L. Dickenson

38 Sex Selection 291
Dena S. Davis

39 Dolly and Polly 296
Alan Petersen

40 Cloning of Animals in Genetic Research: Ethical and Religious Issues 302
Andrew Linzey and Ara Barsam

41 Distributive Justice and Genetics 308
Colin Farrelly

42 Reprogenetics: Visions of the Future 312
Regine Kollek

Index 318

Acknowledgements

As well as the authors, we must also thank the Nature Publishing Group, Macmillan Publishers for their vision in publishing the original Encyclopedia, and John Wiley and Sons Ltd, who now hold copyright in the Encyclopedia and have given us permission to publish this volume of material. Angus Clarke would also like to thank David Cooper for the original invitation to structure and develop the ethical and social areas of the Encyclopedia of the Human Genome.

List of Contributors

Gwen Adshead, Traumatic Stress Clinic, London, UK

Joseph S. Alper, retired; formerly University of Massachusetts, Boston, USA

Einar Árnason, University of Iceland, Reykjavik, Iceland

Paul Atkinson, Cardiff University, UK

Henri Atlan, Hadassah University Hospital, Jerusalem, Israel, and Ecole des Hautes Etudes en Sciences Sociales, Paris, France

Ara Barsam, US Agency for International Development, Armenia

Patrick Bateson, University of Cambridge, UK

Jon Beckwith, Harvard Medical School, Boston, USA

Philip L. Bereano, Department of Technical Communication, University of Washington, USA

Aditya Bharadwaj, University of Edinburgh, UK

Barbara B. Biesecker, National Human Genome Research Institute, Maryland, USA

Jennifer Bostock, J Consultants and South London & Maudsley NHS Trust, London, UK

Hannah Bradby, University of Warwick, Coventry, UK

Timothy Caulfield, University of Alberta, Canada

Ruth Chadwick, Cardiff University, UK

Irun R. Cohen, The Weizmann Institute of Science, Israel

Dena S. Davis, Cleveland State University, Ohio, USA

Donna L. Dickenson, Birkbeck, University of London, UK

Gabriel Dover, University of Leicester, UK

Troy Duster, University of California, Berkeley, USA

Carol Emslie, MRC Social and Public Health Sciences Unit, University of Glasgow, UK

Veronica English, British Medical Association, London, UK

Colin Farrelly, University of Waterloo, Canada

Katie Featherstone, Cardiff University, UK

Mary Ford, University of Nottingham, UK

Michael A. Fortun, Department of Science and Technology Studies, Rensselaer Polytechnic Institute, USA

Sun-Wei Guo, Medical College of Wisconsin, USA

John Harris, University of Manchester, UK

Sheila Hollins, St George's Hospital Medical School, London, UK

Kate Hunt, MRC Social & Public Health Sciences Unit, University of Glasgow, UK

Regine Kollek, University of Hamburg, Germany

Andrew Linzey, University of Oxford, UK

Jonathan Marks, University of North Carolina at Charlotte, USA

Paul A. Martin, University of Nottingham, UK

Jane McCarthy, St George's Hospital Medical School, London, UK

Robyn McDermott, University of South Australia

Tony McGleenan, School of Law, University of Ulster, UK

John J. Mitchell, McGill University, Canada

Bernadette Modell, University College London, UK

Derek Morgan, Faculty of Law, University of Technology, Australia

Dorothy Nelkin (deceased), formerly University Professor at New York University, USA

Alan Petersen, University of Plymouth, UK

Lindsay Prior, Queens University, Belfast, UK

Annie Procter, Institute of Medical Genetics, Cardiff, UK

Philip R. Reilly, Interleukin Genetics Inc., USA

Martin P. M. Richards, Professor Emeritus, University of Cambridge, UK

Jeremy Rifkin, The Foundation on Economic Trends, Washington, USA

Steven P. R. Rose, The Open University, Milton Keynes, UK

Michael Ruse, Florida State University, Tallahassee, USA

Charles R. Scriver, McGill University, Canada

Tom Shakespeare, University of Newcastle Upon Tyne, UK

Valerie Sinason, St George's Hospital Medical School, London, UK

Ann Sommerville, British Medical Association, London, UK

Paul Weindling, Oxford Brookes University, Oxford, UK

Frank Wells, Faculty of Pharmaceutical Medicine, Royal College of Physicians, London, UK

Glossary

abiotic environment The physical surroundings of an organism, excluding other organisms from consideration.

accommodation In theology, John Calvin's belief that scripture is written (by God) in a simple or metaphorical form to make it understandable to simple and uneducated people.

affine A relative by marriage.

allopathic medicine A system of medical practice which aims to combat disease by the use of remedies which produce effects different from those produced by the special disease treated. This term was invented by Hahnemann to designate ordinary medical practice, as opposed to homeopathy in which the remedies produce similar effects to the disease itself.

anthropometric The technique that deals with the measurement of the size, weight and proportions of the human or other primate body.

apoptosis Programmed cell death, the body's normal method of disposing of damaged, unwanted or unneeded cells in the course of disease or development.

appellate Law (of a court) dealing with appeals.

Aryan In Nazi (National Socialist) terminology, Christian Caucasians, especially those of Nordic ancestry. The term can also be used more broadly to describe the peoples speaking Indo-European languages, including those derived from Sanskrit, Greek, Latin and the Norse languages.

Bayesian density discriminant functions Related to Bayesian theory, a theorem in conditional probability named for Thomas Bayes (1702–61). In epidemiology, it is used to obtain the probability of disease in a group of people with some characteristic on the basis of the overall rate of that disease and of the likelihoods of that characteristic in healthy and diseased individuals. The most familiar application is in clinical decision analysis where it is used for estimating the probability of a particular diagnosis given the appearance of some symptoms or test result.

bio-colonialism A political relationship in which economically developed nations or corporations acquire the environmental knowledge or biological resources of indigenous peoples cheaply and transform them for profit, without distributing the gains equitably among those peoples.

bioinformatics The science of collecting and analysing complex biological data such as genetic codes, using advanced computer techniques.

bio-piracy A term, used by critics of biotechnology, to describe a form of neo-colonialization wherein the developed world is seen to exploit the underdeveloped world for its genetic heritage and its traditional knowledge for commercial gain, primarily through the enforcement of intellectual property rights, including patents.

bioprospecting The collection and testing of biological samples and the collection of indigenous knowledge to help in the discovery and exploitation of genetic or biochemical resources, which is seen as having primarily economic purposes (e.g., the development of new drugs, crops, industrial products).

biotic environment All the organisms in a community and their interactions with each other.

casuistry In ethics, this is the careful comparison of similar cases to identify the key features that justify a discrimination between one situation and another – thereby refining ethical judgement. The term can be used disparagingly to suggest that an argument or opponent uses clever but false reasoning.

catecholamines Class of hormones (e.g. epinephrine and norepinephrine) secreted by the adrenal glands and also used as neurotransmitters.

clinical trial A biomedical experiment involving patients that follows very precise rules so that a particular drug or treatment can be evaluated for safety and therapeutic potential.

cloning Producing multiple copies of a molecule (e.g. DNA), a cell or an organism. When referring to a cell or a whole organism, cloning results in multiple copies with the same genetic constitution.

Code of Federal Regulations The corpus of regulations issued by departments of the executive branch of the United States Government. The federal regulations pertaining to the oversight of federally funded research involving human subjects is found in Title 45, Part 46 of the CFR.

commodification To turn into or treat as a mere commodity.

communitarianism A philosophy that stresses social participation and service to the common good, in contrast to liberal individualism. Individuals have responsibilities as well as rights, and their freedom has to be balanced with the good of the community.

community consent An emerging doctrine in the bioethics literature on informed consent asserting that, among certain highly self-defined cohorts of the human population, especially those that have historically been the target of sustained patterns of prejudice from other groups, efforts to enrol members as research subjects ought to be preceded by review and approval by an entity that represents the interests of the body politic of that cohort.

complementary deoxyribonucleic acid (cDNA) DNA produced *in vitro* so that the sequence of bases is complementary to a particular messenger RNA. It is used to study gene expression because it is more stable than RNA and is more amenable to recombinant cloning techniques.

concordance The degree of similarity in phenotype between individuals.

concretization To make concrete instead of abstract; this may sometimes be appropriate, as an explanatory tool, or inappropriate.

consanguinity Related by blood; in genetics the term is usually used to describe a mating or marriage between close relatives.

contiguities The state of being contiguous with, adjacent to, or abutting onto something. Adjacency is the attribute of being so near as to be touching. Continuous genes lie adjacent to each other on the same chromosome.

corporate liberalism A modern variant of classical liberalism borrowing from the fascist emphasis on order, as in Mussolini's 'corporatist state'; a political/social philosophy based on constitutionally limited government that protects the individual (including corporate) rights, a materialistic emphasis on property as the basis for society, especially a free-market system, and a belief in social progress.

craniology The science (now regarded as unfounded) of studying the mind and human behaviour by means of measuring the shape and dimensions of the skull.

creationism A generic term for the belief that organisms were produced in six days, miraculously, as described in the early chapters of Genesis in the Old Testament.

Darwinism The belief that the main cause of evolution is the mechanism proposed by Charles Darwin of variation and selection – 'survival of the fittest' – with selection including both natural and sexual mechanisms.

dendrogram Bifurcating tree-like diagram indicating relationships among items – especially organisms or molecules that are related through evolution.

determinism A philosophical concept arguing that all events, including human behaviour, is predetermined.

dialectically Adverb from 'dialectic' – a way of discovering what is true by considering opposite theories. This was developed in Hegel's philosophy of history into an account of material reality and gave rise to Marx's ideas of historical development.

differentiated cell A specialized cell that has developed from a single (undifferentiated) cell, the zygote.

disability rights The campaign for the extension of civil, social, political and legal rights to all disabled people.

disquisitions Pieces of long or complex discourse.

echocardiography Ultrasound examination of the heart.

endogamy The practice of marriage within a community; the laws, customs or rules which specify this practice of marriage within defined boundaries (such as geographical location, kinship group, religion, language group, etc).

enucleated eggs Female sex cells, which have had their chromosomes (their nuclear genetic material) removed.

ethnic affiliation estimation A strategy for the use of DNA marker analysis to establish probability estimates that a subject has come from a select population group.

ethologists A scientist who studies animal behaviour.

etiology (or aetiology) A branch of knowledge concerned with the causes of particular phenomena, specifically a branch of medical science concerned with the causes and origins of diseases.

eugenics A programme designed to improve the human species by encouraging 'genetically superior' people to have more children and by discouraging, in more or less coercive ways, the 'genetically inferior' people from reproducing. Hence, the cultivation of conditions that may improve the physical and moral qualities of future generations.

evolution The belief that all organisms, including ourselves, are the descendants of one or a few original forms, and that the causes are natural or law bound, rather than miraculous.

***ex vivo* gene therapy** Patient cells are harvested and cultivated in the laboratory and incubated with vectors carrying a corrective or therapeutic gene. Cells with the new genetic information are then harvested and transplanted into the patient from whom they were derived.

Expressed Sequence Tag (EST) A short sequence of the complementary DNA (c-DNA) expressed by a full-length gene. Knowledge of an EST can be used to identify the gene from which it derived. Multiple EST sequences can be used to piece together a full-length gene sequence.

externalities Effects of an action that are not taken into account in deciding whether or not to go ahead with that action.

Factor IX A protein that assists in the coagulation (clotting) of blood, a deficiency of which leads to one form of haemophilia, haemophilia B, a bleeding disorder.

forensics The attempt to deploy the scientific method, usually at the crime scene, to find information that can be used in a court of law.

founder effect A phenomenon through which newly established populations are more representative of their founders than of the population from which the founders came.

free will A philosophical concept arguing that human behaviour is the result of individual free choice.

fundamentalism The belief that truth is to be found in the original articles of faith of a religious tradition, often as recorded in the tradition's holy scriptures. Variants of this are found in several religions, including Judaism, Christianity and Islam. The conflict between the biblical accounts of creation in Genesis and the Darwinian account of evolution led certain fundamentalists to instigate the prosecution (and conviction) of the teacher John Thomas Scopes, in 1925, in Dayton, Tennessee, USA.

gene therapy The treatment of disease by either replacing or supplementing damaged or abnormal genes with normal ones, or by providing new genetic instructions to help fight disease, e.g. cancer. Therapeutic genes are transferred into the patient either through a weakened virus, a non-viral vector, or through direct delivery of so-called 'naked' DNA.

genealogy A lineage of descent from an ancestor; the study of lineages and descent.

genetic defence A shorthand term referring to the use of behavioural genetic evidence in a legal defence to criminal charges.

genetic determinism The ideological belief that genes determine an organism's future development, including a person's physical features and behaviour.

genetic discrimination Discrimination, especially in regard to socio-economic matters including insurance or employment, against an otherwise healthy individual on the basis of genetic information that suggests an increased predisposition to future illness or to reproductive risks.

genetic essentialism 'Genetic essentialism reduces the self to a molecular entity, equating human beings, in all their social, historical, and moral complexity, with their genes' (Nelkin and Lindee, 1995, p. 2).

genetic homogeneity (1) A population may be genetically homogeneous when there are low levels of inter-individual variation (may be in relation to a specific trait or locus or to the whole genome); (2) a genetic disease shows genetic homogeneity when all cases are the result of mutations affecting a single gene locus.

genetic isolate A subpopulation that has become geographically isolated from the rest of the species, so that it may have begun to evolve independently.

genetic–environmental complexity Interactions among genetic and environmental influences that affect a trait or behaviour, so that it is impossible to separate the genetic from the environmental causes of the differences in the trait between individuals; the complexity makes it very difficult to identify the relevant genetic factors in human populations although laboratory breeding programmes do enable this in species such as *Drosophila*.

geneticization A term coined by Abby Lippman to mean 'the ongoing process by which priority is given to differences between individuals based on their DNA codes, with most disorders, behaviors and physiological variations . . . structured as, at least in part, hereditary' (Lippman, 1993, p. 178).

genotype The set of genetic variants (alleles) possessed by an individual organism – usually specified in relation to one or more specific loci.

glycosylation The addition of chains of sugar molecules to the amino acid sequence of a protein, producing a glycoprotein.

Hassidic community A mystical Jewish movement founded in Poland in the eighteenth century, represented today by fundamentalist communities in Israel and NewYork.

hegemony Political dominance; authority supported by power.

h(a)emoglobinopathies Disorders of h(a)emoglobin, which is a protein in red blood cells that carries oxygen around the body.

heritability In population genetics, the proportion of the variance (the statistical measure of differences) of a trait within a specified population or group at a specified time that is attributable to genetic as opposed to environmental factors.

heterozygous A genotype consisting of two different alleles of a gene for a particular trait (Aa), often one normal or wild-type allele and one disease-associated allele. Individuals who are heterozygous for a trait are referred to as heterozygotes.

homozygous A genotype consisting of two identical alleles of a gene for a particular trait. An individual may be homozygous dominant (AA) or homozygous recessive (aa). Individuals who are homozygous for a trait are referred to as homozygotes.

hygienics The science concerned with the prevention of illness and maintenance of health.

hypercholesterolemia Increased concentration of cholesterol in the bloodstream.

***in vivo* gene therapy** A vector carrying the therapeutic gene(s) is directly administered to the patient.

informed consent The process by which a legally competent individual, after being provided with a morally adequate description of the research project in which he/she is being asked to participate, decides autonomously to participate as a subject in that research.

Institutional Review Board (IRB) A group of individuals (especially in USA), acting as a duly constituted body, as described in relevant federal regulations, to monitor proposals to conduct research involving human subjects.

intelligent design The belief that organisms show the evidence of divine planning and direct manufacture. The contemporary term for some fundamentalist accounts of creationism.

internal entropy Entropy: the amount of disorder in a system. The second law of thermodynamics specifies that disorder increases with time in a closed system, so that 'order' can only be sustained through the expenditure of energy.

introns Those parts of a gene that are excised from the primary transcript in the process of RNA editing, so as to produce the messenger RNA molecule (which consists of the exons and the 5′ and 3′ untranslated regions) for export from the nucleus to the cytoplasm.

irreducible complexity The intricate organization of living things, supposedly making impossible a natural (that is, evolutionary) origin (cf. creationism; intelligent design).

karyotype A karyotype is the complete set of all chromosomes of a cell of any living organism. The chromosomes are arranged and displayed (often on a photograph) in a standard format: in pairs, ordered by size. Karyotypes are examined in searches for chromosomal aberrations, and may be used to determine other macroscopically visible aspects of an individual's genotype, such as sex (XX vs. XY pair). The study of karyotypes (also known as karyology) is a major aspect of cytogenetics.

kinship A set of practices, along with associated beliefs, values and categories, which structure social relationships, social perceptions and actions, including biological connectedness.

learning disability This is the current UK term for what used to be called 'mental handicap' and in other countries is called 'intellectual disability'.

macrosomic A body that has grown to be excessively large – in terms of stature, not obesity. A child with macrosomia may be said to have overgrowth.

Malthusian prediction Thomas Malthus (1766–1834) was an economist and clergyman who predicted that the 'natural' unchecked geometric increase in population growth would produce a crisis of the food supply. Only war, famine, disease or other natural disasters would slow this crisis, which he theorized was inevitable.

matrilineal A system of kinship based on the relationship to one's mother and grandmother; this may influence inheritance, site of residence, etc.

medicalization This concept is used by social scientists to describe the processes through which the power of medicine extends to areas of human life previously considered non-medical.

Mendelian trait A Mendelian trait is one determined by a single gene; effectively synonymous with 'monogenic'.

messenger ribonucleic acid (mRNA) RNA that carries information from DNA to the ribosomes, the sites of protein synthesis in the cell. In eukaryotes, once RNA has been transcribed from DNA, it is processed into mRNA and then exported from the nucleus into the cytoplasm, where it is bound to ribosomes and used to direct the synthesis of proteins from amino acids.

micro-phylogeny Recent biological history of closely related populations.

monogenic disorders Describes a single gene or Mendelian trait.

non-directive counselling A key component of ethos of the genetic counselling profession; it indicates that professionals should not impose their value judgements or recommendations for action on clients but should rather support individuals and families in the exercise of their autonomy.

(somatic cell) nuclear transfer The process of transferring the nucleus of an adult cell into an egg cell that has its nucleus removed, as used in the cloning of Dolly.

pathogenesis The origin and development of a disease.

patronymic An individual's name that is derived from that of the person's father or paternal ancestor, usually by the addition of a suffix (as in Russian usage).

penetrance The likelihood that a disease-associated genotype will result in an abnormal phenotype.

pharming The genetic modification of animals to produce therapeutically valuable materials, usually proteins.

phenotype The visible properties of an organism that are produced by the interaction of the genotype and the environment, but a given phenotypic trait is not necessarily genetically determined.

phlebotomy A procedure to remove blood from a vein in the body, also called venesection.

phosphorylation The addition of a phosphate group to a compound by an enzyme. Phosphorylation is an essential step in many cellular processes.

plasmodium falciparum A protozoan parasite that causes the most severe types of malaria in humans.

polity A form or process of civil government or constitution. An organized society, a state as a political entity.

polygenic Refers to characteristics (e.g. measurements) or diseases which result from the interactions between multiple genes, which individually would only have a small effect on the phenotype; environmental factors would often also be involved.

polymorphism A naturally occurring variant in the DNA sequence among individuals in a population. Most polymorphisms are harmless and are part of normal human genetic variation.

positional cloning A technique used to identify genes based on their location on a chromosome.

post-natal genetic screening Analysing tissue or blood samples from asymptomatic populations (usually newborn) to detect a chromosomal abnormality or a genetic condition that is the focus of the screening procedure. Newborn screening programmes can identify many inborn errors of metabolism as well as diseases that may be of non-genetic origin (e.g. hypothyroidism).

predictive genetic testing The use of a genetic test in an asymptomatic person to predict the likely onset of disease in the future. The term 'predictive' is used for testing Mendelian genes for high-penetrance diseases, while 'risk-modifying' or susceptibility testing refers to tests looking at genetic factors of lower penetrance.

presumed consent Consent (e.g. to donate; to participate in research) is presumed unless the person has 'opted out' by expressly indicating otherwise. This may be justified in some contexts but is problematic in others, where a health professional who presumes that a patient or research participant gives their consent may not appreciate their failure to understand what is at stake (cf. informed consent).

probability The probability of an event is defined as the frequency of its occurrence in a long series of 'trials' – such as the frequency of getting 'heads' in 100 tosses of a fair coin. When linked to the notion of hazard or adverse outcome, it may be referred to as a 'risk'.

proband The person whose case is the stimulus for the study of other members of the family to identify the possible genetic factors involved in a given disease, condition or characteristic.

promulgated Publicly announced, especially a new law.

race Formerly, a large natural division of the human species; now, since such groups are recognized as illusory, with continuous variation in many genetic variables across large distances and geographical boundaries so that 'races' cannot be regarded as distinct and separate groups, the term is sometimes now used to designate a politically meaningful category into which people may be allocated on the basis of some aspect of ancestry.

racial hygiene The practice in National Socialist Germany in the early to mid-twentieth century of removing individuals from the population – either on the grounds of disease or degeneracy (including some clearly genetic disorders but also alcoholism and mental illness) or on the grounds of racial admixture (with the goal of keeping the 'Aryan' gene pool of the German population free from contamination by 'inferior' groups.

racialist A synonym for racist (especially in Britain).

racist A person who believes that races can be distinguished, that racial differences are important, that some races are superior to others, and that his or her race is one of the superior ones.

recombinant deoxyribonucleic acid (rDNA) Genetically engineered DNA made by recombining fragments of DNA from different organisms.

reductionism This philosophical term describes the idea that the nature of complex things can always be reduced to (explained by) simpler or more fundamental things, e.g. that the behaviour of a complex organism can be understood in terms of the atoms and molecules of which it is constituted.

reify To treat as concrete or real something that is in fact abstract, e.g. a nation, normality.

reproductive cloning The application of the technology of cloning for reproductive purposes – to create new individuals genetically identical to each other and to an already existing person.

retrovirus A virus that has RNA instead of DNA as its genetic material. When introduced into a host cell, it is used as a template to produce viral DNA, this allows the formation of more, identical viruses.

risk factor An attribute that is associated with an increased probability of a disease occurring. A risk factor is not to be understood as a cause of the disease but may be implicated as contributing towards the disease.

salvage anthropology Research on indigenous peoples requiring immediate action due to the impending destruction of those peoples or their lifeways.

Scientific Creationism A 1960s and 1970s form of Biblical literalism, dressed up in scientific language, in order to avoid the US Constitution's separation of church and state.

sequelae A condition or complication following as a consequence of a disease.

serological Relating to serology: the scientific study or diagnostic examination of blood serum, usually relating to circulating immunoglobulins (antibodies).

single-nucleotide polymorphisms (SNPs – pronounced 'snip') A polymorphism (q.v.) at a particular base site in a coding or noncoding sequence of DNA. The frequency of the less common allele should be at least 1 per cent. This type of polymorphism is extensive throughout the genome, and may be physically close to, and associated with, important genetic variation that contributes to the risk of disease, even when the SNP itself is not the cause of the altered disease risk. The patterns of SNPs vary between population groups and so can also be related to a person's origins. SNPs have the great advantage of being detectable without the need for gel electrophoresis, which opens the way for large-scale automation of genotyping.

Sinic or sino- Refers broadly to anything having Chinese influence. Thus, most East Asian nations can be accurately characterized by this adjective, especially Japan, the Koreas, China (Hong Kong), Taiwan, Singapore and Vietnam.

social relations model of technology A philosophy that holds that technical phenomena are an outgrowth of social dynamics in a society, in particular the distribution of power in the society, so that the technologies produced reflect and manifest that power.

somatic therapy Involves the manipulation of gene expression in the somatic (body) cells of the patient but not affecting the germline so that it will not be transmitted to the next generation.

spatiotemporal Physics and philosophy belonging to both space and time or to space-time.

stochastic Occurring by a random chance process.

technological fix A public policy approach which holds that since modern social problems are too messy and complex to solve by direct social interventions, the use of technical factors should be relied on to produce changes that are easier to make socially and politically acceptable.

telos In philosophy, an ultimate object or aim.

tendentious Expressing or supporting a particular opinion with which many other people disagree.

teratogenic Adjective of 'teratogen', a substance which, when a pregnant woman is exposed to it, can cause birth defects in her foetus.

thalassemia A genetic form of anaemia in which there is abnormality of the globin portion of haemoglobin. Affected individuals cannot synthesize haemoglobin properly, and they produce small, pale, short-lived red blood cells. Widespread in Mediterranean countries.

theistic Relating to theism: the belief in the existence of a god or gods, specifically of a creator who intervenes (or has intervened) in the universe.

therapeutic cloning The application of the technology of cloning for therapeutic purposes – perhaps to grow cells of a particular tissue to replace damaged tissue.

transcription factor A protein that controls gene expression by binding to particular DNA sequences within specific promoter regions.

transgenesis The introduction of foreign genes into an organism to confer new properties on that organism. This method is generally used to study the function of genes in the context of a living organism.

transgenic organism An organism that is the product of the mixing of genes of different types of plants and/or animals.

vectors Gene-therapy delivery vehicles, or carriers, that encapsulate therapeutic genes for delivery to cells. Vectors include both genetically disabled viruses such as adenovirus and non-viral vectors such as liposomes.

XYY syndrome A term used to describe the condition of men who possess two Y chromosomes instead of the usual single Y chromosome (their chromosome constitution is therefore 47, XYY).

References

Lippman, A. 1993. 'Prenatal Genetic Testing and Geneticization: Mother Matters for All'. *Fetal Diagnosis and Therapy* 8, 175-188.

Nelkin, D., Lindee, S. 1995. DNA Mystique. *The Gene as a Cultural Icon.* New York: Freedman.

Introduction

Angus Clarke

Recent developments in the genetics of humans and other animals have generated fascinating insights into our biology. We see more clearly our place in the animal kingdom and the historical details of our relationship with other creatures; we are learning about the origin of different human populations and the early migrations out of Africa that led us to inhabit the rest of the globe; we are studying the processes that control the development of our bodies – both the shaping of our bodies in embryonic life and the shaping of connections between the cells in the brain that underlie our ability to learn and to behave in a meaningful manner; we are coming to understand more about disease and the degeneration of our bodies, and we are just beginning to be able to use this knowledge for the treatment and prevention of disease.

This knowledge and understanding that we claim to derive from genetics is based upon information – an accumulation of facts that are generated and interpreted through very specific social processes. Whatever the appearances, therefore, this knowledge emerges neither from a vacuum nor from an oracle. As a consequence, and alongside the undoubted benefits, there are some troubling questions. How is such knowledge produced and how is it applied? Who controls this generation and application of knowledge?

In the context of medicine, and health care more generally, how do professionals and patients respond to the more detailed understanding of inherited disease that is now becoming available? Along with the new knowledge and understanding come new choices, responsibilities and burdens. How do *we think* of ourselves – and how do *we feel* about ourselves – if we learn that we carry a genetic change that may affect our future health, or the health of our children? If we already have a disease or a disability, how does the knowledge of the genetic basis underlying this affect us 'inside'? – how does it affect our self-understanding, and how does it impact upon our feelings? How will others, or society at large, respond to us? How will we be treated and cared for if our disease or disability is now better understood and could perhaps have been 'prevented'? Indeed, should the termination of a pregnancy likely to be affected by a specific disease or disability be regarded as disease prevention, or as the taking of life? In short, how does knowledge of the biological mechanisms responsible for a disease or a disability impact upon moral, social and legal concepts of responsibility?

Consideration of these issues will itself generate further questions. If genetic information is important, then should it be guarded as confidential? – should society ensure a right to the privacy of personal genetic information, even though the employers, insurers or family members of someone carrying a disadvantageous genetic alteration may wish to know about it? Conversely, is there a danger that we may attach too much importance to genetic influences and forget that other factors also affect our health and well-being? Will we forget that the cultural transmission of language, wealth and ways of life are also forms of (non-genetic) inheritance? Where are the limits to genetics – what else is required, apart from a knowledge of genetics, in order to give a satisfying explanation of the biological and social facts of our existence? Where might genetics take us, collectively, in the future? If we allow the powerful combination of science and commerce to push forward in the direction of anticipated future profits, without the application of any social constraints, how will this alter the future of our local societies, or of global(ized) society? We have become familiar with the military–industrial complex as a powerful force shaping the national and global political economy; how will biotechnology influence our collective futures? At the most intimate and personal level, can we foresee how biotechnology may alter our children and grandchildren?

This book cannot provide final answers to this list of questions. What it can do is to set out some of the best available thinking about this range of topics. We are not presenting an overview or an interpretation of the science for a lay audience – that has been done before. Nor are we presenting our own predictions about the likely social consequences of current or future scientific developments, or our own views about the relevant ethical issues, except in a minor way towards the end of this introduction. Instead, we have chosen 42 articles written by different authors, who present their individual, and sometimes sharply contradictory, positions towards human genetics. Here, therefore, you will find critical thinking about a wide range of the issues set out above. There is no one position adopted or being promoted by the authors or the editors. Without any requirement to make this volume comprehensive in its scope, we have been free to select authors and articles both on the basis of the intrinsic interest of the writing and to ensure that a wide range of topics and views are represented. (The weblinks, listed at the ends of the chapters, at time of writing were all live, but as websites change all the time, readers should be aware that some may disappear during the lifetime of this book.)

The articles, sometimes edited (with the approval of the authors), have all been drawn from the *Encyclopedia of the Human Genome*. One of us (AC) was the editor of the 'ethics and society' section of that major reference work. In this role, he selected the topics to be covered in the *Encyclopedia*, set out the scope of each article and then identified appropriate authors; he was also involved in some of the ensuing discussions and negotiations between authors and publisher. This effort has been well rewarded by the high quality of the resulting product – both the *Encyclopedia* as a whole and the articles on ethical and social issues in particular. The *Encyclopedia*, however, has been more readily accessible to those working or studying in human genetics than to those from the social sciences or humanities or, indeed, to interested members of the public. The publication of this volume will enable a much wider potential readership to access a selection of articles from the *Encyclopedia* on the ethical and social impact of human genetics. This prospect gives us great pleasure, and we are confident that the views expressed here will serve to stimulate and challenge all those who wish to engage with these important issues.

Background: What is genetics?

Genetics addresses two fundamental questions: (a) how do we change, in the course of development and in the course of evolution? and (b) conversely, how do we stay the same? – in effect, how is it that children resemble their parents? Genetics therefore encompasses the systematic study of three related topics:

1. How is the genetic information in each cell copied and transmitted to the daughter cells at each cell division – and especially in the production of the egg and sperm cells that contribute to the next generation?
2. How does the genetic information in the fertilized egg control the growth and development of the embryo and thereby influence the form of the adult individual and their propensity to disease?
3. How do populations of an organism accumulate changes over time so that the species evolves, perhaps giving rise to new species?

These three topics consider 'genes' from very different temporal perspectives – the life of the cell, the life of the individual organism and the history of populations. The overall story has four principal roots in the nineteenth century: Mendel's recognition that inheritance is carried by specific particles, despite his not knowing of what these consisted; the discovery of chromosomes as the possible physical basis of inheritance, with major input from several scientists, including Boveri; Galton's development of statistics and its application to continuous traits (such as height or weight) and to the resemblance between relatives; and Darwin's theory of the evolution of populations through the operation of selection. These roots came together at the start of the twentieth century and led to the development of the field of modern genetics.

Studies of the transmission of genetic information at cell division and reproduction led first to the identification of deoxyribonucleic acid (DNA) as the chemical basis of heredity in micro-organisms (in 1944) and then to modern molecular biology, which began in earnest with the Watson and Crick double-helix model of the structure of DNA (in 1953). The theme of genetic influences upon the development of disease in the individual began with Garrod's ideas of gene action as the production of an enzyme that would carry out a specific biochemical (metabolic) reaction, but it took some decades for this approach to lead to an understanding of metabolic pathways. It was only once the DNA from specific chromosomal sites or genes could be isolated that the mutations disrupting gene function could be examined. This led to an appreciation of the different ways in which mutation could interfere with gene function – indeed, new types of mutation are still being recognized. Especially noteworthy and fruitful in this field has been the work of Weatherall and his team (Weatherall & Clegg 1999, Weatherall 2001) on the mutations affecting the haemoglobin genes and leading to imbalances in production of the several types of globin molecules, especially the beta-globin protein molecule.

The theme of population changes over evolutionary time has also undergone interesting developments. Once Mendel's work had been rediscovered, around 1900, it took three decades for 'the new synthesis' to emerge, in which Mendelian genes were seen as contributing to variation in quantitative traits and permitting evolution through the operation of natural selection. It took longer for the enormous scale of genetic variation between individuals in a population to be recognized and for the relative importance in evolution of natural selection and of sheer random events to be clarified. Random changes in gene frequencies, often referred to as genetic drift, are particularly likely to occur through pure chance when the size of a population is very small, perhaps when disease, famine or war has killed a major part of the population. Such 'bottleneck' effects can alter the genetic constitution of a population through strong selection as well as through pure chance. In humans, the relative importance of natural selection and of genetic drift in accounting for genetic differences between population groups is still unclear. The legacy of genetic differences between populations is useful for tracking patterns of human migration in prehistory but may also have consequences for the pattern of diseases evident in modern societies.

Genetics in medical research

The importance of genetics in medicine has been increasingly recognized over the last three decades. Whereas human genetics was for long seen as being esoteric, with medical geneticists regarded as boffins interested in obscure ailments and hens' teeth, it is now portrayed as being science's chief hope in the fight against the common ailments of developed, and even developing, societies. Genetics was thought of as relevant merely to foetal malformation (such as congenital heart disease, spina bifida), some diseases of early childhood (such as cystic fibrosis, immune-deficiency diseases) and some rare adult-onset diseases, and it was not seen as making any significant contribution to the common disorders of adult life in economically developed societies, such as coronary heart disease, cerebrovascular disease (strokes), Alzheimer's disease (dementia), hypertension, diabetes mellitus type 2 (maturity-onset diabetes), schizophrenia and cancer of the large bowel (colon) and of the breast.

From the 1980s onwards, researchers worked to locate disease genes along the length of the chromosomes, and then to identify and isolate them. The genes that could be approached in this way were those in which mutations transmitted through a family are strongly associated with the particular disease that is also being transmitted as single-gene (Mendelian) traits; a Mendelian trait is one that is transmitted through families: a mutation in one or both copies of the gene present in an individual. Such diseases are mostly uncommon but have been important for our understanding of many pathological processes. Some progress has even been made toward the development of rational, gene-based therapies for some of these rare conditions – not always gene therapies as such, but therapies based on an understanding of the disorder that has been made possible only from a knowledge of the gene(s) involved.

Over the past decade, investigators have largely turned from their efforts to identify single genes in which they hoped to find mutations that could be said to 'cause' genetic disorders to investigate the common, *complex* diseases instead. They have set out to dissect the contribution of genetic variation at many sites across the genome to the individual's propensity to develop these conditions. In contrast to the Mendelian diseases, where the cause can be thought of as any mutation disrupting the action of a specific gene, these much more frequent, degenerative diseases – common in many developed and developing societies – do not have a simple mode of inheritance; while there is some familial clustering, with a tendency for more than one case to arise within a family, the process of disease causation is much more complex than when a particular mutation in a particular gene can be said to be a cause of the disease. There has been a shift in research strategy from those methods appropriate to a simple genetic cause – tracking the genes shared by affected family members – to the use of vastly more elaborate technologies applied to the analysis of samples from hundreds or thousands of patients (and those who are not yet patients). Such efforts seek to identify sites of genetic variation that collectively exert only a modest, merely probabilistic influence on disease susceptibility. This shift in strategy has resulted from the failure of the earlier approach to yield consistent results across different populations, along with the development of the molecular techniques and computational capacity required to generate and analyse the vast arrays of data necessary in this type of study.

Genetics and genomics

'Genome' is a collective noun for all the items of a complete genetic constitution. It is the entire array of genes and chromosomes that make up the inheritance of an individual or even of a population. It may be applied to the single set of genes and chromosomes in an egg or sperm – the 'haploid' (single copy) genome; or to the double set present in each cell in our bodies (except for eggs and sperm) – the 'diploid' (double set) genome. It may also be applied to the genetic constitution of an entire population or species – the human genome, the chick genome, the *Drosophila* (fruit fly) genome – including the variation found between individuals and groups within the species.

Genomics is the study of genes as they are found, or as they function, in this collective, wholesale sense. A gene or set of related genes may be compared across species, to look at the changes that have accumulated in their respective copies of these genes in the course of evolutionary time – since their last, perhaps remote, common ancestor. This can be helpful in tracing the historical relationships between closely related or even distantly related species.

Another aspect of 'genomics' is the study of the hierarchies of organization within the genome – the genes themselves, their constituent elements and the structural and organizational features apparent at a higher level of complexity. Genes usually consist of regions that are transcribed into ribonucleic acid (RNA) – the coding regions – and other, regulatory regions, sometimes remote from the transcribed region, that interact with specific proteins to control when and where the gene will be transcribed. The transcribed RNA is edited within the nucleus, resulting in the excision of intervening sequences (introns) that break up the protein-coding region. The edited RNA, known as messenger RNA (mRNA), leaves the cell nucleus bound for the protein-synthesizing apparatus in the cytoplasm; it consists of a coding region, that will be translated into amino-acids, and untranslated regions in front and following behind, involved in the regulation of mRNA movement, translation and stability. The common features of the regulatory, the intronic and the coding regions of genes can be examined within and between species. The location and function of repetitive DNA sequences, not usually coding for protein, can be studied. The factors that influence gene expression can be investigated. The structural organization of chromatin – the complex of DNA with its associated proteins – can be explored, as well as the relation between chromatin structure and gene expression.

If we wish to focus on disease, the characterization of mutations may help to define the functional importance of different domains within the gene, including both the protein-coding and the regulatory regions of the relevant gene. Evolutionary studies may also be helpful in understanding the importance of specific domains within the protein-coding region of the gene. When genetic variants are found in patients affected by a disease, it may be unclear whether the variation identified in a plausibly relevant gene has actually contributed to the disease: could *that* genetic change be relevant to the disease in *this* patient, or is just a chance finding unrelated to the disease? – the answer is that, if the mutation is in a region of the gene that is strongly conserved (has varied little in the course of evolution) then it is more likely to affect a functionally important region of the protein and perhaps to be a disease-*causing* mutation.

We may look to find sequences from around the genome that are shared more commonly by those affected by a given disease than in the general population. This would not necessarily mean that such a sequence caused the disease – it might perhaps modify the chance of developing the disease, or it might simply happen to be located on a chromosome close to another site of genetic variation that *is* functionally related to the risk of disease. Analysing DNA sequences at many sites around the genome, using variation at some hundreds or more different and unique (non-repetitive) DNA sequences, can be termed genome-wide analysis.

The recent development of microarray technologies – also known as 'DNA chips' – has been reducing the cost of such genome-wide investigations into the DNA sequence composition of an individual or a tissue; looking for the changes in genetic constitution of a tumour can be helpful in understanding the origin of the malignancy, and such information is now becoming clinically useful in guiding the treatment of individuals with specific types of tumour. Similar approaches, looking at tissue-specific RNA content rather than DNA content, can be used to examine the pattern of gene activity in specific tissues at specific stages of development. This is transforming the study of developmental biology and is also beginning to have clinical applications when used to compare the pattern of gene expression in normal and disease-associated (especially tumor) tissues.

One specific area, where there has been much talk of the likely benefits from the new genetics, relates to the development of new drug therapies and the tailoring of drugs – the choice of drug and the dose of drug – to the individual patient. We do not need to dwell on this at length, but it is appropriate to place some emphasis upon the distinction between two types of gene-based research of relevance to clinical practice in the future. The reason why I wish to clarify this distinction is because the type of research required to promote the two types of application may differ, and the ethical and legal issues raised by the research may also differ.

- One type of research aims to uncover the genetic variation relevant to the risk of developing a disease, in the hope that this will identify novel genes – perhaps not previously suspected as being involved in the disease – whose corresponding proteins are then implicated in the disease process. These proteins, or others with which these interact, may then become targets for the development of future drugs that could be used in the treatment of this disease, whether or not the patient has a genetic variation in this gene or protein. This research may be termed *pharmacogenomic research* if it is being conducted with the goal of identifying such novel drug targets, although any disease-related research that studies variation across the entire genome may then be said to fit this bill.
- The other type of research, termed *pharmacogenetics research,* is intended to distinguish between those patients likely to benefit from one drug rather than another, or to predict the drug dose appropriate to the individual. The goal – the implicit promise of such research – is that each patient can be given the perfect remedy for their disease – a drug that works, that does not cause toxicity or trigger an allergic or other idiosyncratic adverse response, and whose minimum effective dose can be predicted from a knowledge of the variants of the genes involved in drug absorption, activation, action, metabolism, inactivation and excretion (in short, drug handling) in that individual patient.

There may be some overlap between these two types of research, because (for instance) the most effective drug to prescribe for a patient's hypertension, asthma or epilepsy may depend upon the genetic basis of the ailment in that individual as well as on their predicted pattern of drug handling, but they will usually be conducted in rather different ways. In fact, there will often be important differences between the ways in which a research study is set up, depending upon which type of research is being conducted, with corresponding implications for the appropriate regulatory framework to be applied. For example, the clinical information that is required will depend on whether the research cohort is (a) a large group of healthy individuals being tracked to see what diseases develop over time, as in UK Biobank, (b) a moderately sized collection of patients already with a specific disease, or (c) a small group of individuals who have had a severe adverse reaction to a particular class of therapeutic medication. Equally, the question of anonymization will differ in these circumstances – anonymization being completely impossible if data are to be collected prospectively, as in UK Biobank – and the question of results being passed back to research participants will also depend enormously on the context. Where data are preliminary and unconfirmed (not yet replicated in other studies) and indicate only a modest association between one allele at a marker and the disease being studied, then feedback of individual results could be very misleading and would clearly not be justified, even when an ethics committee mistakenly declares otherwise. In other circumstances, of course, feedback might be entirely appropriate, as when a Mendelian effect is being studied and a very strong influence of one genetic factor has been identified.

The new genetics as an activity

The shift in research focus from the study of single genes to the study of the common diseases, involving complex gene–gene and gene–environmental interactions, has occurred as a direct result of the accumulation of genetic sequence information about the human genome and the genomes of other species. The accumulating data can be interrogated through the use of increasingly sophisticated programmes that look for common patterns of sequence. These new tools for the investigator allow a wide new range of questions to be asked. A few examples of the types of question that can now be asked (if not yet always answered) can be given:

(a) from genome sequence information, can we identify the sequences that will be transcribed into RNA and translated into protein as functional genes?
(b) what domains within a gene have been heavily conserved over evolution (by comparison with related, homologous genes in other species)?
(c) are mutations in these conserved domains cause more likely to result in severe disease when found in human genes?
(d) what functional domains can be recognized within genes, by homology with the specific domains of known function found in other genes?
(e) what accounts for the pattern of mutations observed in specific genes?
(f) are these patterns similar, or do they differ, depending upon whether we look at mutations that have occurred over the timescale of evolution, over a historical timescale as causes of genetic disease within families or over the life-course of an individual as causes of malignant or degenerative disease within that person's tissues and organs?
(g) given the structural properties of DNA sequences and of proteins, as far as they can be predicted from the raw DNA sequences, what proteins are likely to interact with each other or with particular gene regulatory sequences?

Asking these questions will inevitably involve information technology (IT) and the intelligent interrogation of data, where in the past human genetics research would have consisted predominantly of traditional laboratory work. This corresponds to the shift from the 'wet' (traditional, bench-top, *in vitro*) experimental research to the 'dry' (new, computer-based, *in silico*) research conducted through the keyboard. Of course, such IT-based research itself generates predictions that demand experimental testing – so it would be wrong to imagine that the 'dry' research is ousting the 'wet', but there has certainly been an enormous shift in the pattern of work of the active researcher in genetics – and more generally in the rest of genetics too. It has become harder to categorize an individual researcher as studying one species when the comparative genomic approach may require the focus of a research project to shift from one organism to another over the course of a few months even when the goal is to learn about one particular human disease.

Genetics in the clinic

In order to understand a disease process, it would in the nineteenth century have been 'natural' to dissect the organ affected – to approach the disease through pathological anatomy. In the early and middle twentieth century, it would have been usual to define the physiological or biochemical consequences of the disease process, or to isolate an invading micro-organism. Today, and over the past two decades, the standard approach would be to find the genes involved. In order to understand the development and function of the kidney or of the heart, one might now seek out rare patients of families with a malformation or malfunction of the kidney or the heart and identify the genetic basis of their problems. This will often give important insights into the normal development and functioning of the relevant organ. In relation to muscle disease, for example, there have been enormous strides taken in understanding the function of muscle since the protein product of the Duchenne muscular dystrophy (DMD) gene was identified almost twenty years ago. This has been relevant not just to patients with DMD but also to many others with different types of muscle problem. Indeed, the DMD protein dystrophin was only identified through this genetic work, and until it was identified there was no opportunity to identify the proteins with which it interacts, and which in their turn can be associated with other muscle disorders. But how is such knowledge feeding through into clinical practice? Where will it take us over the next few decades?

Genetic investigations are being applied in several clinical contexts. The most readily appreciated may be the diagnosis of known genetic disorders. Where clinicians suspect that a patient has a particular genetic condition, confirmation of this may be sought by searching for mutations in the relevant gene (or genes). This may be simpler and less invasive or less hazardous than other types of investigation – for example, it may allow the accurate diagnosis of a muscular dystrophy to be made without a muscle biopsy, or a diagnosis of a particular type of dementia may be made without the need to wait for a post-mortem examination of the brain. Such a diagnosis then brings with it the possibility of tests being made available to others in the family. With an autosomal dominant disorder, such as Huntington's disease or some types of familial cancer, predictive testing to see who will develop the disease in the future will be possible if – after appropriate counselling – it is wanted. Similarly, prenatal diagnosis may become feasible for an autosomal recessive or sex-linked recessive disorder once the cause of the condition in a previously affected child has been identified, and it will be up to the family whether or not they wish to use this.

It also becomes possible to screen populations of healthy individuals to identify specific genetic alterations. Most pregnant women in developed societies undergo at least some type of screening for structural malformation or for chromosomal anomalies in their foetus/baby – by a combination of maternal blood tests and foetal ultrasound scanning. Whether or not these women have thought this through in advance, so that only those who would want to know of a potential malformation or chromosome anomaly go ahead with screening, is another matter. From the experience of myself and others in my field of clinical genetics, I am regretfully confident that many women still have these tests performed without really choosing to do so but simply because these tests appear to be the expected behaviour of a responsible woman, one of the normal rites of pregnancy. This does lead to some women becoming the casualties of prenatal screening, when they are faced with burdensome decisions such as whether to continue or to terminate a pregnancy in which the very-much-wanted foetus/baby has been found to have (or probably to have) a malformation or a chromosomal anomaly. These decisions can be heart-breaking for the families concerned and for the professionals too, and the shock of discovering an anomaly can lead to a hasty decision to terminate the pregnancy even when the outlook for the foetus/child (if born alive) might have been medically quite favourable. Such events lead on to subsequent parental regrets and depression.

The very existence of prenatal screening programmes is also seen as deeply offensive by some of those affected by the conditions 'for' which screening is promoted, or at least made available. Such individuals have begun to organize opposition to screening programmes, which they experience as an assault on their right to be alive. The passion underlying some of these arguments from disability-rights groups may appear to remove them from the sphere of reasoned public deliberation, but it is vital that we should engage in the process of public discussion. It is too easy to forget that many of the so-called medical terminations of pregnancy for foetal abnormality carried out in developed societies are performed because of the fear of adverse social consequences of a genetic condition. Much of the suffering of those with genetic conditions is the result of social stigmatizations and discrimination on account of physical difference or learning disability. These terminations are not necessarily carried out because of the severity of the medical consequences of the particular diagnosis but because society stigmatizes those who are physically different in any way and mocks those whose behaviour is at all unusual or who find it hard to cope with schooling and employment. The disability-rights critique of prenatal screening has a strong case when it argues that the termination of pregnancies on these supposedly medical grounds may actually reinforce and promote the very social intolerance that leads parents to fear having a child with such problems. This is not to argue that all terminations of pregnancy on the grounds of foetal abnormality are carried out on spurious grounds, but to point out that (a) there is no clear divide between 'social' and 'medical' grounds for such terminations of pregnancy, (b) that programmes of prenatal screening do in fact, and understandably, cause deep offence to some individuals affected by genetic conditions and other causes of disability, and (c) these real consequences of prenatal screening programmes should be addressed and discussed publicly instead of being sidelined.

Other types of population screening include the screening of newborn infants for a range of disorders, especially where treatment is available, and screening adults to identify healthy carriers of recessive diseases such as cystic fibrosis (CF), the haemoglobinopathies (i.e. sickle cell disease and thalassemia) and Tay–Sachs disease. Newborn screening began with programmes aimed at early treatment for phenylketonuria (PKU) and congenital hypothyroidism, both of which can be treated very effectively if treatment is started early in life, but it has now been extended in many countries to include conditions where the benefits of treatment are less clear. While I think that screening to identify some untreatable disorders can be justified – as with screening for DMD (Parsons et al. 2002) – this is becoming a tangled area and the benefits of screening for a wide range of inborn errors of metabolism is at present unclear. The fact that screening for a wide range of disorders is becoming available because of developments in technology – the introduction of tandem mass spectroscopy in medical biochemistry laboratories – seems to be driving some of the expansion in scope of screening ahead of any evidence that the screening is helpful to the children or families concerned.

Carrier screening has been possible for the haemoglobin disorders for several decades and was introduced in the USA in the 1970s in an unhelpful fashion, leading to unjustified episodes of social (e.g. employment) genetic discrimination against the (mostly Black) carriers of sickle cell disease. Lessons have been learned from that episode, but the adoption of widespread carrier screening for this and other disorders has remained very 'patchy' around the globe. One reason is that the societies that have the technology to offer the testing usually also have the means to make available effective treatments – leading to very real improvements in the prognosis for individuals affected by CF and the haemoglobin disorders over the past few decades. Carrier screening, like prenatal screening, may then be motivated by health services as a way of sparing resources – through the termination of affected pregnancies – rather than because of any public wish to embrace such a programme.

The consensus recommendation of the National Institutes of Health in USA that carrier screening for CF be made available in pregnancy has certainly delivered a strong signal in support of the pro-screening lobby (for a critical view, see Clarke 1997). Given the potential application of 'chip' technology in this area – microarrays capable of detecting many thousands of mutations for carrier detection of the more common recessive diseases – there may well soon be pressure from the biotechnology industry and other interested parties to introduce such testing at the population level for a wide range of disorders, especially in the prenatal clinic, as also occurred in Denmark in relation to CF carrier screening. We could be faced with the impact of such screening on the choice (selection?) of partners as well as on the screening of pregnancies – but we will have to see how these different orientations to screening play out over time.

Another type of population genetic screening for disease that is held up as one of the 'promises' – the laudable goals – of the new genetics, is screening for genetic susceptibility to disease. It would of course take many years to accumulate evidence of benefit from such screening, in part because the dissection of the genetic contribution to the complex diseases is indeed enormously complex, and in part because the behavioural consequences of risk information about health are also complex and may be intrinsically unpredictable. One of the many reasons for this unpredictability may be the instability of social responses to marketing, fashion and other trends. Certainly, there have in the past been unfortunate – paradoxical and unanticipated – consequences of screening for hypertension and hypercholesterolemia, when the pharmacological remedies were unpalatable and/or not very effective (Clarke 1995). As with carrier and prenatal screening, one can see pressures for the introduction of profit-generating technology aligning with other interests to promote testing against the more conservative social forces – some professionals, some faith groups, some green movements and other cultural conservatives. It will be interesting both to take part in these debates and to observe their progress.

Genetics in the (near) future of the clinic

When it comes to the delivery of health care over the next five to ten years, we can think separately about personal health and reproduction. In the area of personal health, there will be continued progress in the diagnostic applications of gene-based technologies to the uncommon genetic diseases. This will generally remain the province of specialist health care services because these disorders are so uncommon. In contrast, the assessment of disease risk in those with a family history of cardiac disease and familial cancers will need to develop in collaboration between primary health care and specialist services.

The first step in this will be the identification of those with a family history of such conditions, so that they can be offered genetic testing either through specialist services or – as for those at risk of familial hypercholesterolemia – through primary health care. This identification of the uncommon 'Mendelian subgroups' of patients through the interpretation of family history information will assume a greater importance in medical and nursing education, so that appropriate services can be offered more systematically to those with an increased risk of heart disease, strokes, dementia or the common cancers as a result of inherited mutations in genes of major effect. We will then need to hope, but not to assume, that those at high risk find it helpful to be so identified, and that they respond with the recommended lifestyle changes, that they comply with screening for disease complications and perhaps with targeted preventive strategies. It is already clear that the stratification of disease risk on the basis of family history can generate anxiety in some individuals that it is difficult to resolve if there is no very useful intervention to reduce the risk. It can also leave some dissatisfied with the health services; for example, those who regard themselves as at high risk of disease but who are assessed as being only at moderate risk and who are not given access either to additional surveillance for disease or to genetic testing, may be dissatisfied. Some patients may feel worse off than they had been before the initial risk assessment.

For those already with malignancies, whether these are the usual sporadic tumours or the much less common inherited forms, therapeutic trials of the potentially toxic anti-cancer treatments are already incorporating genetic studies of tumour material. Such a categorization of tumours is likely to become increasingly helpful in guiding therapy. Knowledge of the genetic basis of some tumours is already being used to devise new specific therapies that target tumour cells and spare the patient's healthy cells. This development of rational therapies, and the ability to select appropriate treatments, are both likely to develop further, especially in the context of cancers, leukemias and lymphomas.

In the sphere of reproduction, there may well be institutional pressures to extend the scope of carrier screening and prenatal screening, from the biotechnology industry and from enthusiastic professionals. As discussed above, it is difficult to predict just how the various social factors at work will play out, and the extent to which those who oppose prenatal screening will be heard in the debate. Outcomes are likely to differ between countries as these factors will operate in different ways.

The scope of the social consequences of modern genetics

The promise of the new genetics is that it will deliver substantial health benefits – the prevention of disease through the identification of those at risk, and then the use of risk-reducing measures and even patient-specific therapies.

The attempt to reach such laudable goals will of course have unintended consequences, and the extent to which these promises can actually be delivered is also unclear and will remain so for many years. It will take a long time for the biological limits to the realization of our hopes to become clear and to be acknowledged. In the meantime – and in the rest of this volume – we need to consider the possible broader social consequences of these developments in genetics.

The conduct of genetic research within a specific social framework – with much research being funded by private investment and by the large, multinational pharmaceutical corporations, and with commercial interests being protected by intellectual property laws and agreements – has consequences for the selection of problems to be investigated. For example, the need for research into diagnostic and therapeutic technologies for tropical diseases and HIV/AIDS has never been greater – but little has been channelled into research where the ability of patients (or their governments) to pay is clearly limited. In HIV/AIDS research, where sufficient funds for the long-term treatment of patients in developing countries cannot be relied upon, access to treatment may be provided in return for agreements to permit research trials to proceed on terms that might not be acceptable within developed countries. Some funding bodies have been notable for their commitment to research into the diseases of developing countries – but they have been in the minority.

Generating genetic information about individuals has the potential to introduce unfair discrimination in several areas of life – in access to health care, insurance, employment and perhaps education; in some societies, there may also be enhanced opportunities for social discrimination in the selection of a spouse or partner. Personal, psychological responses to unwelcome genetic information may also be profound whether the information relates to risk of disease in the individual or in their children.

The search for genetic differences between populations is undoubtedly of real academic interest, and is helping to work out the pattern of migration of different groups across the globe. This allows us to learn much about prehistory – giving information not accessible through archaeology or historical linguistics. Comparative genetic studies do have the potential to reveal helpful information about disease susceptibility, but information about population history may conflict with the traditional accounts of preliterate peoples, so the question of how to obtain valid consent for that type of research is a real one.

We have little idea as to how predictions about personality or intelligence may be used socially; we can anticipate complex and problematic consequences such as feelings of fatalism, impaired self-esteem and stigmatization in some, but we have little idea as to the strength or duration of such responses or the scope of their application in society, especially in reproduction. If differences are identified in such factors between population groups, then it is highly likely that such results will be misapplied so as to foment tensions between ethnic groups, however dubious any such extrapolation might be. These social and emotional consequences of genetic information may of course be realized whether or not the genetic information produced is of any biological validity.

If genetic factors are identified as being implicated in criminality, then again we have the potential for very real problems. Do we excuse an individual for committing a crime because they were genetically predisposed to do so? Or will we incarcerate those with such a predisposition, before they commit a crime, so that the rest of us are protected adequately? Might it be more appropriate to ignore genetic factors influencing criminality because we simply would not know how to use such information?

How can society debate such difficult issues? There are real differences of perspective among knowledgeable professionals, and even these 'experts' are usually knowledgeable about a restricted area within the whole field of genetics. So how can those with little technical knowledge contribute to a sensible shaping of policy on some of these contentious but important issues? Fortunately, the 'deficit' model of the public understanding of science ('let us, the experts, tell you, the public, what you need to know, so that you come to think as we tell you to') has at least in part been ousted by the public 'engagement with' science model, in which an interested public is taken to be capable of reaching appropriate conclusions when given access to appropriate information. I have been enormously impressed over my years in clinical genetics by the ability of so many people to understand very complex information when its relevance is clear; when confronted by genetic disease in their family most people are able to grasp the essence of what is at stake. Similarly, in the two citizens' juries on genetic topics held in Wales over the past decade, diverse groups of citizens have been able to engage most productively with highly complex issues. We do need as a society to decide how we can, collectively, address these and similarly technical issues – as in the debates about genetically modified crops in agriculture (GMOs) or nuclear safety – and we do not yet have a satisfactory resolution. The ideal of a full and direct democracy may not be realistic, but the shortcomings of a representative democracy, whose decisions can be influenced too readily by commerce, the technocrats, the media or other interest groups, are all too apparent.

The generation of genetic information about the population for institutional purposes – for effective policing, accurate insurance and for the planning and delivery of health care – can be presented as clearly necessary and entirely sensible. On the other hand, such potential applications of genetic information can be seen as grave threats to human rights that should be resisted by all upright citizens. A police genetic databank can be regarded as a threat to civil liberties and an open invitation to corrupt officers to frame suspects or worse. Incorporating genetic information into the setting of insurance premiums can be regarded as undermining the very basis of insurance, at least if it is seen as a collective, social agreement to support each other in the event of misfortune. The genetic testing of healthy individuals to identify their carrier status or susceptibility to illness for health care purposes may be regarded as unlikely to be of benefit but rather to raise inappropriate anxieties in large numbers of people who would never come to harm, and perhaps as an opportunity for 'society' to blame individuals for their own ill health or the inherited disorders in their children. The allocation of biological responsibility for disease could all too easily become the imposition of social (legal, financial ...) responsibility for coping with disease, so that genetic information could effectively present a way for those born lucky to escape from their social obligations to less fortunate fellow-citizens.

In the end, the discussions about public engagement and about social responsibility come together. If we have a collective discussion about human genetics, and how it can be applied helpfully to the benefit of society, then respect for the liberal values of justice, equity and human dignity will guide deliberation about these difficult social issues – the need for collective security and a criminal justice system on the one hand and respect for individual freedom on the other, and the balance of the need to reward innovation and the entrepreneurial spirit against the principle that we should share scientific knowledge for the benefit of all. If instead we approach these issues through a series of loud monologues – leaving individuals to make their own decisions, shouting for what (we each think) will be to our own advantage – then we will indeed be respecting autonomy but we will fail to protect the weak and vulnerable both within our societies and internationally, and the autonomy that we protect will amount to a very narrow form of self-interest which will increasingly come to resemble destructive caprice. In other words, the way we choose to conduct these public debates may well determine the type of conclusion at which we arrive. So it is truly vital that the form and terms of debate are established with great care.

Patterns of inheritance

It is appropriate here merely to outline the main ways in which chromosomes, genes and inherited traits (especially diseases), are transmitted between generations. Those familiar with this material will want to pass over the rest of this introduction; those unfamiliar with the material may wish to read something that gives a fuller exposition; this is a compressed, skeletal account that is all we can provide in a volume of this sort.

1 Genes are the coded messages that control the growth, development and function of the body and all its tissues and organs. They generally come in

matching pairs, with one of each type being transmitted from a child's father through the sperm and another of each type from the mother in her egg, to the next generation. There are around 30,000 pairs of genes, which between them convey the information required to specify the amino acid sequence and structure of the body's proteins.

2 The genes are made of the chemical DNA arranged as an interlocking double helix – with two DNA molecules oriented in opposite direction and with the sequence of the four bases A, C, G or T along one helix being matched by the sequence T, G, C or A on the other DNA molecule. The coded genetic message consists of the sequence of the four DNA bases in the protein-coding portion of each gene and is specified twice – once on each of the two, interlocked DNA molecules, although only one strand is used as a template for the copying of this information into the related chemical, RNA. The sequence of bases in the RNA is used as the basis for the synthesis of the corresponding protein molecule, with three bases in the DNA determining the choice of each amino acid in the protein. The fact that the molecule is a double helix, with both chains carrying essentially the same information, accounts for the way in which the DNA can be copied in the process of preparation for the next cell division.

3 The genes are organized along the length of large physical structures, visible under the light microscope and termed 'chromosomes'. The approximately 30,000 genes in the human are packaged onto 23 pairs of such chromosomes. Each chromosome consists of one enormously long double helix of DNA, intimately entwined with structural, enzymatic and regulatory protein molecules. The complex of DNA plus protein is known as chromatin and this is packaged into condensed (relatively inactive) and open (relatively active) regions. One area of chromatin may be active at some stages of the life of the cell or organism and inactive at other times. The differences between tissues and organs arise during development as the cumulative result of the differences in the overall patterns of gene activity in the various cell types.

4 For the autosomes, the 22 chromosomes that are present in pairs in the cells of both males and females, the genes will also be present as pairs. If a genetic alteration occurs – a mutation – that disrupts the function of that gene or the resulting protein, then a genetic disorder may occur. For some genes, particularly those that code for an enzyme, the body will often function perfectly adequately with only a single functioning copy. Someone carrying such a mutation in one of their two copies of such a gene will be perfectly healthy and is known as a carrier of the gene mutation; the mutation is said to be recessive, because it does not manifest in these circumstances. It is only when the function of both copies is disrupted by mutation that the disease occurs. With other genes, including many of those coding for structural and signaling molecules, an alteration may cause health problems

even with a mutation in just one of the two copies. The disease is then said to be transmitted in dominant fashion, where an affected person has a 50 per cent (1in 2) chance of handing it on to any child they have. In contrast, it will be the brothers and sisters rather than the children of those affected by a recessive disease who are most at risk of the same condition.

5 The other pair of chromosomes, the sex chromosomes, is passed on in a different fashion. A woman has two copies of the X chromosome, which carries many genes involved in many different body functions – not just in sex and reproduction. She passes one X chromosome to each child. A man has one X chromosome and one Y chromosome, which is much smaller than the X chromosome and has many fewer genes on it. When he passes his X chromosome to a child, the child will develop as a girl. When he passes his Y chromosome to a child, the child will have one X and one Y chromosome and will develop as a boy. The clinical disorders associated with mutations on the Y chromosome are mostly related to male sexual differentiation and male fertility. A male who has an altered gene on his single X chromosome will usually show signs of it while a female carrying a mutation in one copy of her two copies of an X chromosome gene may show signs but often will not – because she will use the mutated copy of the gene in only some of her cells, and may therefore have enough function to escape without problems. A woman carrying an X chromosome gene defect will transmit it to half of her sons, who will be affected, and to half of her daughters, who will usually be healthy carriers. A man affected by an X chromosome gene defect will transmit it to all of his daughters, who will be carriers, and to none of his sons. A sex-linked recessive condition can therefore be transmitted for generations through the female side of a family and then crop up apparently out of the blue when it is passed on once more into a male.

6 There are other patterns of inheritance but they are less common.

- Disorders of chromosome number usually arise *de novo* (as new, random events) and usually have serious developmental consequences, unless it is one of the sex chromosomes that is involved when the associated condition may be rather mild.
- More complex structural rearrangements of the chromosomes can have a range of different consequences. They can sometimes be carried through a family without causing many problems except recurrent miscarriages, while other rearrangements may have very serious consequences for children who receive an altered set of chromosomal material.
- Mitochondria, the tiny chemical engineering plants located in each cell, were derived from bacteria in the course of evolution and carry some of their own genes. A fertilized egg receives (almost) all of its mitochondria in the cytoplasm of the mother's egg cell, and (virtually) none in the

sperm from its father. Those mitochondrial diseases caused by defects in the mitochondrial genome are therefore transmitted only from a woman to her children, although both boys and girls may be affected.

7 Finally, it is worth mentioning that cancer can be regarded as a genetic disease even when it is not inherited. A cancer usually results from the loss of a series of gene-based cellular controls that restrict the processes of cell division and growth. It is the accumulation of such a series of mutations within one cell line in the body that leads those cells to continue dividing inappropriately, and then spreading locally or further around the body as a malignant cancer. The loss of each cell control mechanism is the result of a new mutation occurring in that tissue. Occasionally, we are born already with a mutation in one of these genes and it is then easier for a full set of cancer-causing mutations to accumulate in one or more sites in the body. Under those circumstances, the cancer can be said to have a head start. This is what occurs in the uncommon, familial cancer disorders.

References

Clarke A (1995) Population screening for genetic susceptibility to disease. *British Medical Journal* (BMJ) 311: 35–38.

Clarke A (1997) Population screening for genetic carrier status. In Harper P S and Clarke A, *Genetics, Society and Clinical Practice* (pp. 77–92). Oxford: Bios Scientific Publishers.

Parsons E P, Clarke A J, Hood K, Lycett E, Bradley D M (2002) Newborn screening for Duchenne muscular dystrophy: A psychosocial study. *Arch Dis Child Fetal Neonatal Ed* 86: F91–95.

Weatherall D J (2001) Phenotype–genotype relationships in monogenic disease: Lessons from the thalassemias. *Nature Reviews: Genetics* 2(4), April: 245–255.

Weatherall D J & Clegg J B (1999) Genetic disorders of hemoglobin. *Seminars in Hematology* 36 (4 suppl. 7), October: 24–37.

Part 1

The Human Genome Project: Genetic Research and Commercialization

Introduction

Angus Clarke and Flo Ticehurst

This first part contains eight articles that emphasize the essentially human nature of genetic research as a social activity. This is not in any sense meant as a challenge to the validity of the science; a proper appreciation of the science is only possible when it is seen as a social activity undertaken within a community of scientists and under constraints of funding, institutional hierarchies, career structures and the micropolitical decisions about how, when or whether to work in a spirit of collaboration or of competition with other individuals or groups. The knowledge produced by science is of course always provisional, tentative and subject to restrictions on its applicability by the potential for future findings that may qualify a current 'fact'. The very production of the knowledge itself, however, is a social process in which researchers seek to persuade competing investigators, who have been trying to understand or illuminate the same or similar phenomena and who are therefore familiar with the practical difficulties of the task, that their observations are accurate and their conclusions are sound.

Fortun (chapter 1) takes us through the events leading up to the successful sequencing of the human genome and the tensions between the public and private efforts to 'win the race'. In particular, he outlines the role of Craig Venter and Celera in this, highlighting Venter's role in the development of high-throughput technologies that provided the infrastructure needed to make sequencing more efficient. This account necessarily includes a consideration of Venter's influence in both the public and the private sectors, through his involvement with 'the hybrid operations known as The Institute for Genomic Research (TIGR) and Human Genome Sciences, Inc. (HGSI)'. The main achievement of TIGR was sequencing the genome of *Haemophilus influenzae* (a disease-causing bacterium) but the method they used was later used by Celera for the sequencing of the human genome.

In 1998, Celera announced their intention to sequence the human genome within three years, far ahead of expectations. This spurred on the public initiative in the UK, which was given a major boost by the injection of funds from the Wellcome Trust, while in the US additional funding was also made available. This allowed those involved in the public effort to announce that they would also complete a working draft in 2001. The race was on ...

Fortun uses metaphors of war to describe the ensuing relations between the Celera team and other scientists involved in the public international consortium, although the two groups were not completely distinct. The main points of contention related to data access and ownership. This culminated in the crashing of the share values of biotechnology companies as the claims and counter-claims flew back and forward between the two camps. Fortun concludes by suggesting that Celera's work, and the impetus it gave to the public effort, was central to speeding up the sequencing of the human genome.

Bereano (chapter 2) explores understandings of science and technology as social phenomena; he outlines several approaches to technology. There is the classical view of technology as progress, the 'corporate liberal' paradigm that is widespread today – the view that technologies are neutral but may be put to harmful use. And Bereano's view, the 'social relations model of technology', that

> technologies developed in class societies ... reflect the underlying relationships of political power, because the powerful sectors of the society are able to articulate the research agendas (i.e. what is interesting to them), secure the funding for these projects, and assure that the programs get up and running, have the science carried out and the technologies developed.

Bereano uses the Human Genome Project (HGP) as an example of the 'contested terrain' between the dominant 'corporate liberal' view and the social relations model. He then explores the values on which the Human Genome Project has rested – biological reductionism, the commodification of biological information, the use of computers to exercise social control and surveillance, and a toleration of exacerbated social inequities. 'Any benefits from the HGP will, because of cost and related social policies, be limited in availability mainly to the richest and most privileged strata of society.'

This account criticizes the ideological elements of corporate liberalism manifested by the HGP – the view that technology increases human options and therefore human freedom, that technology is neutral, objective and value-free, that discussion of how technology changes society needs no theory of social change, and tolerance of the ethical review of genetic science only in so far as it avoids challenging the research.

Caulfield (chapter 3) gives an overview of the ethical and social concerns about the commercialization of genetic research. He outlines the role that commercialization has played in enabling the goals of the HGP to be achieved – particularly in terms of the financial flexibility and responsiveness offered by private funds. He then goes on to present an account of challenges to commercialization, such as the view that patenting and the ownership of genetic material constitute an infringement of human dignity and the commodification of a resource that should belong to everybody, and the view that genetic knowledge constitutes a series of discoveries rather than inventions (and should therefore not be subject to patenting).

It is argued that the commercialization of research has had extensive consequences – an impact on the openness of research environments and therefore higher education, and skewing of research priorities and the undermining of public confidence in the objectivity and independence of genetic scientists. There is also a greater sense of suspicion that genetic-testing services might be introduced and marketed before being properly evaluated, and a concern that the importance of social, political and economic considerations may be downplayed and too much weight given to genetic factors underlying contemporary social problems.

Rifkin (chapter 4) develops these arguments further. He presents his strong personal view about the commercialization of genetic research, in particular the patenting of genetic material. He compares the power that genetic resources will have over the world economy with the power exerted in the recent past by control over fossil fuels, metal and minerals. He is deeply critical of the moves being made by large multinational corporations and biotech companies to 'commodify the gene pool' and is especially keen to point out the inequities evident in this process as most of the scientific and technological expertise and corporate power lies in the Northern hemisphere, and (arguably) most of the genetic resources necessary to fuel a bio-economy lie in the Southern hemisphere.

Rifkin takes up the crucial question as to whether genetic research leads to discoveries or inventions. He believes that the potential value of the HGP should not be brushed aside, but that it is most important for this new knowledge to be applied responsibly and wisely – it should not be allowed to fall into the hands of a small number of powerful multinational corporations. In his view: 'our common biological destiny should be a shared human responsibility.' He predicts that 'genetic rights' will become a central issue of what he calls the 'Biotech Age'.

Marks (chapter 5) describes the aims of the Human Genome Diversity Project (HGDP), that researches the pattern of descent of different human groups, and the criticism that has been leveled at it since its inception as an extension of the HGP. He explains how the focus on the differences between culturally different groups became the main point of controversy. He contextualizes his account by providing insights into the social and political background to the HGDP. He describes the dominant paradigm in the field of anthropology at the beginning of the twentieth century, which sought to understand the relation of blood group to racial categories, drawing attention to the associated assumption that once a substance is out of a person's body it belongs to science: 'It was a classic colonial enterprise, agents of a powerful state acting with little regard for the powerless.'

The author also describes the difficulties encountered by the HGDP in its early days, including the rumors circulating amongst some indigenous populations that there were plots (by 'white people') to steal their bodies, body parts and blood. He goes on to examine the concept of 'group consent' as a means of gaining the voluntary participation of the community from which a representative blood donor comes, and the associated problems relating to representation and identity as well as possible coercion by political entities representing the groups. Another concern has been that this scientific approach to uncovering evidence of migration and evolution might undermine people's folk ideas about their origins.

Marks ends by saying that the main impact of the HGDP has been that it has reinforced worries about biocolonialism and the potential for commercial exploitation of indigenous groups by Western powers. These fears have not been completely dispelled, and remain vigorous in relation to commercial bio-prospecting for pharmaceuticals among disappearing tropical habitats. The active involvement of contributing peoples in the current HapMap project – a contemporary descendant of the original HGDP – does mean that scientists have learned at least to attempt to carry communities with them rather than simply obtaining samples and withdrawing to their laboratories.

Árnason and Wells (chapter 6) use the deCODE Genetics database project in Iceland as a case study to illustrate some of the main concerns about the commercialization of modern genetic research into the common, complex disorders. They explain the rationale/business plan behind the project: in an isolated population gene mapping is considered to be easier and deCODE could identify genetic variation contributing to susceptibility to the common disease and market the information they obtained. Prospective customers were pharmaceutical companies, health maintenance companies and insurance companies. The relative genetic homogeneity of Icelanders, their developed health care system and their excellent genealogical records all make Iceland a favourable context for such research. The types of consent that were sought to attain these goals are considered, and the authors explain how the scheme gained acceptance within the community. They then tell how, despite having gained acceptability among the public and politicians, skeptics began to air their concerns. These related to consent, the interweaving of corporate and government interests, data protection and the 'interconnection' of genealogical data with health records. They elaborate how some of the opposing views have played out and discuss how these experiences in Iceland may have implications for similar projects elsewhere.

Reilly (chapter 7) charts the development of the doctrine of informed consent from the Nürnberg (Nuremberg) Code, through the declaration of Helsinki and the US Belmont Report, bringing us up to date – where there is the potential for 'informational harm' in the light of advances in molecular biology. In particular, he discusses the potential for harm from disclosure of genetic information to third parties, and the issues related to ownership of genetic data. He describes the process of informed consent in human genetics research as it currently works in the US context – particularly the ethical and legal duties of the research team to the research subjects. He then discusses the ownership of research samples (especially DNA), the threat of genetic discrimination and the concept of community consent.

Finally in this section, Martin (chapter 8) defines the two broad types of gene therapy, somatic and germ-line, and describes the development of gene therapy as a method of treatment for genetic conditions. He recalls how experiments in the 1970s and 1980s led to organized opposition to gene therapy and how, after the first full clinical trial, most of this organized opposition ceased.

Martin describes how industry became involved in the development of technologies for gene therapy in the 1980s – initially the main source of commercial interest was from academics who started their own companies to fund their research, only later attracting investment from mainstream pharmaceutical companies. He explains the clinical milestones reached since the explosion in clinical trials in the late 1990s, and some major setbacks. He describes the first death of a research volunteer, but this article was written before some of the recent sad deaths of treated patients. Despite significant investment, no gene therapy has yet reached the market, but Martin views the prospects for gene therapy with cautious optimism, emphasizing the need for researchers and others to make realistic claims.

1
Celera Genomics: The Race for the Human Genome Sequence

Michael A. Fortun

Introduction

Celera – from the Latin *celer* meaning swift, or simply the middle letters of the English word ac**celer**ate. Such a distillation could suffice for the entire entry for this late-twentieth-century genomics corporation, since it crudely encapsulates Celera's scientific and business logic: be faster than the nearest competitor. But since Celera's nearest competitor in the so-called 'race to finish' the sequencing of the human genome that occurred in late 1999 and early 2000 might be most easily referred to as 'the entire state-supported genomics communities of the Western industrialized nations', a story of such apparently epic proportions surely merits further elaboration.

By pushing the logics and strategies of acceleration in the world of genomics as far as possible, Celera became a kind of epic genesis machine, where titanic oppositions were seen to clash: public *versus* private science, sharing *versus* patenting genomic data, smart science *versus* dumb technology, British noblesse oblige *versus* ruthless American capitalism. At least, that is the kind of epic struggle that was transmitted through the television reports, the popular and scientific magazine articles, the newspaper accounts and, increasingly, the internet investor sites that tracked the race of Celera *versus* Everybody Else with almost disapproving interest. But one of the main lessons about biological entities that we are gleaning from the proliferation of genomic mapping and sequencing data, might fruitfully be applied to the social and historical event of Celera: things are much more complicated than such either/or logics would allow.

Intensifying the Human Genome Project

Long before he founded and became president of Celera, J. Craig Venter appreciated the main goal and underlying logic of the Human Genome Project

(HGP): ironically, not so much to fully sequence the genomes of humans and other organisms, but to create an infrastructure of faster, more efficient genomic technologies that would inject new power into the engines of the biotechnology industry. In 1987, Charles DeLisi, one of the earliest and strongest advocates of the HGP and at that time in the US Department of Energy, testified to a US Senate committee that the goal of the HGP was

> to develop technologies that would make sequencing ... a lot quicker than it currently is ... [I]f you want to sequence a hundred thousand bases [in] twenty people and compare their sequences and understand disease susceptibilities, you can't do that, it's not a clinically viable procedure. We can make that a clinically viable procedure. That's the goal, it's not to sequence the human genome, at least initially.
>
> (US Senate, 1987a, p. 12)

Johnson Johnson Corporation's Jack McConnell, who helped draft early US legislation on the HGP, testified:

> If we want the US to maintain its position as a dominant force in the pharmaceutical industry in the world, I cannot imagine our letting this opportunity pass us by ... [T]he group that first gains access to the information from mapping and sequencing the human genome will be in position to dominate the pharmaceutical and biotech industry for decades to come.
>
> (US Senate, 1987b, p. 329)

In a very real sense, then, Venter's history is one not so much of competing with the HGP, but of extending and intensifying some of the original rationales for the HGP.

Working within the US National Institutes of Health (NIH) at the time when the HGP was just officially getting under way, Venter was already developing and advocating high-throughput technological strategies. Venter leapfrogged seemingly endless discussions about the most appropriate strategy for sequencing generic stretch after generic stretch of DNA by assembling and automating a variety of technologies that used messenger ribonucleic acid (mRNA) to focus on expressed genes, churning out large quantities of complementary deoxyribonucleic acid (cDNA) fragments and only working out just enough sequence information to 'tag' them (hence the name for the overall technology as expressed sequence tags or ESTs) and render them useful (Adams et al., 1991). It was a stroke of technological and scientific brilliance on Venter's part, although more than a few scientists considered it all rather mechanical and base. But when the NIH in 1991 applied for patents on thousands of these fragments with practically no understanding of their biological function, but based primarily on their utility as basic tools for further genomic research, many biologists lined up behind HGP director James Watson in their outrage, expressed in various combinations of such terms as 'mindless', 'land grab', 'unscientific', 'monkey business' and 'Gold Rush' (Andrews, 1991; Roberts, 1991, 1992; Veggeberg, 1992; Fortun, 1999).

Human Genome Sciences and TIGR

Venter had already moved on, however. Along with almost his entire laboratory staff, he left the NIH in July 1991 to begin a new venture straddling the public–private divide, the hybrid operations known as The Institute for Genomic Research (TIGR) and Human Genome Sciences, Inc. (HGSI). Here the high-throughput sequencing technologies and faster search algorithms that Venter's group developed did double duty: producing loads of sequence data on a panoply of organisms from bacteria to fruit flies at the nonprofit TIGR, and producing loads of potential human gene candidates and drug targets exclusively for HGSI (which in turn had an exclusive arrangement with the pharmaceutical company SmithKline Beecham). If, as one commentator has evocatively put it, 'Venter had one foot in the world of pure science and one foot in a bucket of money' (Preston, 2000), it was also the case that both the pure science world and the money-bucket world benefited considerably from each of Venter's feet.

With Venter's talents and technologies as key asset, HGSI was one of the first, largest and most successful of the commercial genomics companies, and set the tone and standards for much of this new industry sector. Again, the logics of speed and international competitiveness were already in force, as the *New York Times* made clear:

> Wallace Steinberg, chairman of the board of HealthCare Investment Corp., which is financing Dr. Venter, said he suddenly realized that there was an international race to lock up the human genome. If Americans do not participate, he said they will forfeit the race and lose the rights to valuable genes to Britain, Japan and other countries that are in the race to win. He said the NIH could not afford to invest enough money in Dr. Venter's enterprise to make it truly competitive.
>
> (Kolata, 1992)

Among TIGR's achievements was the first complete sequencing of an organism, *Haemophilus influenzae*, using the whole-genome shotgun sequencing method that would later become Celera's linchpin strategy for the human genome: blast it to fragments, automatically sequence the pieces, enlist computers to look for overlaps and build up contiguities. In both cases, initial widespread disbelief in the viability of such a strategy – mostly stemming from the computational difficulties entailed – would eventually turn to grudging admiration and ultimately adoption. But as interesting and noteworthy as technical triumph is, the real issues and bones of contention lay elsewhere.

In early 1998, Applied Biosystem's Michael Hunkapillar showed a prototype of the ABI Prism 3700 to Venter; within the right context, Venter realized, a few hundred of these representatives of the next generation of automated DNA sequencers could sequence and assemble any large genome in a few years. In May 1998, the PE Corporation (now named Applera) was formed, a parent company with Applied Biosystems and Celera as squalling twins: one for building, selling and servicing sequencers to all customers, 'private' and 'public' alike, and one for sequencing, databasing, annotating and patenting genomes. Celera announced its intention to fully sequence the human genome in 3 years, far ahead of the public project's timetables; everyone else, it was suggested, could do the mouse.

Like a wounded animal, the more public consortium responded immediately and vehemently. With the prodding of John Sulston of its Sanger Centre, the Wellcome Trust in the United Kingdom immediately and dramatically raised its funding levels for human genome sequencing on the order of a £100 million. At the behest of National Human Genome Research Institute director Francis Collins and ex-director James Watson, the US government also pumped money into the sequencing side of its efforts. Reorganization occurred, and a few large laboratories (including the Sanger Centre, the Whitehead Institute, Baylor College of Medicine and Washington University) received the lion's share of the money along with responsibility for large-scale production and quality control; other laboratories that had been sequencing saw their funding evaporate. The international public consortium announced it would complete a 'working draft' sequence of the human genome by spring 2001.

End game

It was the beginning of the end, and as sequencing rates on both sides shot up, civility went downhill. Relations between Celera and its scientists and the international consortium's scientists (the majority of whom had their own involvements in other corporate endeavors) became acrimonious; terms like 'slimy', 'egomaniac', 'dumb', 'schoolyard brawl' and 'grudge match' were employed by all parties and provided embellishment to the comparatively lackluster term 'race'. The fundamental differences were over data access and ownership. At a 1996 meeting in Bermuda, the members of the international consortium had elaborated a set of principles concerning human genome sequencing: all data should be 'freely available and in the public domain to encourage research and development'; 'sequence assemblies should be released as soon as possible', preferably daily; and 'finished annotated sequence should be submitted immediately to public databases' (Editorial, 1996). It was often difficult to discern exactly how Celera proposed to handle the information it was producing, but there were clearly differences: if some Celera data were to be 'freely available' it was clear that this would not happen 'immediately', and Celera would also be selling licenses and different kinds of access to the database and software tools it was creating. By November 1999, Celera announced that it had filed 6500 'provisional patent applications', placeholders for future patents built on and around DNA sequence information.

A new round of talks began between Celera and representatives of the international consortium, attempting to come to some agreement about data access and ownership. Like many diplomatic negotiations in wartime, these started, stalled, broke off and restarted. In early March 2000, the Wellcome Trust enacted a different kind of sortie across the public/private frontlines when it publicly released a confidential letter sent from Francis Collins to Craig Venter, in which Collins laid out a set of (relatively nonnegotiable) terms and set a (relatively unreasonable) deadline for response; 'open warfare' broke out (Gillis, 2000). A week later, a joint statement by US President Bill Clinton and UK Prime Minister Tony Blair concerning the patenting of DNA sequences was widely misunderstood as a blow against gene patents more broadly; stocks across the biotechnology sector crashed, and the high-tech NASDAQ index as a whole followed suit. The genomics bubble of the previous year had burst; Celera's stock lost nearly 20% of its value overnight and continued to slide for months, along with the rest of the sector. (This, of course, did not fully wipe out the nearly US$1 billion in capital that Celera had amassed in the bull genomics market of late 1999 and early 2000.) In May 2000, some of Celera's own shareholders filed a class-action suit against the company, charging that its failure to fully disclose the details of the breakdown of talks with the government in the previous months had exposed its shareholders to unknown risk (Philipkoski, 2000).

Through it all, both public and private databases filled rapidly with sequence information on humans and other organisms, with each side claiming its database more nearly finished or more reliable or more comprehensive. What was indisputable – and for whatever it was worth – was that Celera's work and the goad it proved to the public effort sped up the flow and availability of genomic information far in advance of the original public timetables (the full human sequence was supposed to be completed only in 2005). Celera had released the entire Drosophila sequence on CD-ROM in March 2000. The on-again, off-again negotiations between Celera and the consortium began again, as the machine assemblages of both efforts gushed out information with a patient rapidity, and completion loomed along with the need for some public show of amity, if not unity. Finally, on 26 June 2000, a symbolic end came as Venter and Collins appeared on either side of President Clinton at a White House press conference; Prime Minister Blair appeared via videolink. The more traditionally official print publication would come 9 months later in February 2001, with simultaneous articles in *Nature* (International Human Genome Sequencing Consortium, 2001) and *Science* (Venter et al., 2001) still symbolically marking the differences.

The *Chicago Tribune*, referring to the PE Corporation's clever business practice of sequencing DNA privately to start a 'race' and then selling DNA sequencers to its public competitors, called Celera 'a sales story for the ages' (Gorner and Van, 2000), which is another way of saying 'a tale of genes in the sell era'.

References

Adams M D, Kelley J M, Gocayne J D, et al. (1991) Complementary DNA sequencing: expressed sequence tags and Human Genome Project. *Science* 252: 1651–1656.

Andrews E L (1991) US seeks patent on genetic codes, setting off furor. *New York Times* 21 October: A1ff.

Editorial (1996) International large-scale sequencing meeting. *Human Genome News* 7(6): April–June. Oak Ridge, TN: US Department of Energy (see also Web links).

Fortun M A (1999) Projecting speed genomics. In: Fortun M A and Mendelsohn E (eds.) *Practices of Human Genetics: International and Interdisciplinary Perspectives*, Sociology of the Sciences Yearbook, vol. 19, pp. 25–48. Boston, MA: Kluwer.

Gillis J (2000) Gene-mapping controversy escalates. *Washington Post* 7 March: E1.

Gorner P and Van J (2000) Now, Celera begins job of selling itself. *Chicago Tribune* 28 June.

International Human Genome Sequencing Consortium (2001) Initial sequencing and analysis of the human genome. *Nature* 409: 860–921.

Kolata G (1992) Biologist's speedy gene method scares peers but gains backer. *New York Times* 28 July: C1ff.

Philipkoski K (2000) Investors sue Celera. *Wired News* 19 May (see also Web links).

Preston R (2000) The genome warrior. *New Yorker* 12 June: 66–83.

Roberts L (1991) OSTP to wade into gene patent quagmire. *Science* 254: 1104–1105.

Roberts L (1992) NIH gene patents, round two. *Science* 255: 912–913.

US Senate (1987a) Workshop on Human Gene Mapping. Committee on Energy and Natural Resources, August 1987 (100th Congress, 1st session, S. Prt. 100-71).

US Senate (1987b) Department of Energy National Laboratory Cooperative Research Initiatives Act. Subcommittee on Energy Research and Development, Committee on Energy and Natural Resources, September 15 and 17, 1987 (100th Congress, 1st session, S. Hrg. 100-602).

Veggeberg S (1992) Controversy mounts over gene patenting policy. *Scientist* 27 April: 1ff.

Venter J C, Adams M D, Myers E W, et al. (2001) The sequence of the human genome. *Science* 291: 1304.

Web links

Human Genome News archive edition.International large-scale sequencing meeting http://www.ornl.gov/hgmis/publicat/hgn/v7n6/19intern.html

2
Human Genome Project as a Social Enterprise

Philip L. Bereano

Introduction

Our era is characterized by an increased reliance on science and technology; indeed, these human activities shape our civilization. They must be understood as social phenomena, and their practitioners must exercise commensurate social responsibilities.

Whether the Human Genome Project (HGP) is categorized as 'science' or 'technology' is not material to this analysis. If it is basic science in one aspect, then it also simultaneously fulfills the definition of technology: that is, 'the practical application of knowledge especially in a particular area; a manner of accomplishing a task especially using technical processes, methods, or knowledge' (Merriam-Webster's online dictionary). Although the popular notion of technology can be limited to physical artifacts, increasingly it is being understood that the concept should also include the coexisting social, political and cultural institutions and mechanisms.

Technologies, by definition, are not acts of God or nature; they are acts of human beings. They result from individual and social processes and therefore embody human intention and values. We can identify several distinct models that have been used to describe this interrelationship between technical and social phenomena.

Four views of technology

The classical view, which used to hold unchallenged dominance in the United States, equates technology with 'progress'. (Note that the analysis presented in this essay uses the United States as a reference for several reasons beyond any possible bias of the author; in reality, the HGP has been largely an activity carried out and funded by sources in the United States.) Literally, progress is a statement about 'means', and not 'goals', so this view obscures any exploration of the end values encapsulated in any particular technical venture. Consistent with the general optimism of that bygone era, however, the notion that technology was equal to progress meant that technology was good.

Beginning in the early 1960s, people like Rachel Carson and Ralph Nader began to articulate a discordant view, but one that reflected what most people were beginning to know from their own experiences and observations. That is, there were problems associated with the cornucopia of technological goodies being churned out by the miracle of science and technology. Economists might call them 'externalities' (which suggests that they really need not be central to our concerns), but these side effects – sometimes unintended, often indirect – were beginning to have unacceptable negative consequences. The currently dominant ideology accommodates these unpleasant aspects of technological phenomena but still suggests that the overall unfolding of events promotes the common good. British author David Dickson (1974) has called this the 'use/abuse model'; in other words, technologies are objective and have neutral value. It is true that there may be some negative effects, or that the technology may be put to a harmful use, but these are merely secondary considerations. Thus, even when the problems are acknowledged, technological change is labeled 'progress'; it is also seen as inevitable (thereby obscuring the social 'actors' responsible for it). This is the 'corporate liberal' paradigm within which we are immersed today.

Two views contest this current dominant ideology. One is a collection of opinions – including those on 'greening' or antitechnology (Ellul, 1964) and those on relying on 'appropriate technologies' (Schumacher, 1973) – that says that there is something wrong with the dominant technical systems and that we must back off, make do with less, and/or reconfigure these situations very differently. Although there is significant variation in this category, a point of unity is a focus on the technical devices as the start of the analysis.

The second, very different emphasis is the 'social relations model of technology' put forward by several authors, including Dickson (1974), Marcuse (1964) and myself (Bereano, 1976). In this view, technologies developed in class societies (such as the United States, Canada and the European Union) reflect the underlying relationships of political power, because the powerful sectors of the society are able to articulate the research agendas (i.e. what is interesting to them), secure the funding for these projects, and assure that the programs get up and running, have the science carried out and the technologies developed. There is no better example of the contested terrain between the dominant use/abuse model and this last model than the HGP.

The HGP as a manifestation of values

The HGP was created and nurtured under a claim that it is neutral-value science, but in reality the enterprise crucially rests on several specific value orientations:

- Biological reductionism and an ideology of genetic essentialism (Nelkin and Lindee, 1995) or genetic determinism (Hubbard and Wald, 1993), which is best expressed in the exultation of James Watson, the first director of the HGP, that 'at last we will know what it truly means to be human' (as though Shakespeare and Jane Austen did not have a clue).
- The commodification of biological information, through the unprecedented expansion of the range of patentable subject matter to include living organisms and mere discoveries of naturally occurring structures (such as genes), as well as the demise of the prototypical 'altruistic biologist' in favor of a new archetype – the academic entrepreneur (whose work is often compromised by unrevealed conflicts of interest).
- Control and surveillance, as enshrined in the growth of computer power, because the HGP itself, as well as its applications, requires extraordinary capabilities to measure, categorize and track people and their activities. (Thus, one can argue that genetic discrimination – the subject of considerable legislative activity since the mid 1990s – is not an unintended externality of the HGP but is intrinsic to genetic medicine.)
- Toleration of exacerbated social inequities, because the US $3–5 billion for the HGP, obtained by telling stories to Congress about achieving medical marvels, is spent in a society with a shameful infant mortality rate and a scandalous lack of access to modern medicine by over 40 million citizens. Any benefits from the HGP will, because of cost and related social policies, be limited in availability mainly to the richest and most privileged strata of society.

HGP and the elements of corporate liberalism

The HGP is a vast societal activity conducted within the parameters of the main ideological elements of corporate liberalism. First is the simplistic notion that technology increases human options and hence, by definition, human freedom. In this view, freedom is conceptualized as merely the proliferation of choices (a market-based view of liberty). And the historical reality that technologies also 'foreclose' options is ignored. The HGP purports to increase options, for example, for therapies, for selecting a mate and for influencing the characteristics of your child.

Second, as already noted, is the belief that technology is neutral, objective and value-free. Indeed, any externalities or other problems can be remedied by applying even more technology – the so-called 'technological fix'. So, instead of fostering a culture that respects human diversity (including what we might call disabilities), the technology of the HGP can give us 'designer babies'. Perfection is achievable (for those who can pay). And those who are different will be even more likely to be treated as 'damaged goods'. The HGP has led to a revival of eugenic proposals claimed to be technical fixes for social ills such as criminality, and for eliminating disabilities and other human variations that are currently in social disfavor.

Third is the principle that no theory of social change is necessary to be an expert at discussing how any technology will change our society. So politicians routinely explain their votes for technological systems (be they prisons, rail transportation or computers for schools) without there being any accepted theoretical construct for how the technological phenomenon will produce the claimed beneficial change. Demonstrated failures of the benefits to be produced are routinely ignored, as faith in this ideological principle overpowers facts to the contrary. The mysterious workings of the market, and the assumption that these workings are beneficial for society, comprise a frequently heard mantra. The HGP is touted as providing better healthcare for all on the basis of a child-like model that falls far short of describing US social reality, which is that therapies are private patent monopolies, even when derived from publicly funded research, that drugs are priced exorbitantly, and that millions of people lack access to modern medicine.

Whenever there is any inkling of a relationship between technology and social change, it is usually presented as being unidirectional – that is, technology causes social change. But any hard look at the real world indicates what corporate liberals really know, that is, that the relationship is dialectical because social factors also produce technological changes. Surely the change in the patent laws represented by the Chakrabarty decision by the Supreme Court of the United States allowing the patenting of a new lifeform, a genetically engineered bacterium, as a 'composition of matter' has had an effect on the evolution of genetic technologies; and religious opposition to abortion is currently stalling research based on stem cells in many countries (Chakrabarty v. Diamond, 447 U.S. 303 (1980)). The HGP has been promoted by many social stimuli, such as government funding, new university rules for patenting and the privatization of public knowledge (which allows monopoly claims over the 'knowledge commons' by private corporations).

Fourth, when Watson precipitously announced the 'Ethical, Legal, Social Implications' (ELSI) activity in response to a question from a news reporter, he apparently was thinking that this program would follow on the technical work of the HGP. And so it did for a few years; its activities had essentially no impact on the conduct of the HGP nor the rush to apply its findings. But when the ELSI Working Group began to deal with policy issues that would have influenced the shape of the genetic research itself (e.g. by questioning the ethical status of re-using existing banks of tissue for unrelated studies), the Director Francis Collins terminated it. (Collins had himself engaged in ethically suspect research, in which a Boston repository of blood samples from research on Tay–Sachs disease was screened for genes linked to cancer.)

Last, general discussion about technological and social impacts under the premises of corporate liberal ideology deals with society either as an undifferentiated whole or as a collection of atomized individuals, instead of analyzing organic social groupings. Public speakers routinely refer to 'society' as though it were a single person with one mind and one set of interests or refer to the micro-level of the individual: 'don't you want to give your child every advantage – the best schooling, the best nutrition, the best genes?' Not only does this discourse perpetuate the notion of technological isolation and alienation, but it suggests that market solutions (rather than social ones that have developed, for example, through political processes) are the only ones possible. (Obviously, in nations with rich traditions of social solidarity, such as Scandinavia, this ideological point is far weaker; but it flourishes in the United States.)

In the 1990s there was a substantial growth of civil society organizations (CSOs) in all parts of the world. As if to take the place of the traditional organic associations that are being stressed and sometimes displaced by technological phenomena, these CSOs are intentional organizations formed for specifically defined social purposes. And many of them are directly confronting aspects of the new biology – such as genetically modified foods, genetic discrimination and privacy, genetic surveillance, eugenics and biosafety – and are aided in their work by the growth in other technological realms, notably computing power and the internet. The HGP was mostly a pact made between legislators and prominent academic biologists. Owing to either paternalism or neglect, consumers, the disabled, insurance companies, religious organizations and other social groups that would be profoundly affected by this work were not included in the original discourse. Instead, a vague faith in 'the market' was relied on to suggest that benefits would be shared by all and that there would not be any significant negative consequences. But increasingly the failure of this viewpoint is becoming obvious; for example, when the University of Pennsylvania Medical School tried to give 'breast cancer tests' to women on low incomes (as part of a commendable University program to re-engage with its neighbors), the President of Myriad Genetics quashed the program because he had not licensed this use of 'his' gene.

Such contradictions are inevitably going to proliferate. CSOs will form around these disputes or move into them if they are already in existence. Nothing in the education of young geneticists is helping to prepare them to participate in this type of dialog. The era of 'doctor knows best' is over. The laissez faire of the market will not be the only resolution of these problems. The HGP is a social institution that was founded with insufficient attention being given to its essentially social nature and the changing dynamics of real societies. Will scientists now be able to move it in the new dynamics into a position to be a socially accountable body – one that can engage with other civil organizations in responsible dialog.

References

Bereano P L (1976) *Technology as a Social and Political Phenomenon*. New York: John Wiley.

Dickson D (1974) *The Politics of Alternative Technology*. Universe Books.

Ellul J (1964) *The Technological Society*. New York: Knopf.

Hubbard R and Wald E (1993) *Exploding the Gene Myth*. Boston: Beacon Press.

Marcuse H (1964) *One Dimensional Man*. Boston Press.

Nelkin D and Lindee S M (1995) *The DNA Mystique: The Gene as Cultural Icon*. New York: Freeman.

Schumacher E F (1973) *Small is Beautiful*. New York: Harper Row.

Web links

Merriam-Webster's Online Dictionary http://www.m-w.com

3
Commercialization of Human Genetic Research

Timothy Caulfield

Introduction

The commercialization of human genetic research is necessary to ensure that the public ultimately benefits from the rapid advances occurring in this area. Genetic products will inevitably be refined and delivered by industry. Indeed, it is hard to imagine how the goals of the Human Genome Project can be achieved without significant involvement of the private sector. It has been noted that industry funding for research can flow more rapidly, and is often more responsive and flexible, than public funding mechanisms. In addition, biotechnology is emerging as an important part of the economy and, as such, the commercialization of genetic research is being promoted by both government and universities.

Existing tensions between the public and private sectors were exemplified in the race to sequence the first draft of the human genome. The National Institutes of Health (NIH), a publicly funded entity, sought to sequence the human genome and to publish its sequence data online to make it freely available to researchers. Celera Genomics began competing with the NIH to complete the first draft of the human genome, its intent being to profit from the final compilation of sequence data. In this controversial move, Celera used the freely available NIH data to augment its own sequence data and, arguably, stood to profit at the public's expense.

Despite the benefits of industry involvement (Cockburn and Henderdon, 1996), the commercialization and marketing of genetic research has raised numerous ethical, legal and social issues. What follows is a brief overview of some of the most commonly articulated concerns. The goal is to provide the reader with a sense of the scope and nature of the concerns that surround commercialization. It is important to note that many of these concerns remain speculative and are not necessarily unique to genetic research; however, because of the complex social ramifications connected with the application of human genetics, the concerns related to the commercialization of genetic research have received a significant amount of attention.

Ownership of human genetic material

The commercialization of human genetics has become closely linked to protection of intellectual property. Obtaining a patent is viewed by industry, venture capitalists, researchers and even universities as a crucial step in the commercialization process. A patent is a limited term monopoly – in most cases 20 years – that prohibits others from making, using, selling or importing an invention. The patent system is designed to promote research and innovation. Advocates of gene patents suggest that they are necessary to justify the large amount of money that is required for biomedical research.

Although patent offices worldwide have issued patents on human genetic material since the beginning of the Human Genome Project (Thomas, 1999), many commentators are not comfortable with the idea of gene patents. Indeed, technology critics have gone so far as to suggest that they amount to a form of 'biopiracy' (Rifkin, 1998). In the most general sense, the concern is that allowing a patent on human genetic material amounts to an infringement of human dignity and a commodification of a resource – the human genome – that should belong to all humankind (Knoppers et al., 1999; Nelkin and Andrews, 1998). Also, because of the vast public resources that have been invested into the project, private entities should not be permitted to benefit unjustly at the expense of the citizenry. Although the exact nature, scope and validity of this broadly based issue continue to be debated (Gold, 2000), many national and international bodies have embraced the spirit of the commodification concern. For example, the Universal Declaration on the Human Genome and Human Rights by the United Nations Educational, Scientific and Cultural Organization (UNESCO) suggests that 'The human genome in its natural state shall not give rise to financial gains' (UNESCO, 1997). Similarly, a Working Group for the European Commission on the Ethical, Social and Legal Aspects of Human Genome Analysis has suggested that 'the human genome, or any part of it as such, should not be patentable' (European Community Working Group, 1991).

Despite such statements, the broad concerns about the ownership of human genetic material have had little impact on the policies of most national patent offices. So long as the patent applicant can satisfy the technical requirements of the patent application process (generally, this requires evidence that the invention is new, is not obvious to a person skilled in the art, and has potential industrial application), a patent will be issued. Since the beginning of the Human Genome Project, thousands of patents on human genetic material have been granted.

It has also been argued that overly liberal gene patenting policies have the potential to deter innovation. The worry is that upstream patents on small pieces of human deoxyribonucleic acid (DNA) with unknown functions, such as patents on expressed sequence tags (ESTs) or single-nucleotide polymorphisms (SNPs), might deter downstream patents on useful innovations such as genetic tests (Heller and Eisenberg, 1998). The concerns about the adverse impact of gene patents on useful research was a primary motivator behind the joint statement in March 2000 by Bill Clinton and Tony Blair that called for public access to human genetic information, and the decision by the United States Patent and Trademark Office to apply the patent criteria more stringently in the context of human genes (Willing, 2000).

Research environment

Commercialization and patenting are also associated with many more specific concerns. An increased emphasis on patenting and closer ties between researchers and industry sponsors has changed the nature of university-based research. For example, by encouraging scientists to be more secretive and less open about their research results, it has been suggested that the commercialization process has had an adverse impact on the research environment. Survey work by individuals such as David Blumenthal has shown that the growing ties between academic researchers and industry have created an atmosphere of secrecy among university scientists. Notably, Blumenthal's research suggests that 'investigators in the field of genetics are more likely to engage in data-withholding behaviors' (Blumenthal et al., 1997). Not only does such behavior have an adverse impact on research collaborations, it may also hurt the teaching environment by making it more difficult for students to publish their work.

The commercialization of the research setting is also associated with the skewing of the university research agenda toward initiatives that are likely to have commercial application. Although university researchers have always been influenced by outside pressures, there is concern that the current unprecedented ties between universities and industry will result in a loss of funding and academic enthusiasm for basic research (Varma, 1999).

A related issue is that the commercialization of the university environment may make it increasingly difficult to find a genetic scientist without ties to industry. Academics who do not have a financial interest in the progress of private industry are among the only individuals with the necessary expertise and credibility to assess and to analyze the claims of commercial enterprises. As the lines between university-based and industry research become progressively blurred, public confidence in the objectivity and independence of the academic genetic scientists could be damaged irreparably (Angell, 2000).

Finally, the commercialization of the research environment has created new opportunities for conflicts of interest. The concern is that if the researcher has a financial stake in the outcome of the research, then there may be an actual or a perceived bias. This is a particularly problematic issue when human research subjects are involved (the interests of the patient/research participant should be a paramount consideration). As highlighted by the controversy surrounding the death of a patient involved in a gene therapy experiment at the University of Pennsylvania (Weiss and Nelson, 1999), there is concern that financial interests may contribute to the premature, and possibly dangerous, running of clinical trials. In this case the principal investigator had significant ties to the company that was developing the gene therapy.

Commercialization and the marketing of genetic technologies

More and more genetic technologies are moving from the research setting into a clinical one. There are, of course, many commercialization issues that are relevant to the practical use of these technologies. Commentators have proposed that commercialization pressure may lead to the implementation of genetic services before their efficacy and social ramifications have been evaluated appropriately (e.g. see Holtzman, 1999; Welch and Burke, 1998; Kodish, 1997; Andrews, 1997). It has been argued, for example, that financial interests and professional enthusiasm have led to the premature commercialization of some genetic tests (Andrews, 1997). In this regard, Edward McCabe, Chair of the United States Secretary's Advisory Committee on Genetic Testing, was quoted as stating that 'Many [genetic] tests are in the quasi-research category but everybody wants to charge for them' (Weiss, 1999).

Many commercialization concerns have been connected with the strategies that are used to sell genetic technologies. First, it has been noted that genetic testing companies may, for the purposes of marketing, minimize the clinical uncertainties associated with genetic testing and overemphasize the potential healthcare benefits (Andrews, 1997; Welch and Burke, 1998; Holtzman, 1999). Other than for the monogenic disorders such as Huntington disease, genetic tests usually involve the provision of complex and often ambiguous information about probability, predispositions and health risks. Although there is no doubt that genetic tests can provide important healthcare information for certain 'at-risk' populations, market pressures may lead companies to oversell the clinical value of genetic tests. As noted by Kodish in the context of genetic testing for predispositions to cancer: 'In contrast to the many published statements by professional bodies which call for restraint, industry positions assume that genetic technology will provide clinical and potential public health benefits in cancer control' (Kodish, 1997, p. 253).

A second and related concern is that market pressures will lead commercial testing companies to sell their services to an inappropriately broad sector of the population (i.e. the broader the definition of the at-risk population, the bigger the market) and that industry-generated interest in testing may cause unnecessary anxiety in individuals who may not, but for the marketing, be interested in testing (Holtzman, 1999; Koenig et al., 1998; Nelkin and Lindee, 1995). This concern is illustrated by the debate around the utility of the test for breast cancer susceptibility. Several nonindustry policy statements suggest that testing for mutations in the breast cancer 1, early onset (BRCA1) and BRCA2 genes is not appropriate for most individuals, but commercial companies will generally test anyone willing to pay, whether or not there exists a family history suggestive of a genetic predisposition to breast or ovarian cancer.

These are some of the reasons why many policy groups have concluded that advertising and marketing of genetic tests should be monitored carefully or even prohibited (SACGT, 2000; Koenig et al., 1998; Holtzman, 1998). Likewise, it has been suggested that greater regulatory oversight of the genetic testing industry may be required (SACGT, 2000; Holtzman, 1999).

Geneticization and commercial pressure

One of the overarching and controversial concerns associated with human genetics is that it will lead to a further 'geneticization' of our society. As summarized by Abby Lippman, the scholar who introduced the term, 'Geneticization is a term coined to capture the ever growing tendency to distinguish people one from another on the basis of genetics; to define most disorders, behaviors, and physiological variations as wholly or in part genetic in origin' (Lippman, 1998). The speculation is that an overemphasis on genetic 'essentialism' may inappropriately de-emphasize social, political and economic considerations and drastically alter the way in which we view ourselves and others.

The commercialization of genetic research is closely connected to the geneticization issue. A significant amount of money has been invested in the development of genetic technologies and products. A successful biotechnology industry requires an enthusiasm for genetic technologies and a receptive market. Given this fact, the issue is whether commercial pressure, through marketing to both the public and professionals, will facilitate the geneticization of our society by overemphasizing inappropriately the significance of genetics (Testart, 1995). Although there is as yet little empirical evidence to support the concerns of a market-driven geneticization of society, it is hard to deny that market forces can have a profound influence on social perceptions.

References

Andrews L B (1997) Past as prologue: sobering thoughts on genetic enthusiasm. *Seton Hall Law Review* 27: 893–918.

Angell M (2000) Is academic medicine for sale? *New England Journal of Medicine* 342: 1516.

Blumenthal D, Campbell E G, Andersen M S, et al. (1997) Withholding research results in academic life science: evidence from a national survey of faculty. *Journal of the American Medical Association* 277: 1224–1227.

Cockburn I and Henderdon R (1996) Public–private interest in pharmaceutical research. *Proceedings of the National Academy of Sciences of the United States of America* 93: 12725–12730.

European Community Working Group for the ELSI Aspects of Human Genome Analysis 1991. Report of 31 December 1991.

Gold R (2000) Finding common cause in the patent debate. *Nature Biotechnology* 18: 1217–1218.

Heller M and Eisenberg R (1998) Can patents deter innovation? The anticommons in biomedical research. *Science* 280: 698–701.

Holtzman N A (1998) Bringing genetic tests into the clinic. *Hospital Practice* 33: 107–112.

Holtzman N A (1999) Promoting safe and effective genetic tests in the United States: Work of the task force on genetic testing. *Clinical Chemistry* 45: 732–738.

Knoppers B, Hirtle M and Glass K (1999) Commercialization of genetic research and public policy. *Science* 286: 227–228.

Kodish E (1997) Commentary: risks and benefits, testing and screening, cancer, genes and dollars. *Journal of Law, Medicine and Ethics* 25: 252–255.

Koenig B A, Greely H T, McConnell L M, et al. (1998) Breast cancer working group of the Stanford program in genomics, ethics and society: genetic testing for BRCA1 and BRCA2: recommendations of the Stanford program in genomics, ethics and society. *Journal of Women's Health* 7: 531–545.

Lippman A (1998) The politics of health: geneticization versus health promotion. In: Sherwin S et al. (eds.) *The Politics of Women's Health*. Philadelphia, PA: Temple University Press.

Nelkin D and Lindee M S (1995) *The DNA Mystique: The Gene as a Cultural Icon*. New York: W H Freeman.

Nelkin D and Andrews L (1998) Homo Economicus: commercialization of body tissue in the age of biotechnology. *Hastings Center Report* 28: 30–39.

Rifkin J (1998) *The Biotech Century*. New York: Penguin Putman.

Secretary's Advisory Committee on Genetic Testing (SACGT) (2000) Enhancing the Oversight of Genetic Tests: Recommendations of the SACGT. National Institutes of Health, June 2000: http://www4.od.nih.gov/oba/sacgt.htm

Testart J (1995) The new eugenics and medicalized reproduction. *Cambridge Quarterly of Healthcare Ethics* 4: 304–312.

Thomas S (1999) Intellectual property rights and the human genome. In: Caulfield T and Williams-Jones B (eds.) *The Commercialization of Genetic Research: Ethical, Legal and Policy Issues*, pp. 55–62. New York: Kluwer Academic Plenum Publishing.

UNESCO, Universal Declaration on the Human Genome and Human Rights, 11 November 1997. Available online at: www.unesco.org/human_rights/hrbc.htm

Varma R (1999) Professional autonomy vs industrial control? *Science as Culture* 8: 23–45.

Welch H G and Burke W (1998) Uncertainties in genetic testing for chronic disease. *JAMA* 280: 1525–1527.

Weiss R (1999) Genetic testing's human toll. *Washington Post* 21 July: A1.
Weiss R and Nelson D (1999) Teen dies undergoing experimental gene therapy. *Washington Post* 29 September: http://www.washingtonpost.com/wp-srv/WPlate/1999-09/29
Willing R (2000) Gene patents get tougher. *USA Today* 15 November: 14A.

4

Patenting of Genes: A Personal View

Jeremy Rifkin

The recent announcement of the mapping of the human genome has focused public attention on a revolutionary change taking place in the global economy.

We are in the midst of a historic transition from the Industrial Age to the 'Biotech Age'. Whereas the twentieth century was shaped largely by spectacular breakthroughs in physics and chemistry, the twenty-first century will belong to the biological sciences. Scientists around the world are deciphering the genetic code of life. After thousands of years of fusing, melting, soldering, forging and burning inanimate matter to create useful things, we are now splicing, recombining, inserting and stitching living material into commercial goods.

Genes are the raw resource of the new economic epoch. Molecular biologists around the world are mapping the genomes of many of the earth's creatures, from the lowliest bacteria to human beings, thereby creating a vast genetic library for commercial exploitation. Gene technology is already being used in many fields of business, including agriculture, animal husbandry, energy, construction materials, pharmaceuticals, medicine and food and drink, to fashion a bioindustrial world.

At the heart of any public discussion of the new genetic commerce is the issue of patenting the genetic blueprints of millions of years of evolution. The economic and political forces that control the genetic resources of the planet will exercise tremendous power over the future world economy, just as access to and control over fossil fuels and valuable metals and minerals helped to determine control over world markets in the industrial age.

In the years ahead, the planet's shrinking gene pool is going to become a source of increasing monetary value. Multinational corporations are already scouting the continents to locate microbes, plants, animals and humans with rare genetic traits that might have market potential. After locating desired traits, biotechnology companies are modifying them and seeking patent protection for their new 'inventions'.

Corporate efforts to commodify the gene pool are meeting with strong resistance from a growing number of nongovernmental organizations and countries in the southern hemisphere, who are beginning to demand an equitable sharing of the fruits of the biotechnology revolution. Although the technological expertise needed to manipulate the 'green gold' resides in scientific laboratories and corporate boardrooms in the northern hemisphere, most of the genetic resources needed to fuel the revolution are located in the ecosystems of the southern hemisphere.

On the one hand, Southern countries claim that what Northern companies call 'invention' is really the pirating of local genetic resources and the accumulated indigenous knowledge of how to use them. On the other hand, the life science companies argue that patent protection is essential if they are to risk financial resources and years of research and development to bring new and useful products to market.

Extending patents to life raises the important legal issue of whether engineered genes, cells, tissues, organs and whole organisms are truly human inventions or merely discoveries of nature that have been skillfully modified by human beings. To qualify as a patented invention in most countries, the inventor must prove that the object is novel, nonobvious and useful – in other words, that no one has ever made the object before, that the object is not something that is so obvious that someone might have thought of it using existing tools, and that the object serves some useful purpose.

But, even if something is novel, nonobvious and useful, if it is a discovery of nature, it is not an invention and is, therefore, not patentable. The chemical elements in the periodic table – although unique, non-obvious when first isolated and purified, and very useful – were not considered patentable because they were discoveries of nature, despite the fact that some degree of human ingenuity went into isolating and classifying them.

The United States Patent Office has said, however, that the isolation and classification of a gene's properties and purposes is sufficient to claim it as an invention. The prevailing logic becomes even more strained when consideration turns to patenting a cell, a genetically modified organ or a whole animal. Is a pancreas or kidney patentable simply because it has been subjected to a slight genetic modification? What about a chimpanzee? Here is an animal who shares 99% of the genetic make-up of a human being. Should he or she qualify as a human invention if researchers insert a single gene into their biological make-up? The answer, according to the patent office, is yes.

The patent issue is likely to become a question of increasing public concern as a result of the stunning breakthroughs in the government-funded Human Genome Project. It is expected that in less than 8 years, nearly all of the genes that make up the genetic blueprints of the human race will have been identified and become the intellectual property of international life science companies.

Such firms are also patenting human chromosomes, cell lines, tissues and organs. PPL Therapeutics – the life science company that cloned the sheep named Dolly – has received a patent that includes cloned human embryos as intellectual property.

The increasing consolidation of corporate control over the genetic blueprints of life, as well as the technologies to exploit them, is alarming because the biotechnology revolution will affect every aspect of our lives. The way that we eat, the way that we date, the way that we have babies, the way that we raise and educate children, the way that we work, and even the way that we perceive the world around us and our place in it – all of our individual and shared realities – will be deeply touched by the biotechnology revolution.

What might it mean for subsequent generations to grow up thinking of all life as mere invention, where the boundaries between the sacred and the profane have all but disappeared and life itself is reduced to an objectified status, devoid of any unique quality that might differentiate it from the strictly mechanical?

Life patents strike at our core beliefs about the very nature of life and whether it is to be conceived of as having intrinsic or mere utility value. The last great debate of this kind occurred in the nineteenth century over the issue of human slavery, with abolitionists arguing that every human being has 'God-given rights' and cannot be made the commercial property of another.

Like antislavery abolitionists, a new generation of genetic activists is beginning to challenge the concept of patenting human life, arguing that human genes, chromosomes, cell lines, tissues, organs and embryos should not be reduced to commercial intellectual property controlled by global conglomerates and traded as mere utilities.

No one doubts for a moment the great potential value in mapping the human genome and the genomes of our fellow creatures. But if we are to use this knowledge wisely, we need to begin by ensuring that it will be held as a collective trust, and will not be made the private preserve of a handful of life science companies.

We should consider crafting a great global treaty to make the human gene pool, and the gene pool of our fellow creatures, a 'commons' that is administered jointly by every nation on behalf of all future generations – similar to the treaty that established Antarctica as a commons.

The announcement of the human genome map ought to be regarded as a triumph for the whole of the human race. Similarly, the knowledge that will come from locating all of the genes that make up our common biological destiny should be a shared human responsibility.

The battle to keep the earth's gene pool an open commons that is free of commercial exploitation will be one of the crucial struggles of the Biotech Age. 'Genetic rights', in turn, is likely to emerge as the seminal issue of the coming era, defining much of the political agenda in the years ahead.

5
Human Genome Diversity Project (HGDP): Impact on Indigenous Communities

Jonathan Marks

Introduction

The Human Genome Diversity Project (HGDP) was proposed as an augmentation to the Human Genome Project (HGP; Cavalli-Sforza et al., 1991). Recognizing that the HGP was based on a Platonic design in which the human species was represented by a single ideal specimen (Walsh and Marks, 1986), population geneticists proposed the collection of genetic material from diverse populations of the world. In the years since that initial proposal, the project has come under intense criticism by members and advocates of indigenous peoples.

Why the focus on indigenous communities at all

Many of the problems faced by the HGDP were brought about by its insistence on targeting indigenous, exotic peoples rather than urban populations. If the goal was to study the human gene pool at the beginning of the twenty-first century, it was flawed, because the indigenous peoples only represent a small fraction of that gene pool. The project could be better served, notes the National Research Council's report from 1997, by collecting samples from internally diverse urban populations.

The HGDP was structured around a single research topic: the microphylogeny of the human species – the pattern of descent of different human groups – which had been a principal research issue of the HGDP's leader, Stanford geneticist Luca Cavalli-Sforza. Using statistical analyses of the frequencies of many alleles across many populations, Cavalli-Sforza represented the similarity of human gene pools in a tree-like structure or dendrogram, which resembled the ancestry of species. But these dendrograms grossly oversimplified the historical processes affecting human populations. As a result, these depictions were unstable and were sensitive to the statistics used, the genes analyzed, the particular populations chosen and the demographic history of the groups.

If the HGDP had been principally interested in the structure of the contemporary human gene pool, it might have begun by sampling according to an arbitrary criterion such as geography, as one of its principal organizers, Allan Wilson, suggested at the outset. But its guiding question was the genetic relations among culturally designated groups, and so it adopted the cultural groups themselves as its organizing principle for sampling, which became the root of the controversy.

Historical and political context

Blood has been retrieved in the field by anthropologists since the development of serological technology early in the twentieth century. Carleton Coon was the first to retrieve blood (in 1922) from 'his people', the Rif, in Morocco, to see whether their physical features and their blood-group features would match when allocating them racially.

Blood remained a staple of anthropological collection even as the questions changed and race waned as the dominant anthropological paradigm. Even so, it had special problems associated with it, as Coon himself noted in the 1950s:

> Blood-letting for blood-group analysis falls into the class of blood-letting in general, and evokes the whole ideology of blood-brotherhood, the fear of injury by contagious magic, and that of ritual contamination based on the analogy of menstruation.
>
> (Coon, 1954)

Blood is never, as anthropologists have been known to say, 'just' blood. Nevertheless, its collection had proceeded for decades on a small scale, sometimes as part of a specifically anthropological project or sometimes for medical testing, and it has been retained and 'piggybacked' by researchers interested in other questions. The overarching assumption here has been that once the substance is out of the person's vein, it belongs to the researcher.

This followed tradition in classical anthropology: the great collections of Native American skeletal materials, for example, were 'acquired' in the nineteenth and twentieth centuries through practices that even included grave-robbing. 'It is most unpleasant work to steal bones from a grave,' wrote the great anthropologist Franz Boas to his sister very early in his career, 'but . . . someone has to do it'. Ultimately, the bones came under the control of museum scientists and were used to advance the careers of the many scholars who acquired them or analyzed them, often without regard for the sensibilities of the people whose relatives the bones actually comprised. It was a classic colonial enterprise, agents of a powerful state acting with little regard for the powerless.

The situation changed dramatically with the passage of the Native American Graves Protection and Repatriation Act of 1990 (NAGPRA). This legislation was designed to acknowledge that North American Indian remains were sacred objects, no less sacred than the scientists' relatives' bones and belonged not to science but to the tribes from whom they had been 'acquired'. This was a significant affirmation of the rights of indigenous Americans and came at precisely the same time that the HGDP was being formulated and was naively planning to collect bioanthropological objects of sacred value on a large scale.

A second significant political context involved widespread rumors within indigenous communities, that 'white people' were plotting to steal the body parts, bodies or simply blood of indigenous people. In some cases they were right, as an international trade in body organs later developed. Perhaps it is not surprising that a prominent paleontologist was abducted from an archaeological site in East Africa at gunpoint by local people, who believed mistakenly that he wanted to steal the blood from their babies. Yet the charge was disarmingly close to what the HGDP was naively proposing!

A third political context involved the development of 'biocolonialism' by agribusiness. Availing themselves of 'knowledge' freely given by indigenous people, large agricultural corporations were making considerable profits in which the people whose knowledge they needed were not sharing. In addition, patent law concerning biotechnology strongly favors scientists, as the unsuccessful cases of John Moore (Greely, 1998) and the descendants of Henrietta Lacks (Jackson, 2001) showed – neither was permitted to share in the profits made from cell lines derived from their bodies. In the case of the blood of indigenous people, the National Institutes of Health applied for patents for cell lines derived ultimately from the blood of a Hagahai (Papua New Guinea), a Solomon Islander and a Guaymí (Panama). This seems to recreate the scenario of the rapacious North American capitalist, looting not merely the land or artifacts or knowledge of tribal people, but now their very blood.

Group consent

Since the HGDP was conceptualized around human groups, each individual sample is only interesting to the extent that it is a representative of that group. Consequently, the idea of 'group consent' was devised as a means of securing not only the voluntary participation of the particular blood donor, but also the voluntary participation of the polity represented by the donor.

While introducing group consent was an admirable step in principle, it raised a significant number of ancillary issues, centering on the idea of representation: which blood samples represent which peoples? After all, human groups are fluid and organized hierarchically. Is group consent relevant to someone who is a 'Chiricahua Apache', an 'Apache', an 'Athapaskan' and an 'American

Indian' simultaneously? If the Northern Paiutes decide not to participate in a study and the Southern Paiutes decide to participate as Paiute representatives, are the rights of the Northern Paiutes thereby violated? If the Hopi decline to participate in a genetic study, can geneticists be prevented from soliciting samples from acculturated Hopis living outside their reservation?

Further, the solicitation of permission from a political entity representing the people raises the problem of possible coercion. If the leaders agree to participate, does an individual still have free rein to refuse, or can the leaders now simply act as agents for the scientists and subtly compel compliance? And more specifically, does everyone in the decision-making complex understand fully what the scientists want, why they want it and what they plan to do? For people who do not share scientific comprehensions of blood, cells, deoxyribonucleic acid (DNA), identity, life, illness and medical genetics, the elicitation of full informed consent would seem to necessitate the development of a crash course in local biological idioms.

Ultimately, the issue of group consent would also serve to reify these groups genetically as units of nature, when they are in fact units of social, political and historical manufacture. Perhaps the best statement about the problem of relating human group-assignment to the reality of identity is given by the late Frank Dukepoo (1998, p. 242), a Native American geneticist:

> I call myself a 'full-blood' American Indian of Hopi and Laguna heritage. While constructing my own pedigree, I found this is far from the truth: my father (a 'Hopi') is a mixture of Hopi, Ute, Paiute, Tewa and Navajo; my mother, on the other hand, (a 'Laguna') is a mixture of Laguna, Acoma, Isleta, Zuni and Spanish. Members of other tribes share similar admixture histories as our ancestors raided, traded or kidnapped to ensure survival of their numbers. As it is reasonably safe to surmise the same situation for members of other ethnic groups, what would 'diversity' research reveal?

Pragmatic concerns

In an early attempt to muster interest for the HGDP, its advocates used arguments from 'salvage anthropology' – the impending loss of these peoples, an argument familiar to anthropologists since the middle of the nineteenth century. However, it sounds very cynical to ask for blood from people who are on the brink of extinction. A native of the Solomon Islands wrote:

> The project has very little interest in helping these people to survive, or in addressing the social, the economical, the political, and the exploitation issues that endanger these indigenous groups of people.
>
> (Liloqula, 1996)

Moreover, some groups were simply experiencing the normal historic forces of merging, splitting, reconstituting and forging new identities; they were 'endangered' only as bounded genetic entities.

Participation in the benefits of modern healthcare often requires allowing blood to be drawn. Some insecurity naturally arises about the fate of that blood – a highly symbolically charged substance – once its diagnostic purpose has been served. Many tribal people would feel very uneasy to learn that their blood, or a product derived from it, was sitting in a laboratory in California, and being manipulated exclusively for the benefit of American scientists.

On one occasion, when Cavalli-Sforza was taking blood from schoolchildren in a rural region of the Central African Republic, he was confronted by an angry farmer brandishing an ax. Recalls the scientist, 'I remember him saying, "If you take the blood of the children, I'll take yours". He was worried that we might want to do some magic with the blood.' (Subramanian, 1995, p. 54).

Rather than musing over the ignorance of the ax-wielding farmer, a contemporary reader should instead recognize unfulfilled obligations of disclosure to the participants here. A strong fear of the magic in blood makes it very unlikely that these people could have given their fully informed consent to this research.

The existence of such ideologies about the power of blood could in principle be circumvented by stipulating that the blood, once drawn, can be used only for the medical purpose specified and must then be destroyed. That is commonly the case now with some North American Indian groups, but there is unfortunately no way to enforce it. Blood from indigenous peoples has been a valuable scientific commodity, traded between laboratories and researchers, for different projects, establishing a network of relationships, obligations and coauthorships (Anderson, 2000). The prospect of such a scientific tradition abruptly ceasing is quite unlikely, regardless of the agreements made with tribes, who may also have cynical views on Euroamerican people living up to their agreements at all.

The question of genetic exploitation is of paramount importance. If there is economic value in the blood of indigenous people (as the interest of biotechnology companies might suggest), then what is a fair price? The HGDP's insistence that there are no financial considerations was forcefully undermined by the patent applications, which did not involve the HGDP itself but were obviously relevant. Consequently, any adequate concept of disclosure and voluntary informed consent would necessitate the scientist explaining to participants that there are financial stakes, through which the researcher could get wealthy without precedent for the subject sharing in that wealth.

Since the HGDP's initial interest was to formulate and answer questions of microevolution, another issue is raised, calling attention to science's role in authoritatively contradicting people's ideas of their folk history and identity. One could legitimately ask why anyone would wish to participate in a project designed to undermine their own ideas of who they are and where they came from.

In any case, medical value, which came to be emphasized a few years after the project's inception, would be difficult to establish, as there have not been any plans for collecting detailed medical, phenotype and life-history data to associate with the genotypes. A study of the genetic etiology of diabetes, for example, would require a knowledge of which DNA samples actually came from people who were diabetic. However, those samples could then not be used for a study of the genetics of schizophrenia, because there would be no information about which samples came from schizophrenics. Thus the HGDP samples could only be of exceedingly limited medical use.

The Iceland genome project, on the other hand, in which the national government cooperated with a private biotechnology company, combining full medical records with blood samples in the hopes of sharing in the wealth, calls attention to the economic incentive: a desire to commodify the body and the bioinformation associated with it.

A final salutary goal promoted by the HGDP involved delegitimizing group hatreds by demonstrating the nonexistence of race. Unfortunately, this is unlikely; group hatreds are rarely rooted in biological difference or biological knowledge, and race has long been known to exert a profound impact upon modern lives as a very real set of social categories and statuses.

However, some geneticists have gone so far as to promise a genetic test to determine tribal membership and Native American identity, a false and patently impossible goal, given that tribal membership is a political status, and the history of mixture (Shelton and Marks, 2001). Ultimately, the impact of the HGDP has been to reinforce many communities' worst fears about the avarice of wealthy nations and the residual colonial attitudes of science, appearing more as an instrument for their exploitation than as a fulfillment of the Baconian promise of a better life for all.

References

Anderson W (2000) The possession of kuru: medical science and biocolonial exchange. *Comparative Studies in Society and History* 42: 713–744.

Cavalli-Sforza L L, Wilson A C, Cantor C R, Cook-Deegan R M and King M-C (1991) Call for a worldwide survey of human genetic diversity: a vanishing opportunity for the Human Genome Project. *Genomics* 11: 490–491.

Coon C S (1954) *The Story of Man*. New York, NY: Alfred A Knopf.

Dukepoo F (1998) The trouble with the Human Genome Diversity Project. *Molecular Medicine Today* 4: 242–243.

Greely H T (1998) Legal, ethical, and social issues in human genome research. *Annual Review of Anthropology* 27: 473–502.

Jackson F (2001) The Human Genome Project and the African-American community: race, diversity, and American science. In: Zilinskas R A and Balint P J (eds.) *The Human Genome Project and Minority Communities: Ethical, Social, and Political Dilemmas*, pp. 35–52. Westport, CT: Praeger Publishers.

Liloqula R (1996) Value of life. *Cultural Survival Quarterly* 20: 42–45.
Shelton B L and Marks J (2001) Genetic markers not a valid test of native identity. *Genewatch: A Bulletin of the Committee for Responsible Genetics* 14(5): 6–8.
Subramanian S (1995) The story in our genes. *Time Magazine* 16 January: 54–55.
Walsh J B and Marks J (1986) Sequencing the human genome. *Nature* 322: 590.
Cunningham H (1997) Colonial encounters in post-colonial contexts. *Critique of Anthropology* 18: 205–233.

6
deCODE and Iceland: A Critique

Einar Árnason and Frank Wells

Introduction

The case of deCODE Genetics and its three-pronged database plan for genomics research and commercialization in Iceland has brought into focus the business practices of the genomics industry and a number of ethical issues. These have implications outside Iceland as well.

The company states that deCODE is founded on the assumption that the scarce resource in human genetics as an industrial endeavor is a population that can yield the genetics of common diseases. The company further states that it is taking a revolutionary approach to genetic research by supplementing leading-edge technology with instant access to an allegedly ideal population. An isolated population having expanded in numbers after founder effects and/or inbreeding is considered ideal because the frequencies of deleterious disease genes might have increased, making gene mapping easier. The same effects can also serve to reduce the complexity and heterogeneity that complicate mapping efforts. Access to this population permits deCODE to positionally clone and characterize genes that contribute to the pathogenesis of common disease, and market the information so obtained via the construction and operation of a large database that combines the genetics with genealogies and phenotypic health information.

A business plan

deCODE Genetics Inc. is a for-profit American corporation registered in Delaware and operating under American law. American venture capitalists, Icelandic investors and Icelandic biomedical scientists founded the corporation in August 1996. It is built around the idea of cloning and characterizing the genes of Icelanders and marketing the information so obtained. The prospective customers are pharmaceutical companies, health maintenance organizations

(HMOs) and insurance companies among others. The business plan is based on three important premises. First, a conjecture about the genetic homogeneity of Icelanders is believed to make genes contributing to major disease stand out prominently. Second, an extensive genealogy for Icelanders, presumed to be known or to be knowable, which can be used to link together affected individuals, including even individuals with distant common ancestors. Third, it is said that Iceland possesses a modern health system that can readily provide an accurate ascertainment of disease from its health records (Gulcher and Stefansson, 1998). Add to this modern information technology and the idea was born of combining these various data sets of genotypes, genealogies, health records and utilization of medical resources and services into an inclusive database of the entire Icelandic nation. Such a database could be used for nonhypothesis-driven data mining by 'systematically juxtaposing various data in the search for best fit' (Gulcher and Stefansson, 1999a).

According to the ideology behind this GGPR (genealogy, genotype, phenotype, resource use) database, a revolution in medicine is being promised if only phenotypic data can be obtained and used via 'presumed consent' and if genotypic data can be obtained and used under an 'open' or 'broad consent' (Gulcher and Stefansson, 1998; 1999a). If 'informed consent' had been used, participation was expected to be much less and, according to the proponents, 'without broad consent [for genetic data], the database would be only an extraordinarily effective tool for classic gene mapping, rather than a revolutionary method for studying the interplay between genetics and environment in human disease and health' (Gulcher and Stefansson, 1999a).

Phlebotomy of a nation

In 1995, Dr Kári Stefánsson, deCODE's chief executive officer (CEO), seeking funding for his biotechnology venture, submitted a business/research plan detailing the gene hunting part of the plan to the Icelandic Research Council. It met with an unfavorable review and was not supported (Zoëga and Andersen, 2000). At this time, Stefánsson also expressed the idea of negotiating with the Icelandic government to win exclusive rights to the patient population for the business venture. At the time, Stefánsson was in discussion with Kevin Kinsella, the then CEO of Sequana, a genomics company operating in La Jolla, California, who wrote:

> As we discussed, Iceland is perhaps the ideal genetic laboratory since there has been virtually no immigration (lots of emigration, of course); it is of manageable size (200,000 inhabitants), is an island expected to have many founder effects, has high quality national healthcare – from which

> we can expect excellent disease diagnosis, has formidable genealogies and the population is Caucasian – of most interest to pharmaceutical companies.
>
> What we would propose to do, in partnership with you, is sponsor a massive program to identify probands in the entire Icelandic population, collect blood from the entire Icelandic population, establish pedigrees for important complex genetic diseases, genotype members of the pedigrees, conduct linkage studies on these diseases and positionally clone the gene(s) of interest.
>
> We would require the full support of the Icelandic government and National Health Service. If all the ascertainment, enrollment, and phlebotomy can be furnished by the Icelandic authorities, then the costs of doing this can be manageable.

Although Kinsella and Sequana are out of the picture now, the pillars of deCODE's business plan of massive genotype/phenotype correlation and exclusive commercial rights were taking shape in 1995 (Zoëga and Andersen, 2000).

Implementation of the plan, mingling of interests

How did this bloodletting plan come to be realized? A constructive aspect of nationalism, pride, selfreliance and the stubbornness of independent people (Laxness, 1946) has been an important element in the building of modern Iceland, as for so many modern nation states. In a small nation, there is a never-ending battle for independence and fear of engulfment. In Iceland, the 'Cod wars' with the British and opposition against the American military base are examples.

Telling the Icelanders that they and their contribution are unique and can help make a revolution in medicine that may cure much of mankind's ailments played persuasively on nationalistic pride. Add to that promises of employment of skilled workers, reversal of the brain drain, net influx of money to the Icelandic economy, a catalyst for a biotechnology industry, support for the research community as well as for the healthcare industry, and the promise of free drugs. Furthermore, the popular former president, Mrs Vigdís Finnbogadóttir, was on the board of directors of deCODE Genetics, and the popular Prime Minister Davíð Oddsson visibly supported the company. With deCODE setting up shop, hiring several dozen people using US$12 million in venture capital, and then in February 1998 landing what was stated to be a US$200 million research contract with the pharmaceutical company Hoffmann-La Roche, important elements were in place to convince politicians and public alike and gain their goodwill. A good image was created and expectations were high. However, a few sceptics began to express their deep concerns.

In the summer of 1997, deCODE wrote a draft of a bill on a health sector database (HSD), the first element of the business plan, and presented it to the government, during the summer and autumn of that year. The idea was kept secret by both deCODE and the government, until the government introduced the bill to the Alping (the Icelandic parliament) in April 1998 planned to be passed in a matter of 4 weeks (Zoëga and Andersen, 2000). The bill was meant to secure the phenotypic data of the plan through no consent at all. It was also meant to establish exclusive rights and eliminate all competition. With the plan made public, the sceptics grew in number and the bill met with heavy opposition. In the ensuing debate, a demand for a priori informed consent for participation was rejected. The government conceded, however, that a 'defence' was needed for the population and an opt-out clause was inserted: those who objected to participation could register with the Director General of Public Health. Others were 'presumed' to consent. The company's two demands, presented as foundations of the business plan, the presumed consent and exclusive rights, were agreed on a priori as dictated by the business plan. Under 'warnings' of great urgency that Iceland would lose billions of dollars if the bill was not passed, it became an Act in December 1998. In a last minute change, the Act (Art. 10) specifically permits the interconnection of data from the HSD with a database of genealogical data and a database of genetic data, thus effectively creating the single three-pronged GGPR database originally planned but not debated in parliament or processed democratically.

Both corporate and government interests are interwoven by the Act. The Ministry of Health and its Director General of Public Health shall have free access to data from the database for statistical processing and planning and policy-making and other projects. The prospect of having such access to the database provided further impetus for government support for the corporate interests.

To build the genealogical database, which forms the second element of the business plan, deCODE relies on an exemption made in law. Icelanders have a long-standing tradition of interest in genealogy. As a result of this interest, the recording of family relationships of Icelanders for supporting research in anthropology and genealogy (operating according to Act no30/1956) was specifically exempted from the Data Protection Act (Act no. 121/1989, on recording and processing of personal information, Art. 2). Based on this exemption, deCODE has built a genealogy database to interconnect with the health records and genotype databases.

The exemption made in Act no. 121/1989, however, is specific and exhaustive and expressly does not apply to 'scientific research'. Genealogical information acquires a new meaning under a different context. Thus a family tree may be innocuous by itself, but annotated with disease the context changes and the tree acquires a new meaning. Whether the wholesale 'interconnection' of genealogical data with health records and genetic data will be considered 'genealogical research' (thus exempt from Act no. 121) or 'scientific research' (and thus subject to Act no. 121) remains to be seen and has not yet been challenged. Act no. 77 (23 May 2000), on personal protection and processing of personal information, which took effect on 1 January 2001, superseded Act no. 121. This new Act, which implements provisions of EU Directive 95/46, does not expressly exempt 'genealogical research' as the previous Act did; in fact there is no mention of that exemption in the new Act. This could lead to the Data Protection Commission making stricter requirements under the HSD Act for the interconnection of data.

deCODE is building the third database by searching for genes, disease by disease, in collaboration with various groups of physicians (Gulcher and Stefansson, 1999b). Some physicians are known to have stock options in the company. The collaborators conduct the research with the consent of participants after review of individual research projects by the National Bioethics Committee and the Data Protection Commission. The individual consent obtained in these studies was a broad consent allowing the researcher's discretion in the further use of the DNA samples. In 1999, while the National Bioethics Committee was rewriting guidelines for informed consent which would have limited the researcher's discretion and required an informed rather than a broad consent, the Minister of Health abruptly issued a new regulation changing the method of nomination to the committee from academic departments and professional organizations to government ministries and departments, ousting the previously established independent committee (Zoëga and Andersen, 2000; Abbott, 1999b). The minister was in such a great hurry to expel members of the committee that a new committee was not ready for appointment until a month later, effectively disrupting ethics reviews for nearly 2 months. Under the auspices of the new committee, deCODE operates with broad consent and not informed consent in its genetic research (deCODE, 2000).

The HSD Act stipulates that the interconnection of genetic, genealogical and health data shall be done using 'methods and protocols that meet the requirements of the Data Protection Commission'. The process for obtaining consent for such an interconnection has not been determined and therefore it is not clear at the moment to what extent the broad consent which deCODE has operated under for studying diseases will be applicable in this interconnection (Gulcher and Stefansson, 1999a). There has certainly been no explicit mention of use of the data for the HSD and of potential interconnections on the consent forms signed by patients participating in the disease-by-disease studies so far. It is, however, clearly the intent of the corporation to use the DNA and genotypic data obtained in studies of individual diseases to make the third pillar in its three-pronged business plan. With the National Bioethics

Committee effectively silenced, the Data Protection Commission remains the key regulatory agency controlling this aspect.

Opposing views

deCODE and the Icelandic authorities maintain that one-way encryption of personal identifiers constitutes depersonalized information and therefore that patients' informed consent is not required. They further maintain that various mechanisms such as access control and continuous monitoring by the data protection authorities will adequately protect the privacy of participants. They maintain that an exclusive license for creating and operating the database is necessary to protect the financial investment of the licensee, and that this arrangement does not constitute unfair business practices because all the original data continue to reside in the institutions where they were generated, accessible to scientists as before.

Mannvernd, the Association for Ethics in Science and Medicine, is the organized opposition to the HSD Act. Mannvernd considers the Act to be fundamentally flawed and is mounting a legal challenge to test the constitutionality of the Act. The database is a longitudinal record linkage and the data are not depersonalized. Therefore, Iceland's international commitments and constitution require patients' a priori consent for use of their data. They further maintain that the exclusive license effectively creates a monopoly of research in human genetics whereby, for example, the licensee can 'listen in' on research by other scientists and appropriate the resources. Also the HSD Act 'permits' health institutions that hold the data to hand them over to the licensee without consent of the patients. They are, however, not authorized to hand over data to other scientists because scientists outside of the licensee are subject to the Act on patients' rights that requires informed consent. This creates an inequity.

The Icelandic Medical Association, backed by the World Medical Association, while recognizing the scientific value of health records data, considers that the HSD Act violates basic principles established to allow the use of such resources and at the same time uphold patient autonomy and dignity (Zoëga and Andersen, 2000). The Society of Pharmaceutical Medicine, based in the UK, has also supported this view and has stated that whereas such databases can provide an important source of information, it is essential that confidence in them is maximized and that they are not inappropriately exploited (Clarke et al., 2001).

Doubts have also been raised about the premises of deCODE's business plan. For example, the conjecture of genetic homogeneity of Icelanders has been found wanting, and the onus is on deCODE to provide evidence for their statements about the usefulness of the Icelandic population (Gulcher and Stefansson, 1998; Árnason et al., 2000). The conjecture that the large-scale correlation studies made possible by the database will revolutionize medicine is also debatable (Gulcher and Stefansson, 1999a; Holtzman and Marteau, 2000; Árnason, 2000).

Other corporations

The case of deCODE in Iceland has raised fears about the possible future conduct of other biotechnology corporations. deCODE and its followers in essence are arguing that a 'new ethics' is needed for what they claim will be a 'revolution in medicine' (Gulcher and Stefansson, 1999a; Zoëga and Andersen, 2000). Informed consent, as stipulated for example by the Helsinki declaration, is seen as an impediment to progress (Gulcher and Stefansson, 1999a).

Several other companies are planning work with other genetically isolated groups for accelerated gene mapping. These include Australia Wide Industries working in Tasmania, UmanGenomics in Sweden, Gemini Genomics in Newfoundland, Eesti Geenikeskus in Estonia and SharDNA in Sardinia, to name but a few. These companies have not received the notice that deCODE has, and many say that they intend to avoid the pitfalls of the Icelandic case (Abbott, 1999a).

deCODE Genetics Inc. is an American corporation competing in the international market. Its largest shareholder, the international pharmaceutical corporation Hoffmann-La Roche, backs it. It has successfully instigated legislation in its favor and has been accused of weakening regulatory agencies. By operating under the relaxed ethical rules of Iceland, the corporation could be seen as gaining a competitive advantage. Similarly, attempts by competitors to relax standards by influencing and lobbying legislation and practices are likely to be made in both Western countries and developing countries. 'It seems likely that many countries, not protected by established democratic and legal systems, and possibly also burdened by poverty, low educational and medical standards and corrupt politicians, may be exposed to similar approaches' (Edwards, 1998).

References

Abbott (1999a) Sweden sets ethical standards for use of genetic 'biobanks'. *Nature* 400: 3.

Abbott A (1999b) 'Strengthened' Icelandic bioethics committee comes under fire. *Nature* 400: 602.

Árnason E (2000) The Icelandic Healthcare Database. *New England Journal of Medicine* 343: 1734.

Árnason E, Sigurgíslason H and Benedikz E (2000) Genetic homogeneity of Icelanders: fact or fiction? *Nature Genetics* 25: 373–374.

Clarke A, English V, Harris H and Wells F (2001) Ethical considerations. *International Journal of Pharmaceutical Medicine* 15: 89–94.

deCODE (2000) Decode Genetics, Inc. S-1 Registration (8 March 2000) (see Web links for details).

Edwards J H (1998) Decoding genes in Iceland (see Web links for details).

Gulcher J R and Stefansson K (1998) Population genomics: laying the groundwork for genetic disease modeling and targeting. *Clinical Chemistry and Laboratory Medicine* 36: 523–527.

Gulcher J R and Stefansson K (1999a) The Icelandic Healthcare Database and informed consent. *New England Journal of Medicine* 342: 1827–1830.

Gulcher J and Stefansson K (1999b) The Icelandic Healthcare Database: a tool to create knowledge, a social debate, a bioethical and privacy challenge. *Medscape Molecular Medicine* 1, August (see also Web links).

Holtzman N A and Marteau T M (2000) Will genetics revolutionize medicine? *New England Journal of Medicine* 343: 1496.

Laxness H (1946) *Independent People*. New York: Vintage International.

Zoëga T and Andersen B (2000) The Icelandic Health Sector Database: deCODE and the 'new' ethics for genetic research. Who Owns our Genes? Proceedings of an International Conference (October 1999, Tallinn, Estonia), pp. 34–65. Copenhagen: Nordic Committee on Bioethics.

Web links

Association of Icelanders for Ethics in Science and Medicine. Decoding genes in Iceland: text of an article by J H Edwards http://www.mannvernd.is/english/articles/04.12.1998_decoding_genes_john_edwards.html

Bill on a Health Sector Database (submitted to Parliament at 123rd session, 1998–99) http://www.mannvernd.is/english/laws/HSD.bill.html

DeCODE Genetics:Home page of deCODE Genetics, Inc., a genomics company http://www.decode.com

Mannvernd: Association of Icelanders for Ethics in Science and Medicine.Home page of Association of Icelanders for Ethics in Science and Medicine http://www.mannvernd.is/english/index.html

Ousting of the National Bioethics Committee. A series of articles from the *Icelandic Medical Journal* http://www.mannvernd.is/english/news/imj.nbc.html

The U.S. Securities and Exchange Commission database of company filings http://www.sec.gov/Archives/edgar/data/1022974/0000950123-00-002076-index.html

The DeCODE Proposal for an Icelandic Health Database. An article by Ross Anderson http://www.cl.cam.ac.uk/users/rja14/iceland/iceland.html

The World Medical Association Inc. World Medical Association Declaration of Helsinki: Ethical principles for medical research involving human subjects http://www.wma.net/e/ policy/63.html

7
Informed Consent in Human Genetic Research

Philip R. Reilly

Historical note on the doctrine of informed consent

Although concern for the moral dimensions of the relationship between a scientist and a human subject who agrees to participate in a research project under the scientist's direction was evident as far back as the nineteenth century, sustained social interest in that relationship emerged in the two decades after the Second World War. The Nuremburg Code, which was compiled in response to the horrors of Nazi experimentation, is famous for its declaration that 'The voluntary consent of the human subject is absolutely essential.' This sentence crystallizes the principle that all persons who participate in research are autonomous agents whose interests must be respected. Of course, consent can only be voluntary if the individual has sufficient knowledge about a proposed experiment to permit a reasoned decision (Reilly, 1998).

In the ensuing decades, as biomedical research involving human subjects grew rapidly, physicians, researchers, lay people with a special interest in research and governments recognized the need to expand and codify the doctrine of informed consent. Among the more important milestones are the 1964 World Medical Association Declaration of Helsinki and its subsequent amendments, the 1978 United States Belmont Report, which articulated the fundamental principles guiding research with federal funds that involved human subjects, and various United States regulations governing the conduct of research with human subjects that were implemented during the late 1970s and 1980s. During the 1980s, many European nations adopted principles and practices that were generally congruent with the policies implemented in the United States.

In the United States, the regulations (which were implemented just before the advent of revolutionary advances in molecular biology) were introduced at a time when most participants in policy discussions thought that the principal ethical issues in research with human subjects were (1) the protection of vulnerable populations (especially pregnant women and fetuses, children and prisoners) who were unable or potentially unable to give a morally acceptable consent, and (2) ensuring that a potential human research subject was aware of the nature and level of risk of the physical harm involved in participating in a study. The content of Subpart B of the federal rules, which focuses on the protection of pregnant women and human fetuses, reflects the deep divide in the United States about the use of aborted fetuses in research. Subpart C sets down strict guidelines for research involving prisoners. Subpart D sets forth the regulations that govern research involving children, including the requirement that the child must assent to participating in research that offers no direct possibility of therapeutic benefit.

The revolution in molecular biology that began in the 1970s, as scientists developed powerful new tools to study deoxyribonucleic acid (DNA), set the stage for the Human Genome Project (formally launched in 1990) and created a new set of concerns about risk to the research subject of 'informational' harm (Holtzman and Watson, 1998). As this great international scientific undertaking began to unfold, it became apparent that our ability to study genetic variation in people meant that we would routinely acquire and store data about them, some of which could, if disclosed to third parties, potentially cause psychological and economic injury.

Specific concerns about human genetic research

The doctrine of informed consent in research is dynamic; it evolves in response to scientific and clinical advance, as well as to ethical analysis by many different stakeholders. Because there are relatively few regulatory guidelines and because most decisions about genetic research involving human subjects are made by local institutional review boards (IRBs), or their equivalents in Europe and other nations, much of the current thinking about the topic is reflected in the formal actions taken by those reviewing bodies. In addition, there is a large body of literature that explores these issues.

Currently, after two decades of discourse about consent in human genetic research, there is consensus about a few issues and sustained debate about several others. Although gene mapping studies that are conducted anonymously (thus, in principle without risk of causing informational harm) are, according to Code of Federal Regulations (CFR) 46.101 (b)(4), technically 'exempt' from the 'common rule' and therefore from the need to obtain informed consent, few (if any) proposed studies today attempt to bypass the consent process. There is in the United States a general consensus that the decision to participate in research intended to generate genetic data (such as determining whether mutations in a certain gene cause a particular disorder) involves more than a 'minimal risk' – a view that means that all such federally funded research must undergo full (as opposed to expedited) review by an IRB (CFR 46.110 (b)(1)). During 1995–6 this consensus was influenced significantly by discourse and debate in which concerned researchers and ethicists argued that disclosure to third parties of results that, for example, indicate that a person is at risk for a psychiatric illness posed the same threat of 'genetic discrimination' in a research setting as it would in clinical practice (Clayton et al., 1995). It is interesting that this policy shift was not accompanied by a formal revision of existing regulations.

The basic information that must be disclosed to a potential human subject who is considering participating in genetic research, which he/she in turn must have the opportunity to evaluate (a process that includes having access to a member of the research team), is listed in the federal regulations (United States Government, 1991). This information includes a general description of the nature of the study (including what part is experimental and how long the study will last), a description of reasonably foreseeable risks, a description of any potential benefits to the subject, a disclosure of alternative treatments (if it is a study involving treatment), a statement of how the confidentiality of records will be maintained, a statement of whether there will be compensation for injury, a designation of a contact person who is a member of the research team, and a declaration that participation is voluntary and that there is no penalty for withdrawal. Depending on the research, the consent process may also require a statement of whether participation by a pregnant woman would put her fetus at risk, a description of the circumstances under which a subject's participation could be terminated by the investigator, costs that the subject might bear, the consequences of withdrawing, and a statement that 'significant new findings developed during the course of the research which may relate to the subject's willingness to continue participation will be provided to the subject', the identification and the approximate number of subjects in the study (45 CFR 46.110 (b)1–6).

In addition to these required elements, bioethicists, researchers, attorneys, consumer groups and others, strongly influenced by the rapid rise and globalization of the biotechnology industry, have raised many other unresolved issues. Among the main ones are the nature of the duty to tell subjects and/or their relatives about potentially important clinical information discovered during the research, ownership of tissue provided by the human subject as part of the research, the conduct of genetic research in nations that do not embrace an informed consent doctrine and the idea of community consent.

The federal regulation that recognizes a potential obligation to disclose data to the human subject (45 CFR 46.116 (b)(5)) was implemented to ensure that if researchers learned that the risks of a particular study were greater than had been thought initially, they would inform the subjects. With the exception of the relatively few individuals participating in trials involving gene therapy, subjects who participate in genetic research currently do so because they have a personal stake in knowing whether they carry a gene variant that is linked to risk for a disease. Researchers involved in such studies face two 'knowledge' problems: what to do about a tentative initial finding that does or does not suggest an association between a gene and a disease, and what to do about an inadvertent finding (such as nonpaternity). It is the nature of most research that early findings must be replicated before the researcher can reach the level of certainty that permits the transfer of new knowledge into the clinical arena. When researchers make a clinically relevant finding, they should (and often do) seek guidance from oversight committees about informing subjects. No ethical or legal duty runs from the research team to relatives of the subjects who are not participants in the project. The discovery of inadvertent findings is not usually disclosed. Institutional review boards vary in whether they require investigators to inform potential subjects that genetic research could discover nonpaternity, but most do not.

For more than a century the practice has been that tissue samples that are taken in the course of clinical care or research become the property of the clinician or the researcher, who may use them for research and teaching purposes but who has an ongoing duty to protect the privacy of the donor. The only appellate court opinion on this issue supports this position (Moore v. Regents, 1989). In the mid-1990s, some people began to argue that tissue obtained in the course of research should not necessarily become the property of the researcher or that, if there were to be a transfer of property interests, it should be acknowledged formally in the consent process. This debate raised complicated issues involving intellectual property rights to products growing out of such research – issues that threaten the core of the pharmaceutical and biotechnology industries.

The current situation is that the researcher has a duty to disclose to the potential subject that he/she intends to collect, analyze, store and possibly reuse a tissue sample or its DNA, and that the ultimate goal may be to create a product that is based in part on the use of this tissue and may generate profits in which the subject will not share. The individual is free to decide whether or not to accept this condition as part of his agreement to participate. Too often ignored in the debate over ownership of DNA samples is the fact that the principle of autonomy, which is the foundation of informed consent doctrine, must surely include the right to decide to donate tissue to a third party. Of course, the same principle could support an individual's decision to make it a condition of his/her participation in a study that his/her share in any future economic benefit flowing from knowledge derived from that research. Currently, it is highly unlikely that an entity sponsoring a research project would agree to that demand.

A related problem involves determining the appropriate scope of genetic research that may be carried out on important sets of archived tissue for which consent for genetic analysis was not originally obtained. Because it is often impossible or impractical to attempt to obtain consent for a large cohort of people long after a study has ended, current thinking is that DNA analysis of archived samples may be conducted only if the samples are anonymous. This policy permits the pursuit of a social benefit – important new knowledge – without causing risk of harm to the individual from whom the tissue was obtained. Some argue that this policy causes a 'dignitary harm', to which the counterargument is that the individual's decision to consent to the original research indicates some measure of altruism and suggests that, if asked, he would consent to the DNA analysis.

During the 1990s, worry about the threat of 'genetic discrimination' (Billings et al., 1992) led some to argue that genetic research involving certain well-defined human subpopulations should be preceded by community evaluation and community consent. Geneticists are especially interested in studying relatively inbred populations that are known to be at unusually high risk for particular diseases. Such populations have yielded important findings, but some of them – for example, particular groups of native Americans – have historically suffered from serious discrimination. Could not the discovery, for example in the Oklahoma Apache, of a gene variant that predisposed such persons to diabetes or alcohol abuse be perceived by some to lend support to specious arguments of inferiority (Foster et al., 1998)? If so, should not permission be sought from the community (through its recognized leaders) before approaching individuals or families to be research participants? Although this concern is not covered by federal rules or guidance documents, it is not uncommon today for it to be raised by reviewing bodies. By introducing a preceding layer of paternalistic review, the doctrine of community consent contradicts the principle of autonomy that underlies informed consent (Reilly, 1998).

Arguments that favor 'community consent' parallel a growing concern outside the United states and Europe that genetic research in the developing nations constitutes a form of 'biocolonialism' that amounts to an unrequited transfer of assets to the West. This has led many countries to forbid the export of DNA research samples. It has also led several countries, including Iceland, Estonia and Latvia, to recognize that, in combination with excellent health histories, a national DNA repository could be a substantial commercial asset.

In summary, the key issues concerning informed consent in modern genetic research arise from two distinct ideas. First, many people think that genetic information about individuals, especially because it relates to risk for disease, may cause harm to individuals and groups if disclosed to and misused by third parties. Second, because of its value in the development of new diagnostic tests and new drugs, genetic information is a potentially valuable commodity that may, ultimately, force a re-evaluation of traditional views of tissue ownership.

References

Billings P, Cohn M, de Cuevas M, et al. (1992) Discrimination as a consequence of genetic testing. *American Journal of Human Genetics* 50: 476–482.

Clayton E, Steinbuerg K, Khoury M, et al. (1995) Informed consent for genetic research on stored tissue samples. *Journal of the American Medical Association* 274: 1786–1792.

Foster M, Bernsten D and Carter T (1998) A model agreement for genetic research in socially identifiable populations. *American Journal of Human Genetics* 63: 785–794.

Holtzman N A and Watson M S (eds.) (1998) *Promoting Safe and Effective Genetic Testing in the United States: Final Report of the Task Force on Genetic Testing*. Baltimore, MD: Johns Hopkins University Press.

Moore v. Regents of the University of California (1989) 793 P2d.479, cert denied, 499 US 936 (1991).

Reilly P R (1998) Rethinking risks to human subjects in genetic research. *American Journal of Human Genetics* 63: 682–685.

United States Government (1991) Code of Federal Regulations Title 45, Part 46; Subpart A: Federal Policy for the Protection of Human Subjects. Source: 56 Federal Register 28003, June 18, 1991. Cf. 46.116 et seq. Washington, DC: Government Printing Office.

Web links

The Institutional Review Board – Discussion and News Forum. [A website devoted to promoting discussion about ethical issues in research with human subjects.] http://www.irbforum.org

8
Gene Therapy: Expectations and Results

Paul A. Martin

What is gene therapy?

There are two broad types of gene therapy, somatic and germ line. Somatic gene therapy can be defined simply as the delivery of functional genes to somatic tissue for the treatment of disease. In somatic gene therapy, a therapeutic gene is administered to the individual in order to make changes to the somatic cells in the body but not to the germ cells that are involved in reproduction. The therapy therefore affects only the person to whom it is given.

Germ-line gene therapy is aimed at genetic alteration of the germ-cells for the treatment of disease in future generations. In germ-line gene therapy, the therapeutic gene is inherited by the progeny of the treated individual and becomes a stable part of their genetic make-up. For ethical reasons, germ-line therapy is not being developed in humans at present.

There are two discrete ways in which somatic gene therapy can be applied:

- *Ex vivo* gene therapy involves modifying the individual's cells outside the body (*ex vivo*). In *ex vivo* therapies, cells are removed from the body, genetically altered in culture and returned to the individual, for example, by blood transfusion or bone marrow transplantation.
- *In vivo* gene therapy involves genetically altering the cells by direct administration of the therapy to the individual (*in vivo*). *In vivo* therapies are administered mainly by injection or in an aerosol spray to the lungs.

Early development of gene therapy

The idea of gene therapy is almost as old as that of molecular biology itself and predates the Human Genome Project by almost three decades. The early vision of how gene therapy might develop was most clearly set out by Tatum in 1966, when he predicted that viruses would be used 'in genetic therapy', that cancer would be cured 'by modification and regulation of gene activities', and that 'the first genetic engineering will be done with the patient's own cells, for example, liver cells, grown in culture' (Tatum, 1966).

In 1970, just before the advent of recombinant deoxyribonucleic acid (DNA) technology, the first attempt at human gene therapy was made by Rogers. Three young German sisters suffering from a rare genetic enzyme deficiency were treated using an unmodified rabbit virus that was thought to contain the missing enzyme and so might correct their metabolic disorder. This unsuccessful experiment drew criticism from other scientists at the time, but it was only when a more serious attempt was made in 1980 that the ethical and scientific issues surrounding gene therapy gained widespread public and government attention.

In this second experiment, Cline tried to treat individuals suffering from thalassemia by using a recombinant globin gene to transfect bone marrow cells, which were subsequently transplanted back into the individual. His failure to get final ethical approval for the trial provoked a storm of protest and resulted in the United States National Institutes of Health (NIH) deciding that all future trials would have to be approved ethically by the NIH Recombinant DNA Advisory Committee (RAC). This was, in effect, a government moratorium, and it was ended only in 1985 when the NIH RAC finally agreed to guidelines for research on gene therapy. Throughout most of the 1980s, there was organized opposition to the development of gene therapy and a high level of public anxiety.

Despite the clinical moratorium following the Cline controversy, further scientific progress ensured continuing interest in the development of gene therapy. The early 1980s saw significant advances in the use of modified viruses, particularly retroviruses, as vectors for gene transfer. This, and the first correction of a genetic defect in a transgenic animal in 1984, inspired a large amount of basic research on the transfer and expression of foreign genes in mammalian systems. Initially, the focus was on the *ex vivo* treatment of rare monogenic disorders, but by 1986 serious attention was also being given to the use of these gene transfer techniques in the treatment of cancer and other acquired diseases.

In the late 1980s, leading researchers started to argue the case for human clinical trials. After much negotiation with the RAC over the design of experimental protocols, the first full clinical trial for gene therapy commenced in 1990. After this landmark trial, organized opposition to the technology largely ceased.

Creation and growth of the gene therapy industry

It was not until the mid-1980s that industry first started to become involved in commercially developing technologies for therapy. There was very little interest shown by established biotechnology and pharmaceutical companies; therefore, it was left to the academic pioneers of gene therapy to create their own firms to finance research, support clinical development and start the process of exploiting the technology.

The first firm, Genetic Therapy Inc., was founded by the leading advocate of gene therapy, French Anderson, in 1986 to commercialize *ex vivo* gene therapy, and by the end of 1990 there were nine firms in the United States (Martin, 1999). Subsequently, there was a steady growth in the number of companies dedicated to gene therapy, such that at the end of 1995 there were 16 in the United States and 10 in Europe (Martin et al., 2000). By 2000, this had increased to 22 firms in the United States and 26 in Europe (Martin et al., 2000).

During the mid-1990s the mainstream pharmaceutical industry started to invest heavily in gene therapy, mainly through acquisitions and research collaborations with dedicated firms (Martin, 2001). By 1996, about 34 technology alliances of this sort had been formed, representing a commitment of over US$1.1 billion to gene therapy (Martin, 2001). This period marked the start of the integration of the technology into the pharmaceutical sector and coincided with increasing interest in the development of *in vivo* gene therapies, or so-called 'genes in a bottle'. These products were designed to be much more like conventional small-molecule medicines; they could be stored at room temperature, given a standard formulation, produced at scale and sold through conventional distribution channels. In the next 5 years, substantial investment was also committed to developing internal research and development programs in large firms such as Rhone-Poulenc Rorer (now Aventis), Novartis and Genzyme. It is estimated that by 2000 a total of US$3 billion had been invested by industry in trying to develop gene therapy, even though, as of December 2002, no gene therapy product has actually reached the market.

Clinical development of gene therapy

The growth of the gene therapy industry and investment in the sector also coincided with a rapid increase in the number of clinical trials. Within 5 years of the first approval by the RAC, about 150 other trials also commenced. Notably, 75% of those first trials were for the treatment of cancer, with only 12% aimed at treating classic monogenic disorders (Martin and Thomas, 1996, p. 67). By December 2002, about 636 trials involving 3500 people had been initiated internationally in 20 countries (Journal of Gene Medicine, 2002). The diseases being treated included a range of cancers, monogenic disorders, infectious diseases and vascular diseases. During this period four trials have reached Phase II/III and four have reached Phase III (*Journal of Gene Medicine*, 2002).

Although no product has successfully gone though phase III trials to reach the market, several important clinical milestones have been achieved. First, several genes have been transferred successfully into many different human tissues. Second, there have been very few significant side effects in most clinical experiments. Third, many important trials have shown preliminary demonstration of a positive biological response and/or efficacy (e.g. treatment of cystic fibrosis, some types of cancer and peripheral arterial disease). Last, in April 2000 the first clear case of a successful treatment using gene therapy was demonstrated when two French children suffering from adenosine deaminase deficiency (ADA) were 'cured' (Cavazzana-Calvo et al., 2000). Thus, much has been achieved, and those first results suggest that there are no fundamental barriers to the successful clinical development of gene therapy.

Along with these positive findings, there have been some major setbacks for the field. In particular, many of the most promising early clinical products proved to be either unsafe or ineffective in late-stage trials, significant side effects have occurred in a few trials, and many technical problems, in terms of the efficiency of gene transfer *in vivo*, still have to be resolved. Overcoming these difficulties will take time and patience – something that has been often lacking in a field marked by hype and unrealistic expectations.

Rise and fall of public expectations

The early growth of the industry and clinical trials was based on high hopes that gene therapy would prove to be an effective treatment for many diseases for which existing therapy is poor or nonexistent. After the first clinical trial, public perceptions of gene therapy quickly changed from suspicion or hostility, to seeing it as holding great promise. In the early 1990s, the media reported a series of 'breakthroughs'; trials were inundated with people wanting to receive experimental treatments, and the prospect of a cure for diseases like cystic fibrosis seemed to be just around the corner (Martin, 1998).

By 1995, however, these high expectations began to evaporate and instead turned into disillusionment and criticism of the field after the failure of a few early stage clinical trials and a general lack of progress (Martin, 1998). In response, the National Institutes of Health undertook a review of the development and funding of gene therapy and concluded that some investigators had created unrealistic expectations of gene therapy and that the scientific community needed to be both more realistic and responsible in its public discussion of the prospects for the technology.

Despite this significant loss of confidence, investment continued to flow into the industry and an increasing number of trials took place, although the focus of activity started to shift from North America to Europe. In the rapidly rising stock market of 1999, gene therapy firms raised hundreds of millions of dollars from the capital markets to continue their research and development activities.

Confidence in the industry remained strong until the first death to be caused directly by gene therapy occurred in a clinical trial in September 1999.

The death of Jessie Gelsinger as a result of being given an adenovirus-based gene therapy caused major public controversy, because the investigators running the trial at the University of Pennsylvania had not adhered fully to Food and Drug Administration and RAC guidelines (Nelson and Weiss, 2000). All trials using this vector system were halted in the United States, and a review of gene therapy clinical trials by the NIH indicated that there was a high level of regulatory noncompliance. In the immediate aftermath of this tragedy, the media portrayed gene therapy as potentially dangerous, public confidence in the technology declined further, and regulators tightened the rules governing clinical development.

This incident marked a turning point in the fortunes of the sector and helped precipitate a hasty retreat by both investors and large companies from gene therapy. In the following 18 months, several of the companies with the biggest investment in the field largely withdrew from direct involvement, including Novartis, Chiron and Aventis. Other indicators of decline included a fall in the number of technology collaborations and a shortage of finance. The latter caused big problems for previously highly regarded companies, such as Transgene in France. By December 2002, the industry was in the weakest state that it had been in for many years.

Prospects for gene therapy

So what are the prospects for the successful development of gene therapy? For new products to be introduced into the clinic, several things will be required including significant technical progress in improving gene transfer technology, positive results from late-stage clinical trials, further investment in the leading gene therapy firms and continuing public support. At present each of these prerequisites is missing. Although progress in improving the efficiency of gene transfer vectors is likely to remain painfully slow, however, it seems realistic to expect that one of the products in late-stage development will get regulatory approval in the next 2–3 years. When this happens, public confidence and investment will return to the sector. In the longer term, there are considerable grounds for optimism that gene therapy will provide a series of important new cures in the coming decades.

One of the most important lessons to be learnt from the chequered history of gene therapy is that public confidence and support are essential. Without it the technology is unlikely to ever realize its full potential. The scientific community therefore needs to be more realistic in the claims that are made about gene therapy and more responsible in its conduct of clinical research. Gene therapy cannot afford another setback.

References

Cavazzana-Calvo M, Hacein-Bey S, de Saint Basile G, et al. (2000) Gene therapy of human severe combined immunodeficiency (SCID)-X1 disease. *Science* 288 (April 28): 669–672.
Journal of Gene Medicine (2002) Gene Therapy Clinical Trials. http://www.wiley.co.uk/wileychi/genmed/clinical/
Martin P (1998) From eugenics to therapeutics: the impact of opposition on the development of gene therapy in the USA. In: Wheale P, von Schomberg R and Glasner P (eds.) *The Social Management of Genetic Engineering*. Aldershot, UK: Ashgate Publishing.
Martin P A (1999) Genes as drugs: the social shaping of gene therapy and the reconstruction of genetic disease. *Sociology of Health and Illness* 21(5): 517–538.
Martin P A (2001) Great expectations: the construction of markets, products and user needs during the early development of gene therapy in the USA. In: Coombs R, Green K, Walsh V and Richards A (eds.) *Technology and the Market: Demand, Users and Innovation*. Cheltenham, UK: Edwards Elgar.
Martin P and Thomas S (1996) *The Development of Gene Therapy in Europe and the United States: A Comparative Analysis*, STEEP Special Report no. 5. Brighton, UK: Science Policy Research Unit, University of Sussex.
Martin P A, Crowther S, Corneliussen F, Jaeckel G and Reiss T (2000) *Gene Therapy in Europe: Exploitation and Commercial Development*, Final Project Report. Brighton, UK: Science Policy Research Unit, University of Sussex.
Nelson D and Weiss R (2000) Gene researchers admit mistakes, deny liability. *Washington Post* February 15: A03.
Tatum EL (1966) Molecular biology, nucleic acids and the future of medicine. *Perspectives in Biology and Medicine* Autumn: 19–32.

Web links

Human Genome Project Information. Introduction to Gene Therapy http://www.ornl.gov/hgmis/medicine/genetherapy.html
Office of Biotechnology Activities. Recombinant DNA and Gene Transfer. Information on regulations and current developments. http://www4.od.nih.gov/oba/Rdna.htm
National Heart, Lung and Blood Institute. Programs of Excellence in Gene Therapy. Links to leading research centres. http://pegt.med.cornell.edu/

Part 2

Genetic Disease: Implications for Individuals, Families and Populations

Introduction

Angus Clarke and Flo Ticehurst

In this section, we look at clinical services and how they are delivered. For those who have not encountered genetic disease within their own families, the area of clinical genetics may be relatively unfamiliar; it is therefore often feared. Indeed, the suggestion that a child or adult be referred 'to genetics' can cause great distress. So – just what are genetic services all about?

Biesecker (chapter 9) outlines the goals of genetic counselling as defined by its practice within the US, and particularly from a psychological perspective. She looks at the influence of a client's background on their response to genetic counselling, especially the educational, economic and socio-cultural aspects of their background and their personality traits. Another important aspect she considers is the prior experience of the condition that the client has, not only in terms knowledge about it but also in terms of their experiences of the disorder in question and how this has affected the individual and/or the family. She then considers psychological reactions to genetic counselling as dependent on the interactions of the client and counsellor and the circumstances outlined above.

As a relatively new clinical practice, the effectiveness of the process of genetic counselling is not well understood. Although there have been a good number of studies that have attempted to understand the process further from a variety of standpoints, measures for the evaluation of outcomes have proved difficult to define.

Procter (chapter 10) introduces the controversy that surrounds both predictive and carrier genetic testing of children. She describes the balance that must be struck between the benefit gained by carrying out a genetic test and the potential harm that may be caused as a result of this investigative process. Not only does the child have rights to be protected – the right to make their own decision about testing when they are older, and the right to confidentiality and respect for their genetic privacy – but there is also the question of the social or emotional harms that may result from testing in childhood. Other factors relevant to performing a predictive test on a child include whether there is a treatment for the disease/disorder in question, and whether it is the parents of a young child who are seeking the test or whether the 'child' – as a mature, competent, legal minor – is seeking the test in his/her own right. She considers how testing for carrier status differs from predictive testing – the main difference being that this type of test has no direct implications for the health of the individual but rather for the possible health of their own future children. However, carrier testing still requires careful consideration as a knowledge of carrier status for a genetic condition can have a significant impact on the individual and also on reproductive decision-making in adulthood.

Procter describes the consensus reached in the UK Clinical Genetics Society's working party report on these issues in 1994. She suggests that this area is likely to remain controversial, however. A response to the working party's consensus from the Genetic Interest Group agreed with the cautious stance on predictive testing, but disagreed with the views on carrier testing, promoting the view that such decisions should be left to individual families who may have good reasons to want to know about carrier status. Procter suggests that the clinician should be thought of as a person with the responsibility to help families to consider the issues around genetic testing in childhood carefully, balancing the possible benefits and harms, rather than as the gatekeeper to genetic testing.

Prior's article (chapter 11) explores the recognition of risk perception as a social activity. He begins by giving us some background to risk assessment and the application of this to clinical genetics now that technology has advanced to the point that risk assessments can be applied to individuals on the basis of genetic test results as well as family history. He describes the uncertainty associated with risk assessments in the context of a family history of breast cancer. This entails an exploration of how the process of risk assessment translates into a personal meaning for the patient in question. Prior describes some of the biases in the models used for estimating risk on the basis of family history information and highlights the problems inherent in moving from an assessment of population risk to individual risk. He considers the implications this has for clinicians and individuals undergoing investigation, recognizing the important role genetic counselling has in addressing them.

Emslie and Hunt's article (chapter 12) focuses on genetic susceptibility as one factor contributing to the causation of the common diseases, in the context of research that is exploring the genetic basis of such conditions as coronary artery disease and the environmental factors that interact with these genetic factors. They use coronary artery disease as a case study to examine lay perceptions of genetic susceptibility to disease as well as considering some of the issues that could arise if genetic testing for CHD is introduced.

The dynamic nature of beliefs about inheritance and family history is discussed as well as perceptions of the role of luck, lifestyle and working and living conditions as factors contributing to the individual's overall susceptibility to disease. The authors suggest that new genetic knowledge will be assimilated into, rather than overturning, these existing beliefs. They then introduce the issues around testing for hypercholesterolemia to illustrate some of the complexities that may arise with the possible future development of genetic tests for susceptibility to disease, including an understanding of the limited sensitivity and specificity of testing and the consequent false-negative and false-positive test results, and the complex relation between risk awareness and behaviour. They suggest in conclusion that genetic susceptibility tests may make it more difficult to predict future health rather than offering any certainty.

McDermott (chapter 13) introduces the proposition of the 'thrifty gene hypothesis' as an explanation of the emergence of type II (maturity-onset) diabetes (DM2) in certain populations, and how this has led to populations being labelled as genetically susceptible. She demonstrates that ecologic and migrant studies indicate that non-genetic factors can account for the variation in disease incidence between populations, so asks what evidence there is for diabetes being a genetic condition. She explains that, as there is no gene identified as the cause of DM2, the main evidence comes from observed patterns of familial clustering.

He then describes alternative hypotheses that have been put forward to account for DM2 – especially Barker's demonstration of a link between low birth weight, poor nutrition in childhood and DM2 later in life, especially where excess weight was put on in adulthood; Hattersley's 'fetal insulin hypothesis', which brings together both genetic and foetal environmental factors; and Freinkel's hypothesis that a mechanism for the heritability of diabetes rests in a form of 'fuel-mediated teratogenesis' during pregnancy.

This leads into a consideration of the relationship between population phenomena and the disease affecting specific individuals. Gene frequencies will not usually change much over just one or two generations, so rapid alterations in disease incidence are likely to have some other – social, biological or environmental – explanation. McDermott offers a critique of the news surrounding the completion of the HGP. He suggests that the complexity of multifactorial diseases will limit the ability of science to deliver on the promises that have been made to cure disease, arguing that it is more important 'to address urgently the social and environmental determinants of the epidemic of DM2 at a population level, rather than being distracted by the chimera of genetic causes and costly "cures" for individuals.'

Modell (chapter 14) introduces us to some of the inherited haemoglobin disorders in the context of population screening as a part of community health programmes, using case studies from both Cyprus and the UK to illustrate the 'implications of carrier screening and prenatal diagnosis for inherited disorders in different cultural settings'. She describes how the prevalence in some populations of sickle cell and thalassemia disorders have resulted in pressure from families and communities to prevent these disorders and the 'burdensome' management of the clinical and social implications.

Communities can be consulted on the ethical applications of genetic knowledge, as in research carried out in the 1970s and 1980s into the various approaches to carrier screening for, and prenatal diagnosis of, thalassemia, which examined how genetic information was received and acted on by the relevant communities. Modell goes on to describe the role of the WHO Working Group on Hemoglobin Disorders, which was convened to work out the principles of genetic population screening. As discussed in this article, the WHO group recognized that delivering a population screening programme to diverse minority groups (as found in the UK) is a greater challenge than introducing such a programme to a small community (like Cyprus).

Scriver and Mitchell (chapter 15) close this section of the book by exploring the issues that have arisen in providing a programme of genetic education, counselling and carrier testing for Tay–Sachs disease and β-thalassemia within a high-school setting in Montreal. There was a high uptake of screening among adolescents, and the incidence of these diseases fell by over 90 per cent in these areas (where these diseases were highly prevalent).

The authors note the distinctions between testing individuals and screening populations for disease and the rationale behind the adoption of screening as a method for the prevention or avoidance of genetic disease. They describe the process involved in setting up the programme in response to the wishes of the community, and the outcomes of both the Tay–Sachs and the β-thalassemia screening programmes. The programmes were judged successful in terms of decreasing incidence of the diseases and in communicating relevant information to individuals and communities. The authors note that other communities may have different views about the appropriateness of genetic testing in adolescence in this way.

9 Genetic Counseling: Psychological Issues

Barbara B. Biesecker

Goals of genetic counseling

Clients seek genetic counseling for a variety of reasons that pertain to health risk and reproduction. Genetic counseling addresses their affective, cognitive and behavioral needs. A recent definition states that:

> Genetic counseling is a dynamic 'psychoeducational' process centered on genetic information. Within a therapeutic relationship established between providers and clients, clients are helped to personalize technical and probabilistic genetic information, to promote self-determination and to enhance their ability to adapt over time. The overarching goal is to facilitate clients' ability to use genetic information in a personally meaningful way that minimizes psychological distress and increases personal control.
>
> (Biesecker and Peters, 2001)

In the pediatric or adult genetic counseling setting, the goal is to facilitate client understanding and acceptance (Biesecker, 2001). Most often a child or relative is affected with a genetic condition or birth defect and the family is struggling to understand and adapt to their circumstances. Genetic counseling aims in this setting are:

1. to discuss client understanding of cause as it relates to a scientific explanation and the client's interpretation;
2. to explore the role of client illusions (personal beliefs) and their role in adaptation;
3. to promote feelings of personal control and mastery over a genetic condition.

The diagnosis or risk of a genetic condition or birth defect stimulates adjustment reactions in parents, siblings, grandparents and other relatives and, for later diagnoses, the affected individual. At the time of diagnosis or risk identification, a crisis may ensue if the information is unexpected. 'Crisis intervention' defines the goal of genetic counseling within this circumstance. With time, or in the situation where a 'problem' has previously been identified and a diagnosis is sought, clients may experience loss and disorientation rather than an overt crisis. Whether the news of a diagnosis or risk is new information or eagerly sought after information, clients in large part adapt to it over time. Shelley Taylor's theoretical and empirical work suggests that there are three components to adapting to a health threat (Taylor, 1983). These include: search for meaning from the event (diagnosis or risk), reestablishment of personal control over the threatening event, and restoration or enhancement of self-esteem. Clients largely adapt successfully without the help of healthcare providers. Yet genetic counseling can serve to enhance or expedite the adaptation process.

One component of the search for meaning is attributing a cause to a condition or risk (Taylor, 1983). Genetic counseling offers a scientific explanation about why a condition has occurred but can also addresses clients' metaphysical questions about why the condition has happened or may happen. Clients assign their own explanations (causal attributions) to conditions or risk that may bear little resemblance to scientific or logical explanations. These positive illusions often serve to restore feelings of personal control over the likelihood that the condition might recur. Thus, genetic counseling must embrace clients' nonscientific explanations as well as their own scientific ones in an effort to support clients' needs to restore control over the genetic risk.

In the context of reproductive genetic counseling, the goal is to promote self-determination in exercising choice about the use of prenatal tests and the information gleaned from them. Reproductive genetic counseling aims include:

1. delivery of personalized genetic information to the client in a useful way;
2. exploration of the meaning of the information with the client in the light of personal values and beliefs;
3. promotion of the client's preferences for reproductive options with consideration of alternatives, consequences and barriers; and
4. preparation of the client for adapting to the outcomes of the choice(s).

Genetic counseling that includes the offer of a genetic test for increased risk, such as that for carrying a fetus with Down syndrome due to advanced maternal age, addresses the needs of clients who most often have no prior experience with the condition. The client or couple participates in genetic counseling as a means toward making a decision about whether to take up a test. Clients need to understand what the test can identify and how that information may be useful to them in the light of their own values and attitudes toward parenting. Thus, the focus of the counseling is on the decision-making process. In the infrequent circumstance when an abnormality is identified, clients face yet another and more difficult decision about whether to continue the pregnancy.

In the setting of genetic counseling for common diseases, such as provided by cancer genetics or neurogenetics clinics, the goal is to maintain health in at-risk individuals, and the aims include:

1. enhancement of accurate risk perception;
2. facilitation of adaptation to genetic risk;
3. promotion of health-enhancing behaviors; and
4. prevention of disease.

These health-promotion or disease-prevention settings differ from the previous two in that the counseling focuses on the at-risk individual. Risk assessment is based upon family history and increasingly on the availability of predictive genetic testing. While the personal choice of genetic testing is generally upheld, specific recommendations for health maintenance are made based on risk status. Strategies for changing lifestyle behaviors and promoting disease screening are implemented to increase the success of prevention or early intervention aims. Risks to relatives are discussed as well as risks to future children.

Background and traits of genetic counseling clients

Within this framework for considering the psychological consequences of a genetic diagnosis, risk or test, there are also client traits and characteristics to consider. Clients who participate in genetic counseling come from a particular sociodemographic and ethnocultural background and vantage point that influences how they react to and use the genetics information. Further, they harbor family and other social experiences, beliefs, core values, attitudes, personality traits, perceptions and psychological distress that interact to influence their cognitive and affective interpretations of genetic information. These attributes need to be assessed by the genetic counselor to anticipate clients' reactions and to provide the most therapeutic environment in which the client can understand the information.

Clients' educational and economic statuses contribute to their success at grasping abstract concepts such as probability. The more difficulty clients have in understanding a scientific explanation for a condition in the family, the more likely they are to feel frustration and anxiety. Without some alignment with the healthcare providers' views, clients are more likely to feel marginalized and less in control. Such reactions confound already existing fears stemming from the threat to one's health or the well-being of an affected relative. Thus, the educated, scientifically literate and economically well-situated clients come better equipped to navigate the threat of personal genetic information. Those who are not scientifically literate benefit from making the information more concrete. Special consideration needs to be given to those without access to health insurance or medical resources.

Ethnocultural diversity among clients leads to a variety of world views, health and spiritual beliefs that define a lens through which clients view the genetic information. Loss and human suffering are shared universal experiences, but how they are given meaning varies greatly. The emotional reactions of clients are framed by their beliefs in concepts such as the existence of an afterlife, reincarnation or fate. Clients' ethnocultural background and resulting beliefs provide the context in which genetic information is delivered. The process of genetic counseling is modeled on Westernized notions of health education and psychological counseling and may be limited in working with clients from Eastern countries and clients from the developing world. The success of the counseling depends on mutual respect and shared underlying values such as the potential benefit of learning genetic information.

Clients' personality traits such as dispositional optimism and tolerance for uncertainty also contribute to emotional management of the information. Optimists, for example, may be more willing to take chances such as learning more precisely about one's genetic risk status. If a client is dysfunctional due to an active psychiatric disorder, their reactions to genetic information are greatly altered and impede the counseling process.

Prior experiences of genetic counseling clients

Superimposed upon personality traits are clients' prior experiences with the condition at hand. Whether the family has openly shared genetic information and communicated their feelings about it, whether the condition has led to social isolation or stigmatization, and whether resources have been mined, all contribute to clients' assimilation of genetic information. For instance, if the existence of Huntington disease in a family has been kept a secret, a client's fears escalate, as they understand that it was viewed worthy of concealing. Anger and resentment toward conspirators may result. In such a case, the client is deprived of family models of healthy adaptation. On the other hand, when an affected relative speaks openly about their experience and fears, is supported and cared for in the family, and is eventually moved to a nursing facility according to prior wishes, relatives may perceive the condition to be difficult but survivable. Healthy adaptation may help at-risk relatives to manage their fears.

Differences in perceptions of a condition also depend upon the degree to which clients are physically and/or cognitively impaired, whether the condition is progressive and whether there is chronic pain or frequent medical interventions. Yet illness perception is also substantially subjective and depends on implications of a condition such as disruption of daily living, independence or close relationships to friends and family. One conceptual framework for understanding this phenomenon in the context of genetic conditions is the self-regulatory model of illness (Peters et al., 2001). According to the model, illness perception is comprised of five components: identity (What is the problem?); timeline (How long will it last?); cause (What caused it?); consequences (What will it do to me?); and cure or control (Can it be cured or controlled?). An individual's perception of a specific condition is the result of the assessment of each of the components, both at the concrete level (e.g. chest pain) and emotional level (e.g. fear). There is evidence that these components remain consistent across illness groups; but the content of each component may vary over time, as well as between groups, individuals and even healthcare providers (Leung et al., 1997). Affected individuals tend to perceive their condition as less disabling or disruptive than do the providers. This difference is probably due to adaptation by clients and medicalization of conditions by providers. Yet it begs the question of whose perceptions are most accurate and has implications for genetic counseling with clients who have had no prior experience with a medical condition.

Beyond illness perception are health and spiritual beliefs, personal control and attitudes such as feelings of responsibility toward one's children and other relatives that are associated with the psychological sequelae of genetic diagnosis and risk. Beliefs about the importance of health maintenance and control are pertinent to motivating clients to pursue medical screening or to facilitating decision-making. Characteristics of family and expectations of how members contribute to the family system will influence reactions. For instance, parents who have had a healthy child may be less devastated by a subsequent pregnancy loss than parents who have never had a child and question their ability to have one. Further, clients who believe that one's devotion to God or a higher power is primarily responsible for avoiding cancer recurrence may be more likely to increase their church attendance than their mammography screening. Beliefs about having control over becoming ill or over the ability to bear healthy children are associated with both the meaning and the usefulness of genetic information. Those who believe procreation is in God's hands are less likely to use prenatal testing.

Psychological reactions to genetic counseling

Psychological reactions of genetic counseling clients will depend on the complex interactions of their personality traits and prior experiences. Clients express a wide array of feelings such as fear, worry and anxiety. These are common emotional reactions to any health threat, yet the ambiguity in genetic diagnosis and the uncertainty of genetic risk contribute further to psychological distress. An unexpected diagnosis typically leads to a state of crisis in which clients subconsciously employ defense mechanisms to protect themselves from the full emotional impact of the information. Clients are in shock and therefore it is difficult for them to interpret the information and consequently to make any decisions. Clients initially seek explicit guidance. As the information seeps into their consciousness, there are accompanying feelings of disorientation, being out of control, and feeling lost and alone. Over time, genetic information and counseling provide a venue for restoring some feeling of control, easing fears and employing resources.

Parents feel responsible for their children's health and well-being in general. When a genetic diagnosis or risk is raised for a child, parents may feel particularly guilty that they passed on a gene mutation or risk. Genetic counseling addresses the uncalculated nature of such occurrences, yet may do little to assuage guilt. Guilt serves to bind parents to their children in a state of shared genetic burden (Kessler, 1984). It complicates future childbearing decisions where parents may knowingly choose to take a chance. If a subsequent child is born affected, there may be more substantial guilt involved in living with the consequences of the decision. When nongenetic causes are attributed to the event, such as medication or a household exposure (to something like a cleaning compound), it provides parents with an opportunity to avoid the agent in the future and thus to have control over recurrence of the event. For this and related reasons, most parents have their own strongly held beliefs about nongenetic causes of their children's conditions; yet many clients successfully manage multiple explanations for why genetic events occur.

Shame may be encountered often in genetic counseling as clients struggle with feeling stigmatized or judged less worthy than others (Kessler, 1984). Genetic conditions can lead to physical differences and disabilities that alienate clients from others in our society. Shame is an internal reaction to real or perceived social stigmatization. If society further accepts diversity and differences, there should be less-frequent occasions for clients to experience feelings of shame. Genetic counseling addresses these feelings, but more extensive psychological counseling may be necessary to understand the source and extent of such feelings.

When clients seek genetic counseling, they come with expectations. Their expectations for participating in the service are shaped by their prior experiences and feelings. Yet expectations are also based upon what clients learned prior to attending. If they are referred to genetic counseling to 'have a genetic test', rather than to consider whether a test may be relevant for them and to make a personal decision about testing, there are different expectations. Providers who refer clients for genetic counseling thus play a significant role in setting expectations (Bernhardt et al., 1998). Further, clients may have read about or heard others' experiences with the practice in ways that sets expectations. Genetic counseling should address client expectations in order to engage clients as active participants.

Evaluation of genetic counseling

The therapeutic encounter in genetic counseling addresses clients' most immediate cognitive and affective needs. Active listening and establishment of an empathic connection provide the basis upon which feelings about the decision or the need to adapt to a diagnosis are explored. Counseling strategies employed in genetic counseling depend in large degree on the situation at hand: problem-solving strategies can facilitate decisions about genetic testing; crisis intervention serves to help clients through the initial phases of a new diagnosis; a family-systems approach addresses the impact of a condition on relatives. Together with one of these counseling approaches, genetic information is translated into a useable form. Clients are coached to consider the information during their decision-making or their search for meaning. Coping strategies may be suggested, as well as support resources in the community. The success of genetic counseling depends in large measure on the receptivity of clients to learn, share their feelings, interact and move forward in their lives following a genetic event.

Genetic counseling is a relatively young clinical practice and little research has been conducted demonstrating its effectiveness. 'Outcome studies' have been limited to client recall or understanding of genetic information, client satisfaction and decisions about reproduction or testing (Ever-Kiebooms and van den Berghe, 1979; Sorenson et al., 1981; Wertz et al., 1988; Shiloh et al., 1990; Michie et al., 1997). The results of a small, qualitative study of genetic counselors and their clients argued for expanding the list of outcomes for genetic counseling to include both short-term outcomes such as an increase in the client's sense of being heard, encouraged, valued, supported and attended to, and long-term outcomes such as improved family communication, anticipation of future feelings and experiences and clarification of personal beliefs and values shaping decisions and attitudes (Bernhardt et al., 2000). Of note, the client participants in this study also cited the interpersonal relationship with the genetics counselor as another important outcome. These findings expand upon past genetic counseling outcomes, but need to be measured prospectively and assessed for their generalizability. In response to what has been described by Clarke as the 'manifest inadequacy of outcome measures in genetic counseling', Kessler has called for studies on the process of genetic counseling (Clarke et al., 1996; Kessler, 1992).

'Process studies' are designed to assess practice interior in a manner that captures not only the experiences of, but also the interaction between client and counselor. Aspects of practice that can be assessed include the communication between the participants, the interventions used by the counselor and the needs expressed by the client. The complexity of genetic counseling calls for the careful design of studies. Data from the process can then be used to establish more rigorous clinical practice norms across indications, subspecialties and providers.

Studies of the process of genetic counseling have also been considered with measures of outcomes in an effort to evaluate what variables within the practice affect its outcomes (Michie et al., 1996). Michie and colleagues have assessed how genetic counseling contributes to client satisfaction and mood. They include 'input' in the analysis and define it as the nature of the genetic problem and characteristics of both the client and the counselor. Raters have coded the session transcripts for whether the purpose of the session was explained, patient knowledge and understanding was assessed, counselor uncertainty was expressed, reassurance was given, and whether concerns or questions, emotional issues or social or family issues were raised or addressed. No process predictors of outcome were identified; yet both client and counselor input were found to influence the outcome measures. The investigators called for the importance of including input as a process variable in future studies.

Future process and outcome studies of genetic counseling will likely identify the role of client background (socioeconomic and ethnocultural), personality traits, prior experiences, illness perception, psychological reactions and expectations in shaping client needs and interactions in genetic counseling. Further, such research will enhance our understanding of provider expectations and counseling interventions in meeting those needs. The efficacy of genetic counseling depends upon understanding this professional interaction in greater detail.

References

Bernhardt B A, Geller G, Doksum T, et al. (1998) Prenatal genetic testing: content of discussions between obstetric providers and pregnant women. *Obstetrics and Gynecology* 91: 648–655.

Bernhardt B, Biesecker B and Mastromarino C (2000) Goals, benefits and outcomes of genetic counselling: client and genetic counselor assessment. *American Journal of Medical Genetics* 94: 189–197.

Biesecker B (2001) Mini review: goals of genetic counseling. *Clinical Genetics* 60: 323–330.

Biesecker B and Peters K (2001) Process studies in genetic counselling: peering into the black box. *American Journal of Medical Genetics* 106: 191–198.

Clarke A J, Parson E and Williams A (1996) Outcomes and process in genetic counseling. *Clinical Genetics* 50: 462–469.

Ever-Kiebooms G and van den Berghe H (1979) Impact of genetic counselling: a review of published follow-up studies. *Clinical Genetics* 15: 465–474.

Kessler S (1984) Psychological aspects of genetic counseling. III. Management of guilt and shame. *American Journal of Medical Genetics* 17: 673–697.

Kessler S (1992) Process issues in genetic counseling. *Birth Defects:* Original Article Series 28: 1–10.

Leung S S, Steinbeck K S, Morris S L, et al. (1997) Chronic illness perception in adolescence: implications for the doctor–patient relationship. *Journal of Paediatrics and Child Health* 33: 107–112.

Michie S, Axworthy D, Weinman J and Marteau T (1996) Genetic counselling: predicting patient outcomes. *Psychology and Health* 11: 797–809.

Michie S, Marteau T and Bobrow M (1997) Genetic counselling: the psychological impact of meeting patients' expectations. *Journal of Medical Genetics* 34: 237–241.

Peters K, Kong F, Horne R, Francomano C and Biesecker B (2001) Living with Marfan syndrome I: perceptions of the condition. *Clinical Genetics* 60: 273–282.

Shiloh S, Avdor O and Goodman RM (1990) Satisfaction with genetic counselling: dimensions and measurement. *American Journal of Medical Genetics* 37: 522–529.

Sorenson J R, Swazey J P and Scotch N A (1981) Genetic counseling and client reproductive plans: reproductive pasts, reproductive futures, genetic counseling and its effectiveness. *Birth Defects*: Original Article Series 17: 219–231.

Taylor S (1983) Adjustment to threatening events: a theory of cognitive adaptation. *American Psychologist* 1161–1173.

Wertz D C, Sorenson J R and Heeren T C (1988) Can't get no (dis)satisfaction: professional satisfaction with professional-client encounters. *Work and Occupations* 15: 36–54.

10
Genetic Testing of Children

Annie Procter

Introduction

There are two main situations in which the testing of a child for an inherited genetic disorder might be considered. First, a child who is sick may be tested to make a medical diagnosis and plan treatment. Second, an apparently healthy child may be tested for a genetic disorder that usually, but not invariably, has already manifested in another member of that child's family. The second approach to testing includes both predictive testing, in which the test is being used to predict the likelihood of an apparently healthy child developing a particular disorder at some time in the future, and carrier testing, in which the test is being used to determine the likelihood of an apparently healthy child passing a genetic disorder on to his or her offspring. The predictive and carrier testing of apparently healthy children is the area of genetic testing that causes most controversy and debate between and among the families and professionals concerned. This article discusses the principal issues involved in undertaking such testing.

What is a genetic test?

Genetic testing comes in many forms and is not limited to the analysis of chromosomes or DNA. The scope of genetic tests has broadened enormously and now various technologies are used to identify those individuals who are at risk of developing an inherited disorder. These technologies include, for example, renal ultrasound scanning, which is used to detect renal cysts in otherwise healthy individuals with polycystic kidney disease, and echocardiography, which is used to identify changes consistent with a diagnosis of hypertrophic cardiomyopathy.

Predictive testing

For a child to benefit truly from predictive genetic testing during childhood, a balance must be struck between the 'enabling' aspects of such testing and the potential harm caused by carrying out such investigations. The idea that predictive testing might be performed inappropriately on children was recognized in 1986 in the context of the neurodegenerative disorder Huntington disease (HD) (Craufurd and Harris, 1986). Since then, concern and debate has expanded to encompass other disorders (Harper and Clarke, 1990).

The predictive testing of apparently healthy children may be justified if the results of such testing can be seen to be in the child's best interests, for example if some beneficial medical intervention or lifestyle change can be introduced during childhood. Hence, in some of the familial cancer disorders, screening for tumours in early or mid-childhood may be warranted for those children known to carry disease-associated gene mutations. But for those genetic disorders for which there are as yet no useful interventions, such as HD and prion-associated dementia, the benefits to the child of genetic testing during childhood are much less clear.

Genetic testing during childhood deprives a child of the right to make that decision for themselves when they are older. Any given child, when of an age to make such decisions, might wish not to know their genetic status with regard to a particular condition; testing them in childhood will have robbed them of the right to say 'no' and the right to say 'yes' with confidentiality assured (Michie and Marteau, 1996; Michie et al., 1996). In a family in which a disorder such as HD is well recognized and part of normal life, to be tested for a disorder may seem the obvious thing to do. But it is important to ask 'obvious to whom?' and for whose benefit is the test being performed: the doctor, the parent, the family or the child?

Adults who are considering genetic testing generally do so to remove uncertainty about their risk of developing a particular disorder. It is usual for such an individual to take part in an extensive counseling process before making the final decision about whether to undergo genetic testing or to decline it. An aim of the counseling process is to ensure that the individual recognizes the possible effects of receiving a positive or a negative result to testing, either of which could have enormous and unexpected consequences for the individual. There are complex psychological and social implications associated with the receipt of genetic test results. There are also practical considerations, such as the ability to obtain insurance, buy a house and make certain career decisions. The results of genetic tests may remove one uncertainty, 'will I get the disease?', only to replace it with another, 'when will I get the disease?'. Genetic counseling offers individuals who are considering genetic testing the opportunity to confront all these possibilities and use the experience to aid their decision-making. A child who is subjected to genetic testing during childhood may be denied the opportunity to benefit from this vital process.

The children for whom predictive testing might be considered will often have experienced the 'family disease' at first hand. One could opine that testing in an environment in which a disorder is known might be the most supportive situation in which such testing could occur, and indeed not to test might even, by some individuals, be considered abnormal. However, grave concerns have been expressed about the desirability of predictive childhood testing. These concerns relate particularly to the potential harm of labelling a healthy child as sick (Grosfeld et al., 1997) and causing them to live their lives under the shadow of the uncertain timing of the onset and development of the disorder. That said, and in large part because the opportunity to perform childhood testing is a relatively recent phenomenon, there is still very little clear evidence as to what form this potential harm might take (Codori et al.,1996).

Carrier testing

In general, carrier testing, by definition, will have no implications for the health of the child being tested, but the results may have an enormous effect on the adult life of that child and in particular on the reproductive decisions that such an individual feels able to make. Childhood carrier testing is an area in which families and professionals can come into particular conflict. A family in which a child has suffered and perhaps died of a debilitating genetic disorder may see carrier testing of that child's siblings as hugely advantageous. The professional to whom the request for testing is made is frequently the pediatrician who may have worked with the family, sometimes for several years, to care for the sick family member. That pediatrician can feel that to be seen to refuse such testing could undermine the relationship with the family to such a degree that, despite any misgivings that they might have on behalf of the child to be tested, they feel impelled to do the carrier testing as requested.

The issues relating to autonomy and confidentiality as outlined above apply equally to predictive and to carrier testing in childhood. As with predictive testing, the advantages afforded by 'advance notice' must be balanced against the potential disadvantages of being labelled as a 'carrier', frequently of a very disabling disorder, many years before such information can be used to make the relevant, usually reproductive, decisions.

Consent to testing

The autonomy of the child must not be forgotten in the consideration of genetic testing in childhood. Many professionals therefore feel that it is important to involve older children in the discussion and decision-making processes. The degree of involvement of any given child depends on their capacity to contribute to such decisions. This contribution will be greatly influenced by the child's experience of the disease in their family, as well as their unique personality and their experience of personal and interpersonal conflict (Reder and Fitzpatrick, 1998). Although the Gillick standard is an accepted legal test of competence to consent, there is a need to recognize the abilities of younger or less mature children to consent to a test or to deny their consent (*Gillick versus West Norfolk and Wisbech Health Authority* (1985)).

Even young children have the capacity to contribute to important decisions about their healthcare (Alderson, 1992, 1993), and there is a powerful case for responding sympathetically to requests from adolescents for genetic testing (Binedell et al., 1996; Binedell, 1998). It is important that the views of children with regard to assessing their own genetic status should be explored whenever feasible and that active requests for testing from adolescents should be considered carefully. The adolescent may benefit from both the opportunity to explore the possibilities confronting them and the respect afforded them by the professional who is prepared to consider their views.

General issues

In 1994, the Clinical Genetics Society formed a working party that carried out a survey of professional attitudes to childhood genetic testing (Clarke, 1994). There were very substantial differences of opinion both within and between different professional groups in Britain, but the working party was able to arrive at a consensus: (1) that predictive genetic testing of children for adult-onset disorders, in the absence of a useful medical intervention, would generally be regarded as inappropriate and (2) that the same considerations would apply, albeit less forcefully, to genetic carrier testing of purely reproductive significance to the future adult.

This survey was repeated in 1998 on behalf of the Advisory Committee on Genetic Testing (Procter et al., 1998) and a very similar consensus view emerged. It is evident with regard to predictive testing, however, that differences between professional groups are increasing. The 1994 survey revealed considerable division between paediatricians and geneticists in particular, with regard to the desirability of childhood genetic testing. There is now strong evidence from the 1998 survey indicating that the view that 'it is more important to have the discussions than to do the tests' is increasingly widespread in professional circles, which probably reflects the increasing and expanding availability of genetic tests for several disorders. Carrier testing in childhood remains an area of concern and controversy for the reasons highlighted above and also because many professionals find it very difficult to reconcile, on the one hand, an enthusiastic approach to antenatal testing with, on the other hand, a much more conservative approach to testing once a child is born.

Families may have very different views regarding the desirability of genetic testing in childhood from those expressed by some professionals. This is highlighted by the response of the Genetic Interest Group (GIG), the United Kingdom umbrella organization of lay genetic disease support groups, to the 1994 document of the Clinical Genetics Society. GIG agreed with the very cautious approach to predictive genetic testing of healthy children for late-onset disorders; however, they disagreed sharply with the working party in relation to its similarly cautious attitude toward the carrier testing of healthy children. The GIG response was that there will often be good reasons for carrier tests on children and that such decisions should be left to individual families (Dalby, 1995).

It is important that the 'availability' of childhood testing does not translate directly into the 'advisability' of childhood testing without due consideration. This is never more pertinent than the situation in which a family member is diagnosed with a condition that is recognized to be inherited. Every member of that family, including the children, has a right to make their own decision about the desirability of testing for such a disorder. To avoid the potential for considerable distress and confusion, such decisions are best made in an informed and supportive manner at a time that is right for the individual concerned. Such considerations must also apply when 'new' or 'research' tests become available. The offer of genetic testing of this nature may cause extreme distress within a family and may indeed add very little to the understanding or management of a disorder for the individuals concerned. These issues will become increasingly important as the availability of genetic testing continues to outstrip our abilities to treat or cure many genetic disorders.

As with many things in life, if an individual or family is determined to proceed with an investigation, then it is likely that they will find the means to have their wishes fulfilled. It is therefore vital that the issues surrounding childhood genetic testing are discussed in a fulsome and nonconfrontational manner. This will allow professionals to assess the individual merits of every request for testing and it will allow families the freedom to consider and discover the optimal way to do what they believe will benefit their child most. It must be remembered that, should a decision be made to delay genetic testing until a child is considered old enough to make a personal, informed choice for themselves, then it is vital that provision is made for the required information to be relayed to that child at the appropriate time and in the appropriate manner.

Anyone considering predictive and carrier testing during childhood should approach such investigations with caution. It is important to ensure that the benefits of the information gained and the opportunities presented to that child outweigh the possible disadvantages. With this in mind, families and professionals need to work together, with the interests of the child held jointly as their highest priority. In such circumstances, there is a need for the professional to regard themselves not as the gatekeeper to genetic testing with the absolute right of veto to testing, but rather as the person with responsibility for helping families to consider and think through all the implications of childhood testing with the greatest care (Hoffman and Wulfsberg, 1995).

References

Alderson P (1992) In the genes or in the stars: Children's competence to consent. *Journal of Medical Ethics* 18: 119–124.

Alderson P (1993) *Children's Consent to Surgery*. Milton Keynes, UK: Open University Press.

Binedell J (1998) Adolescent requests for predictive genetic tests. In: Clarke A (ed.) *The Genetic Testing of Children*, chap. 11, pp. 123–132. Oxford, UK: BIOS Scientific Publishers.

Binedell J, Soldan J R, Scourfield J and Harper P S (1996) Huntington's disease predictive testing: the case for an assessment approach to requests from adolescents. *Journal of Medical Genetics* 33: 912–918.

Clarke A (1994) The genetic testing of children. Working Party of the Clinical Genetics Society (UK). *Journal of Medical Genetics* 31: 785–797.

Codori A-M, Petersen G M, Boyd P A, Brandt J and Giardello F M (1996) Genetic testing for cancer in children. *Archives of Pediatric and Adolescent Medicine* 150: 1131–1138.

Craufurd D and Harris R (1986) Ethics of predictive testing for Huntington's chorea: the need for more information. *British Medical Journal* 293: 249–251.

Dalby S (1995) Genetics Interest Group response to the UK Clinical Genetics Society report 'The genetic testing of children'. *Journal of Medical Genetics* 32: 490–491.

Gillick versus West Norfolk and Wisbech Health Authority (1985) 3 All ER 402.

Grosfeld F J M, Lips C J M, Beemer F A, et al. (1997) Psychological risks of genetically testing children for a hereditary cancer syndrome. *Patient Education and Counseling* 32: 63–67.

Harper P S and Clarke A (1990) Should we test children for 'adult' genetic diseases? *Lancet* 335: 1205–1206.

Hoffmann D E and Wulfsberg E A (1995) Testing children for genetic predispositions: is it in their best interest? *Journal of Law, Medicine and Ethics* 23: 331–344.

Michie S and Marteau T M (1996) Predictive genetic testing in children: the need for psychological research. *British Journal of Health and Psychology* 1: 3–16.

Michie S, McDonald V, Bobrow M, McKeown C and Marteau T (1996) Parents' responses to predictive genetic testing in their children: report of a single case study. *Journal of Medical Genetics* 33: 313–318.

Procter A M, Clarke A J and Harper P S (1998) Survey of Genetic Testing in Childhood. Unpublished report on behalf of the Advisory Committee on Genetic Testing and the Department of Health.

Reder P and Fitzpatrick G (1998) What is sufficient understanding? *Clinical Child Psychology and Psychiatry* 3: 103–113.

11
Genetic Risk

Lindsay Prior

Risk and populations

The notion of risk implies danger or hazard plus uncertainty. Linkage of these elements can be achieved only through human decision-making processes. Consequently, the recognition of risk is, essentially, a social activity. In the medical sciences as a whole, the analysis of risk developed substantially in the last quarter of the twentieth century. In fact, Skolbekken (1995) has argued that the late twentieth century medical sciences have been witness to an 'epidemic' of risk. That is not to say that the earlier part of the century was devoid of hazard, only that such hazards were rarely regarded as objects of medical concern. An underlying reason for an accentuated interest in risk is that modern medicine has tended to draw into its orbit of care people who are currently asymptomatic, but who are regarded as being 'at risk' of future harm. This is certainly so in matters of clinical genetics.

In the first half of the twentieth century, genetic risk was understood primarily in terms of populations rather than individuals. For example, according to the Hardy–Weinberg law of genetic equilibrium, in any large population in which discrete generations reproduce by random mating, the distribution of a gene with two alleles can be modeled by the function $(p + q)2$. In other words, AA, Aa and aa would be distributed in the proportions $p2$, $2pq$ and $q2$. Assuming that a recessive and 'undesirable' allele has a low probability (say, $p = 0.01$) then it is clear that only a small proportion of individuals will express the undesirable characteristics ($p2 = 0.0001$). But the recessive allele will be carried by ($2pq = 0.0198$) of the population. It was this law that enabled L. S. Penrose to understand why the proportion of people with forms of inherited mental handicap failed to diminish over generations, despite the fact that the affected individuals rarely reproduced.

The Hardy–Weinberg law is of course still valid, but as a result of developments in genetics that took place after the 1950s it is now possible to zoom in from the broad population risk and make assessments about which specific individuals might be carrying unwanted mutations. In the movement between assessments of population risk and individual risk, however, there is considerable room for uncertainty, and it is this uncertainty that often causes problems for clinicians and patients alike.

Genetics of uncertainty

It is commonly thought that deoxyribonucleic acid (DNA) technology has enabled genetic risk to be translated into certainty, but this is not the case. Even with single-gene disorders there is considerable room for uncertainty in the way of outcomes. Thus, a DNA test for Huntington disease, for example, should enable a person to judge with almost 100% accuracy whether they will succumb to the disease, but it will still not provide any information about the age of onset. A similar ambiguity pertains for cystic fibrosis (another recessive single-gene disorder), in which there are possibly more than 400 mutations of the gene, not all of which are harmful. Similarly, for the breast cancer, early onset 1 (BRCA1) gene that is linked to breast cancer, over 300 mutations had been logged by the year 2000, not all of which lead to adverse outcomes.

Such uncertainties naturally multiply when one considers multigene disorders. Thus, research into Alzheimer disease currently focuses on several mutations. A key candidate is the apolipoprotein E (APOE) gene, which we all carry and which encodes three variant proteins, apoE2, apoE3 and apoE4. apoE2 seems to protect against Alzheimer disease, whereas apoE4 seems to make it more likely. If we inherit one apoE4 variant, we have an increased chance of developing the disease. If we inherit apoE4 from both parents – as do about two in every 100 people – it is suspected that we are much more likely to develop the disease by the age of 80. Given such degrees of uncertainty, however, the assessment of genetic risk is necessarily a multistage process. To illustrate some key features of the process, the remainder of this article focuses on risk relating to cancer in general and to breast cancer in particular.

Assessing cancer risk

Cancer predisposing mutations on the two genes associated with breast cancer (BRCA1 and breast cancer, early onset 2 (BRCA2), which are located on chromosome loci 17q21 and 13q12 respectively) are thought to be carried by around 1–2 individuals per thousand of the population. But such mutations are probably responsible for only about 5–10% of all breast cancers (Evans et al., 2001). BRCA1 is also associated with ovarian cancer. The incidence of cancer with a familial predisposition as a whole is probably in the range of 1 in 30 individuals; thus, about 10% of all cancers have a familial subgroup (McPherson et al., 2000). The penetrance of some of these genes is estimated to be as high as 0.8, which implies that an individual possessing the mutation is at a very high mathematical risk of succumbing to cancer.

Although many families can demonstrate a history of cancers, the decision about whether such histories are significant or not is a complex matter. At present, there are no commonly agreed standards (either in the United Kingdom or otherwise) for deciding when an individual falls prima facie into a category of being at high risk (of developing a familial cancer). In most clinical genetics services, however, the allocation of people into low-, medium- and high-risk categories is initially decided by the use of referral criteria. The latter may be used by general practitioners and secondary care specialists to form a preliminary judgment about the suitability of an individual for detailed investigation.

As far as breast cancer is concerned, inclusion into a high- or moderate-risk category would probably require at least one of the following criteria to be met: one affected, first-degree female relative younger than 40 years of age; two first-degree relatives (on the same side of the family) affected at 60 years of age or younger; three first- or second-degree relatives of any age (on the same side of the family); one first-degree female relative with bilateral breast cancer; or one first-degree male relative with breast cancer. There are numerous published guidelines about who should and who should not be allocated to the various risk categories (Eccles et al., 2000). Different criteria (the Amsterdam criteria) would be used for colorectal cancers (see Vasen et al., 1991). For ovarian cancers, the requirement would be for two or more such cancers in the family with at least one affected first-degree relative.

But referral criteria are flexible. So it would be possible for a clinical service to alter the above rules and abolish, for example, the age barrier of 40 years in the first-degree relative, or to reduce the requirement for three first-degree relatives at any age to two such relatives. Indeed, in setting the boundaries between categories, numerous practical concerns can come into play, including financial and other judgments about the allocation of health service resources (Prior, 2001).

Having gained entry to a clinical service, individuals who fit the criteria are invariably subject to further forms of risk assessment. In assembling such assessments, two types of evidence are commonly sought. The first relates to a family history and the second to the results of molecular tests.

The composition of a family history (a pedigree) depends on information gleaned from individuals and other sources such as hospital records and cancer registries. It should be noted that the manufacture of a pedigree is fraught with ethical and social difficulties. For example, it is not at all clear who properly owns the information in a pedigree, or who should be privy to it. Nevertheless, pedigree data enable a clinician to form an assessment of risk that any individual faces with respect to the onset of a familial cancer. For breast cancer, pedigree data may be fed into a dedicated cancer risk assessment program such as CYRILLIC, which draws a family tree and provides a numerical risk estimate for an individual. Even at this stage, however, estimates of being at risk of breast or of any other cancer can be flexible. Thus, the average lifetime risk for female breast cancer is estimated to be around 1 in 8. But for any individual, the risk may be higher or lower, and will alter according to such things as age, age at menarche and so on. Thus, a 50-year-old woman who has not had breast cancer has a lifetime risk of 11% instead of 12% (McPherson et al., 2000).

In the realm of genetics, risk estimates are commonly calculated according to particular statistical models, such as the Gail model (Spiegelman et al., 1994) or the Claus model (Claus et al., 1991). The role of family history, and of other factors, differs in each model and so a given person's risk of cancer is dependent not simply on items drawn from a personal biography, but also on the relative weight that is given to different factors. Not surprisingly, perhaps, in a comparison of risk assessments applied to 200 women attending a breast cancer clinic in the United Kingdom (Tischkowitz et al., 2000), it was noted that the proportion of such women allocated to a high-risk (of hereditary breast cancer) category varied markedly, from 0.27 using one method to 0.53 using a second method. A third method allocated only 0.14 to the high-risk category. So there are some women for whom risk is systematically 'underestimated' by the very nature of the models. This is because the populations on which the risk models are based are themselves biased. For example, they contain only (North American) women, women who predominantly work in the professions and women with documented family histories. In addition, the samples under-represent ethnic groups known to be at high risk, such as women of Ashkenazi Jewish descent. Men affected by breast cancer are usually absent from the sample populations.

Such biases in the models naturally create difficulties for clinicians, and these difficulties underline the problems inherent in the movement from population to personal risk assessment. When risk prediction models indicate that a person falls into the high-risk category, the possibility of DNA testing arises. For breast cancer, it becomes feasible to consider the use of blood samples and subsequent laboratory analysis to determine the presence of mutations in the CAGT sequences of BRCA1 and BRCA2. It is estimated that the presence of relevant BRCA1 and BRCA2 mutations suggests a lifetime risk of such cancer in the range of 60–85% and a risk of ovarian cancer of about 15–40% (Armstrong et al., 2000). A woman belonging to a high-risk family in whom no mutation is detected would, of course, require a different risk assessment, although there are currently no models available for adjusting the prediction. Negative results in such cases are uninformative.

Given that there are over 300 deletions, insertions and duplications associated with BRCA1 alone, laboratory evidence has to be coupled with family history to form robust and reliable estimates of the genetic risk that any individual might face. In addition, the use of laboratory data in the absence of pedigree information could be dangerously misleading. This is especially so in relation to colorectal cancer, where over a third of affected individuals with a distinct family history of hereditary nonpolyposis colorectal cancer (HNPCC) show no evidence of any abnormality in the DNA sequence. (For colorectal cancer, computer calculations of risk are unavailable, and so categorization into the low- and high-risk bands is achieved on the basis of family history alone.)

This method of using a family history as a gateway to the search for mutations enables geneticists to exert some degree of professional control over the risk assessment process. It also creates a space in which counseling and advice sessions might be directed toward those who need them, because there are enormous potential implications for individuals who undergo such assessments – psychological, social and financial. It is therefore very important that genetic testing be accompanied by counseling on the meaning and the implications of the test result.

References

Armstrong K, Essen A and Weber B (2000) Assessing the risk of breast cancer. *New England Journal of Medicine* 342(8): 564–571.

Claus E B, Risch N and Thompson W D (1991) Genetic analysis of breast cancer in the cancer and steroid hormone study. *American Journal of Human Genetics* 48: 232–242.

Eccles D M, Evans D G R and Mackay J (2000) Guidelines for a genetic risk based approach to advising women with a family history of breast cancer. *Journal of Medical Genetics* 37: 203–209.

Evans J P, Skrzynia C and Burke W (2001) The complexities of predictive genetic testing. *British Medical Journal* 322: 1052–1056.

McPherson K, Steel C M and Dixon J M (2000) Breast cancer – epidemiology, risk factors, and genetics. *British Medical Journal* 321: 624–628.

Prior L (2001) Rationing through risk assessment in clinical genetics. *Sociology of Health and Illness* 23(5): 570–593.

Skolbekken J-A (1995) The risk epidemic in medical journals. *Social Science and Medicine* 40: 291–305.

Spiegelman D, Colditz G A, Hunter D, et al. (1994) Validation of the Gail et al. model for predicting individual breast cancer risk. *Journal of the National Cancer Institute* 86: 600–608.

Tischkowitz M, Wheeler D, France E, et al. (2000) A comparison of methods currently used in clinical practice to estimate familial breast cancer risks. *Annals of Oncology* 11(4): 451–454.

Vasen H F A, Mecklin J-P, Merakhan P and Lynch H T (1991) The international collaborative group on hereditary non-polyposis colorectal cancer. *Diseases of the Colon and Rectum* 34: 424–425.

Web links

Genetic Health http://www.genetichealth.com
Apolipoprotein E (*APOE*). LocusID: 348. EntrezGene: http://www.ncbi.nih.gov/entrez/query.fcgi?db=gene&cmd=Retrieve&dopt= summary&list_uids=348
Breast cancer 1, early onset (*BRCA1*). LocusID: 672. EntrezGene: http://www.ncbi.nih.gov/entrez/query.fcgi?db=gene&cmd=Retrieve&dopt=summary&list_uids=672
Breast cancer 2, early onset (*BRCA2*). LocusID: 675. EntrezGene: http://www.ncbi.nih.gov/entrez/query.fcgi?db=gene&cmd=Retrieve&dopt= summary&list_uids=675
Apolipoprotein E (*APOE*). MIM number: 107741. OMIM: http://www.ncbi.nlm.nih.gov/entrez/dispomim.cgi?id=107741
Breast cancer 1, early onset (*BRCA1*). MIM number: 113705. OMIM: http://www.ncbi.nlm.nih.gov/entrez/dispomim.cgi?id=113705
Breast cancer 2, early onset (*BRCA2*). MIM number: 600185. OMIM: http://www.ncbi.nlm.nih.gov/entrez/dispomim.cgi?id=600185

12
Genetic Susceptibility

Carol Emslie and Kate Hunt

Introduction

The mapping of the human genome opens up the possibility that individuals will be able to gain information about their genetic susceptibility to a wide range of diseases. At present, tests are used to detect single-gene disorders such as Huntington disease. In the future, it is suggested that genetic profiling will be used to advise people about their individual susceptibility to common chronic illnesses such as coronary heart disease (CHD), diabetes and Alzheimer disease. However, these illnesses are known to have multifactorial etiologies, of which genetic susceptibility is just one part. These illnesses develop because of complicated interactions between many different genes, and there are equally complex interactions between genes, behavioral and environmental factors. Thus, there is some debate about when (or indeed, if) it will be possible to develop reliable predictive tests for the polygenic forms of common diseases (Richards, 2001).

Knowledge about how people will interpret and react to this sort of information about likely genetic risk is very limited; however, it is possible to extrapolate from existing research in related areas in order to draw some tentative conclusions about perceptions of and reactions to this new genetic knowledge. This article uses coronary heart disease (CHD) as an example to outline research in two areas. First, it summarizes research on people's perceptions of their genetic susceptibility to heart disease. Second, it highlights some issues that areassociated with cholesterol testing that may well also be relevant if genetic testing for heart disease and other common diseases is introduced.

Perceptions of a 'family history' of heart disease

Lay understandings of the causes of ill health generally encompass many factors, including luck, lifestyle, working and living conditions, and inheritance. Perhaps it is not surprising that inheritance and patterns of illness within families are particularly salient to people's understandings of health and disease (Richards, 1996), given that similarities and differences between relatives are pivotal to the construction of personal and family identity.

When people are asked whether any aspects of poor health 'run' in their family, heart disease is the most commonly reported disease (Hunt et al., 2000). Qualitative research has identified several factors that influence whether people think that heart problems run in their family (Hunt et al., 2001). Lay perceptions of having a 'family history' of heart disease (like medical judgments) generally take account of the number of relatives who have experienced heart problems, whether these relatives were close relatives (such as parents) or more distant relatives (such as cousins), and the age at which family members were affected. But whereas some people have detailed and more or less complete knowledge about both 'sides' of their family, others have less complete knowledge about their family (health) history; family disputes or losing touch with one branch of the family are often cited as reasons for limited knowledge about patterns of illness in current and previous generations.

Two particularly relevant findings have emerged from a study by Hunt et al. (2001). First, whereas some people were certain that heart disease ran (or did not run) in their family, other people were undecided and continued to 'weigh' evidence from their family health history. Men from less affluent backgrounds were more likely to be ambivalent about whether they had a family history than were women, or men from more affluent circumstances. Thus, beliefs about family history can be dynamic rather than fixed, and may be subject to re-evaluation in the light of new illness in the family.

Second, some people made a clear distinction between a family risk of heart disease and their own personal risk of heart disease. Thus, a recognition of heightened risk of heart disease in the family (a family history) was not necessarily translated into increased personal risk, particularly when people either identified differences in their lifestyle as compared with affected relatives (e.g. in diet or smoking histories) or believed that they 'took after' the other side of the family.

New information about genetic susceptibility will not exist in a vacuum, but will be incorporated into these existing beliefs about family history. However, it is difficult to predict how this new information will be regarded: will it be seen as belonging to the realm of the family or to that of the individual? If there is a change from understanding genetic susceptibility as being based on family history to understanding it as being based on a personal blood test, then it could follow that existing knowledge about family health history (and judgments made on the number of relatives with heart disease, their age and their relationship to an individual) will become less important. Heart disease may then be seen as less preventable and more difficult to control. Existing research suggests that information from genetic tests is perceived by people as extremely accurate and more deterministic than information gained from other tests (Senior et al., 2000). However, this new genetic information may still be regarded as family information rather than personal information, and thus understandings may change very little. Genetic susceptibility may still be perceived as just one strand in the etiology of heart disease.

Potential lessons from cholesterol testing

In many respects the identification of genetic bases for raised risk is simply a special case within the general field of screening for risk but this may not be apparent to the lay public nor indeed to many geneticists (Davison, 1996, p. 344).

In the excitement that surrounds discussions about the 'new' genetics, it is sometimes forgotten that there are similarities, as well as differences, between the existing ways of predicting risk of disease and the proposed tests of the future. It is likely that predictive genetic testing for diseases with multifactorial etiology will involve taking samples from asymptomatic individuals and analyzing these samples to provide information about the potential for future disease. A similar process operates, for example, when testing for hypercholesterolemia as a measure of risk of heart disease. Below we draw out key findings from research on diagnosing and treating hypercholesterolemia that seem most relevant to future lay understandings of the meaning of predictive genetic tests.

Tests provide probabilistic risk information about health, not certain outcomes

Hypercholesterolemia is a risk factor for CHD, but there is no certainty that the discovery of high serum cholesterol in an individual will mean death from heart disease. Similarly, predictive genetic testing will provide information about the probability of an individual's risk of future illness, rather than an infallible prediction. Although such genetic tests may increase precision, they will not provide certain forecasts of the future. Davison (1996) argues that uncertainty is likely to remain a central feature of predictive genetic testing for multigene diseases because of the difficulty in estimating the interaction of different genetic components and their modes of inheritance and their variable expression, and because of the problems in predicting how genes and environment will interact.

Tests are unlikely to be completely accurate

There are issues of measurement with diagnostic tests. For example, the cut-off point for 'high' or 'borderline high' cholesterol is always arbitrary to some extent. The population distribution of cholesterol is determined culturally in part and varies between countries and over time. Cholesterol levels also fluctuate over time within an individual: on one test a patient may be classed as having 'high' cholesterol, and a few weeks later they may be below this threshold. This may lead to anxiety in some individuals when their cholesterol seems to climb (Brett, 1991) and to skepticism in others (Troein et al., 1997). Accuracy of measurement may also be affected by laboratory conditions and techniques, and by human error if measurements require any element of subjective evaluation. The consequences of informing an individual that they have a genetic susceptibility to CHD when this is not the case (a 'false-positive' result), or informing them that they do not have genetic susceptibility when they do (a 'false-negative' result), are likely to be serious.

Labeling asymptomatic people as being 'at risk' can cause problems

Hypercholesterolemia is generally symptomless, so understanding this 'diagnosis' and the accompanying higher risk status for CHD may be difficult. Some people find it hard to accept the 'diagnosis', and the ensuing suggestions about behavioral change, when they do not feel unwell (Troein et al., 1997). Those who accept it may question their identity as healthy people and take on a social identity that is neither sick nor well, but 'at risk'; such people have been described as the 'worried well'. This group may be very anxious about their health and behave as though they were sick. Similar problems can apply to people who feel completely well but, as a result of a predictive genetic test, are told that they have a high risk of having CHD in the future. Indeed, they may interpret this high level of risk of future disease as a 'disease' in itself (Gifford, 1986). Problems may be intensified because predictive genetic testing can indicate susceptibility to heart disease much earlier than current markers of risk, such as high serum cholesterol.

Knowledge about increased risk of disease does not automatically lead to behavioral change oriented toward risk reduction

It is often assumed that one reason for informing people that they are at increased risk of a disease (whether as a result of markers for CHD such as high cholesterol or as a result of a predictive genetic test) is to motivate them to make behavioral changes. Davison (1996) has argued, however, that being told that one is at high risk of heart problems because of inherited susceptibility is as likely to lead to a 'fatalistic' attitude (there is 'no point' trying to make behavioral changes) as to lead to a conviction to minimize behavioral risk (that one should go on a low-fat diet, stop smoking or take more exercise).

People may become more skeptical about health promotion and the benefits of behavioral change

Health education often stresses that altering one's diet will reduce serum cholesterol. However, the success of efforts to lower cholesterol vary according to genetic make-up. Brett (1991) reports that individuals who made marked changes in their diet but did not 'achieve' lower cholesterol levels felt disappointed, confused and a sense of personal failure. People may become skeptical about health education and the benefits of behavioral change when they observe that some people eat an 'unhealthy' diet and have low cholesterol but others do 'all the right things' and still have high cholesterol. Existing skepticism about advice on behavioral change (particularly advice about dietary change) may become more pronounced with the introduction of predictive genetic testing.

Conclusion

A more precise and detailed knowledge about individual genetic susceptibility to common, polygenic diseases such as CHD may become available in the future. Owing to the complex interactions between genes and environmental and behavioral factors in the etiology of such diseases, new information about genetic susceptibility is likely to be understood and experienced by people as another risk factor rather than as certain information about future ill health. Gifford argues that risk is experienced as a symptom of a 'hidden or future illness' and 'will always possess an inherent quality of unmeasured ambiguity and uncertainty as a central characteristic ... it must be understood as a dynamic experience of personal uncertainty about one's future' (Gifford, 1986, p. 231). Rather than forecasting the future, predictive genetic tests may intensify our sense of the difficulties of predicting what lies ahead.

References

Brett A S (1991) Psychologic effects of the diagnosis and treatment of hypercholesterolemia: lessons from case studies. *American Journal of Medicine* 91: 642–647.

Davison C (1996) Predictive genetics: the cultural implications of supplying probable futures. In: Marteau T and Richards M (eds.) *The Troubled Helix: Social and Psychological Implications of the New Human Genetics*, pp. 317–330. Cambridge, UK: Cambridge University Press.

Gifford S M (1986) The meaning of lumps: a case study of the ambiguities of risk. In: Janes C R, Stall R and Gifford SM (eds.) *Anthropology and Epidemiology: Interdisciplinary Approaches to the Study of Health and Disease*, pp. 213–245. Dordrecht, Netherlands: D Reidel.

Hunt K, Davison C, Emslie C and Ford G (2000) Are perceptions of family history of heart disease related to health-related attitudes and behaviour? *Health Education Research* 15(2): 131–143.

Hunt K, Emslie C and Watt G (2001) Lay constructions of a 'family history' of heart disease: potential for misunderstandings in the clinical encounter? *Lancet* 357: 1168–1171.
Richards M (1996) Families, kinship and genetics. In: Marteau T and Richards M (eds.) *The Troubled Helix: Social and Psychological Implications of the New Human Genetics*, pp. 249–273. Cambridge, UK: Cambridge University Press.
Richards T (2001) Three views of genetics: the enthusiast, the visionary, and the sceptic. *British Medical Journal* 322: 1016–1017.
Senior V, Marteau TM and Weinman J (2000) Impact of genetic testing on causal models of heart disease and arthritis: an analogue study. *Psychology and Health* 14: 1077–1088.
Troein M, Rastam L, Selander S, Widlund M and Uden G (1997) Understanding the unperceivable: ideas about cholesterol expressed by middle-aged men with recently discovered hypercholesterolaemia. *Family Practice* 14: 376–381.

13 Thrifty Gene Hypothesis: Challenges

Robyn McDermott

'Thrifty gene' hypothesis

In 1962, the population geneticist James Neel proposed the existence of a 'thrifty gene' to explain the apparent paradox of the emergence of diabetes (a disease with a 'well-defined genetic basis') in some populations despite its manifestly adverse effects on reproduction. This thrifty gene made its owners exceptionally efficient in the utilization of food, conferring a survival advantage during times of famine. As hunter–gatherer societies moved from subsistence through agriculture to urbanized Western lifestyles, those carrying the thrifty gene became more susceptible to obesity and diabetes (Neel, 1962).

Since then, numerous studies have tended to label affected populations as 'genetically susceptible', simply on the basis of higher prevalence compared with other ethnic or geographically separate groups. However, ecologic and migrant studies indicate that nongenetic factors acting at various stages in the life course account for a large part of the variation in chronic disease outcomes, including type 2 diabetes mellitus (T2DM), between genetically similar populations in different geographic areas (Cruickshank et al., 2001).

Even in the face of the rapid appearance of T2DM in most populations in the space of one or two generations, including the doubling of diabetes prevalence in the USA and Australia in the last decade, some reviews and textbooks on the subject still claim that T2DM is strongly genetically influenced. What is the evidence that T2DM is a genetic condition?

In the absence of the culprit genes for T2DM, most evidence supporting the genetic theory comes from observed familial susceptibility, documented differences between certain ethnic groups, a high concordance among monozygotic twins and the discovery of rare maturity-onset diabetes of the young (MODY) genes.

Leaving aside the MODY pedigree (which is clearly genetic, accounts for fewer than 1% of diabetics, and is an entity distinct from T2DM), the high prevalence in certain societies, strong family history and observations in twins can be explained equally plausibly by other, nongenetic mechanisms. This is not to argue that genetic differences in susceptibility do not exist, but that the rapid evolution of the epidemic means that contributions from genotypes are probably insignificant compared with changes in phenotype.

Barker hypothesis: 'thrifty phenotype'

In the 1980s the dominant model for the etiology of chronic conditions and T2DM concentrated on identifying adult risk factors. However, these explained only a fraction of observed disease. Barker and colleagues published a landmark series of large cohort studies in the early 1990s in the UK which demonstrated a clear link between low birth weight and poor nutrition in childhood and a range of adult chronic diseases, including abnormal glucose tolerance and T2DM. This effect was amplified among those who were born small and who subsequently gained excess weight as adults (Barker, 1998). These findings have since been confirmed in numerous cohort studies in different populations. Barker proposed a physiological mechanism for this phenomenon, the thrifty phenotype, where poor nutrition in fetal life and early infancy affects the development and function of pancreatic β cells which in turn predisposes the individual to the development of the metabolic syndrome and T2DM (intrauterine programing).

Recent challenges to simple interpretations of Barker's hypothesis come from twins which, despite being lighter at birth than singletons, do not as a group have higher blood pressure or other manifestations of the metabolic syndrome. More recently, Hattersley has proposed the 'fetal insulin hypothesis' to explain the association of low birth weight with later diabetes and vascular disease. In this model, low birth weight and subsequent insulin resistance are merely phenotypes of the same insulin-resistant genotype which impairs insulin-mediated growth in the fetus. Thus, 'the predisposition to NIDDM (noninsulin-dependent diabetes mellitus) and vascular disease is likely to be the result of both genetic and fetal environmental factors' (Hattersley and Tooke, 1999).

Fetal hyperinsulinemia and intergenerational risk (Freinkel's hypothesis of fuel-mediated teratogenesis)

The observation that T2DM runs in families is used to support genetic inheritance theories. However, in 1980 Norbert Freinkel hypothesized that a mechanism for the heritability of diabetes might rest in a form of 'fuel-mediated teratogenesis' during pregnancy, where the fetus exposed to hyperglycemia in utero has long-term anthropometric and metabolic effects, including increased susceptibility to obesity and T2DM (Freinkel, 1980). Supporting this were data from the large Pima Indian cohort studies which showed that the offspring of women with T2DM were more obese and had greater prevalence of T2DM than offspring of nondiabetic women or of women who developed diabetes after the pregnancy. These effects persisted after controlling for age, fathers' diabetic status and the age of onset of the mothers' diabetes.

These studies also suggested that the very early age of onset of T2DM in the female offspring of these diabetic mothers set up a vicious circle where this generation was exposing the next to a diabetogenic intrauterine environment, thus amplifying the risk through the generations (Pettit et al., 1993). Subsequent cohort studies in (mainly Caucasian) nonindigenous women and their offspring in Chicago showed a greatly increased risk of glucose intolerance in the offspring of diabetic mothers by 10 years of age (Silverman et al., 1995). The risk of impaired glucose tolerance (IGT) in this cohort was linked to fetal hyperinsulinemia (correlated with maternal hyperglycemia) rather than to the mothers' type of diabetes, since most of these mothers had type 1 diabetes while their offspring were showing a T2DM syndrome (obesity and IGT).

Animal studies looking at artificially induced fetal hyperinsulinemia showed a similar phenomenon, where offspring developed abnormal glucose tolerance during subsequent pregnancies, and their offspring in turn were macrosomic (Susa et al., 1993). This suggests a teratogenic effect which can persist into the third generation, but which is metabolic, not genetic, in origin.

Causes of cases versus causes of incidence: role of energy imbalance at a population level

Clinicians search for the cause of disease in an individual. Epidemiologists search for the causes of incidence in a population. In 1993, Rose proposed that populations with complex patterns of social history, current exposures and risk behaviors gave rise to 'sick' individuals (Rose, 1993). Complex determinants operating at the population level can shift the distribution of risk in a population, so that 'caseness' (the number of people reaching a diagnostic threshold, for example, for T2DM or hypertension, which is in fact based on probability of risk for certain outcomes, rather than strict 'objective' and immutable diagnostic criteria) can increase suddenly. This is largely what we are observing with the T2DM 'epidemic'.

Populations as a whole are getting fatter, by small increments per year, owing to changes in lifestyle affecting energy balance. Large cohort studies show that body mass index is the single most important predictor of incidence of T2DM, even in a 'low-risk' population of mostly Caucasian middle-class nurses in the

USA (Colditz et al., 1995). The epidemic of obesity and T2DM is appearing in populations (adults and now children) of every genetic type exposed to these social conditions (Visscher and Seidel, 2001). This is challenging for the genetic hypothesis: genotypes cannot change in two generations, so does that mean all populations carry the thrifty gene?

Nutrition, risk behaviors, intergenerational effects and the social gradient

Recent studies of the social epidemiology of chronic diseases in Western societies, including T2DM, demonstrate a clear social gradient in the distribution of risk, particularly obesity, poor micronutrient intake, smoking, low birth weight and poor outcomes.

'Systemic stress', neuroendocrine pathways and hyperglycemia

Another line of argument for the genetic hypothesis is the very high prevalence of T2DM among indigenous people, particularly those affected by 'coca-colonization' in the Pacific and North America. However, new lines of research have begun to address if and how social conditions can affect health outcomes by way of responses mediated by the central nervous system to the social environment. There is evidence that indigenous societies (and some other racial minorities) are subject to ongoing stresses ('systemic stress') in everyday social experience which are manifest in chronically elevated levels of catecholamines which, added to risks accumulated through the life course (low birth weight, adult obesity, poor nutrition) eventually lead to the metabolic syndrome and T2DM, as demonstrated by higher population-level distributions of glycated hemoglobin (Kelly et al., 1997). The subjective experience of racism in daily living has been linked, among African-American men, to higher rates of raised blood pressure, when controlling for other known risk factors (Krieger, 1990).

A 'life-course' approach to type 2 diabetes mellitus and other chronic diseases

A review of the epidemiological, biological and sociological evidence from different settings over the last 100 years has led to a 'composite' model for explaining the incidence of chronic diseases in different populations, where risk accumulates over the life course: poor maternal nutrition and fetal development, poor childhood nutrition, obesity and inactivity in adulthood, smoking and other risk behaviors, adverse physical and social environments and low socioeconomic status all act to increase the likelihood of T2DM and other chronic diseases in adults (Kuh and Ben-Schlomo, 1997).

'Genohype' and the promise of individualized therapies

The announcement of the deciphering of the human genome in February 2001 was accompanied by much fanfare and exaggerated claims that the genome will lead to the unraveling, not just of single-gene disorders, but also of the polygenic forms of common diseases like T2DM, and even 'will eventually tell us ... what we are'. The Human Genome Project has become 'the new power base for big business and big money, as well as the platform for the launch of inflated hopes and monstrous hyperbole' (Radford, 2001). The truth will probably be much less exciting, and the spectacle so far resembles stage one of the 'hype cycle', a term coined by the information technology consulting firm Gartner following the 'dot.com' collapses.

The early phase of the cycle is characterized by a technological breakthrough that excites much media interest. This is followed by a marketing effort that inflates expectations with unrealistic forecasts, and the exaggerated publicizing of some successes, but then more shortfalls as the technology fails to deliver on promises. Much of the money made in this first stage is made by conference organizers and magazine publishers. This phase is followed by disillusion, and finally by a 'plateau of productivity' where the real, much more modest, benefits of the technology are demonstrated. The Human Genome Project has so far discovered over 1000 single-gene disorders that affect less than 4% of the world population. These discoveries may lead to better predictive tests and treatments for some of those individuals.

However, incidence and clinical outcomes in the metabolic syndrome and T2DM are determined by complex interactions throughout the life course. The simplistic reductionist paradigm of the thrifty gene hypothesis is wholly inadequate to describe the complexity of T2DM. The quest for the T2DM genotype has been compared to searching for the susceptibility genes for cholera in the middle of an outbreak.

The other promise, that the Human Genome Project will eventually benefit almost everyone in the world, is also improbable, as the Project proposes to deliver, at best, tailored drug treatments and gene therapy for affected individuals. However increasingly, sufferers of T2DM are poor people, and the roots of their health problems lie in the same poverty which will exclude them from individualized treatments. The logical imperative now is to address urgently the social and environmental determinants of the epidemic of T2DM at a population level, rather than being distracted by the chimera of genetic causes and costly 'cures' for individuals.

References

Barker D J P (1998) *Mothers, Babies and Health in Later Life*, 2nd edn. Edinburgh, UK: Churchill Livingstone.

Colditz G A, Willett W C, Rotnitzky A and Manson J E (1995) Weight gain as a risk factor for clinical diabetes mellitus. *Annals of Internal Medicine* 122: 481–486.

Cruickshank J K, Mbanya J C, Wilks R, et al. (2001) Sick genes, sick individuals or sick populations with chronic disease: The emergence of diabetes and high blood pressure in African-origin populations. *International Journal of Epidemiology* 30: 111–117.

Freinkel N (1980) Of pregnancy and progeny. *Diabetes* 29: 1023–1035.

Hattersley A T and Tooke J E (1999) The fetal insulin hypothesis: an alternative explanation of the association of low birthweight with diabetes and vascular disease. *Lancet* 353: 1789–1792.

Kelly S, Hertzman C and Daniel M (1997) Searching for the biological pathways between stress and health. *Annual Reviews of Public Health* 18: 437–462.

Kreiger N (1990) Racial and gender discrimination: risk factors for high blood pressure? *Social Science and Medicine* 30: 1273–1278.

Kuh D and Ben-Schlomo B (eds.) (1997) *A Life Course Approach to Chronic Disease Epidemiology*. Oxford: Oxford University Press.

Neel J V (1962) Diabetes mellitus: a 'thrifty' genotype rendered detrimental by 'progress'? *American Journal of Human Genetics* 14: 353–362.

Pettit D J, Nelson R G, Saad M F, et al. (1993) Diabetes and obesity in the offspring of Pima Indian women with diabetes during pregnancy. *Diabetes Care* 16: 310–314.

Radford T (2001) Cracking the genome (Book review). *Lancet* 357: 1537.

Rose G (1993) *Strategy of Preventive Medicine*. Oxford: Oxford University Press.

Silverman B L, Metzger B E, Cho N H, et al. (1995) Impaired glucose tolerance in adolescent offspring of diabetic mothers. *Diabetes Care* 18: 611–617.

Susa J B, Sehgal P and Schwartz R (1993) Rhesus monkeys made exogenously hyperinsulinemic in utero as fetuses display abnormal glucose homeostasis as pregnant adults and have macrosomic fetuses. *Diabetes* 42(supplement): 86A.

Visscher T L S and Seidel J C (2001) The public health impact of obesity. *Annual Reviews of Public Health* 22: 355–375.

Web links

Social Science and Medicine. McDermott R (1998) *Ethics, Epidemiology and the Thrifty Gene* http://www.elsevier.com/locate/socscimed

14
Carrier Screening for Inherited Hemoglobin Disorders in Cyprus and the United Kingdom

Bernadette Modell

Introduction

The hemoglobin disorders, thalassemia and sickle cell anemia, are the most common of the severe inherited disorders in humans and are inherited recessively. More than 5% of the human race are carriers without symptoms, and couples who are both carriers have a one in four chance of an affected child in every pregnancy. Worldwide more than 1 in 500 children that are born have a hemoglobin disorder. The disorders are so common because carriers are naturally protected against death from malaria caused by *Plasmodium falciparum* (Weatherall and Clegg, 1981). Carrier prevalence therefore ranges from about 0.1% in northern Europe to over 30% in sub-Saharan Africa, depending on the past (or present) local prevalence of malaria. Global migrations have disseminated the hemoglobin disorders widely, and they are now common in the multiethnic populations of many formerly nonendemic areas, including northwest Europe.

It has been possible to detect carriers by relatively simple blood tests for over 40 years, and prenatal diagnosis has been available for the main disorders for over 20 years (Alter et al., 1976). 'Control' programs for these disorders, which combine best possible individual care with community information, carrier screening and the availability of prenatal diagnosis, were initiated in Cyprus, Italy and Greece in the late 1970s (WHO, 1983). The World Health Organization (WHO) recognized the importance of these pilot genetic screening programs and ensured the organized collection of relevant information.

Prenatal diagnosis is by far the most powerful medical application of genetic knowledge. When it is feasible, most informed couples at risk for a severe inherited disorder use it to ensure a healthy family (Modell et al., 1980). Its application for many inherited disorders is still limited by the difficulty of detecting at-risk carrier couples before they have an affected child, but new diagnostic possibilities arising from the Human Genome Project are likely to resolve this issue in due course. Other important problems revolve around the social and ethical implications of the large-scale use of genetic knowledge within the population. The data obtained from the community-based screening programs for hemoglobin disorders provide the best available guidance on the broad implications of carrier screening and prenatal diagnosis for inherited disorders in different cultural settings.

This article addresses some of the issues involved in establishing and maintaining an equitable program of genetic population screening and genetic counseling; it shows how statistical information can cast light on how a population uses genetic information, and how the community can be consulted on the ethical applications of genetic technology.

Thalassemias and sickle cell disorders have different clinical and social implications. Sickle cell disorders are unpredictable. Typically they cause anemia and attacks of very severe pain, called 'painful crises', which can occur anywhere in the body and sometimes lead to serious organ damage. There is also a risk of sudden death. Diagnosis in the newborn period (by neonatal screening for sickle cell disorders) and simple supportive measures can greatly improve the outlook for affected individuals.

β-Thalassemia major results in a severe anemia that requires treatment by monthly blood transfusions. Blood contains iron that cannot be excreted naturally. If it is allowed to build up, it causes death from overload in early adult life. This is prevented by nightly infusions of a drug that removes iron (desferrioxamine) from a small portable syringe driver. However, this management is very burdensome for individuals, families and the community, and it is too costly for developing countries where the disorders are common (WHO, 1983). As a result, there has been consistent pressure from families and communities to prevent thalassemias. This article focuses on the experience of population screening for thalassemia, with particular emphasis on the issue of how people use genetic information once they have it.

Consulting the community

The three basic ethical principles of genetic counseling are, first, the autonomy of the individual or couple; second, their right to full information; and third, the highest level of confidentiality (Fletcher et al., 1985). Ethical principles that apply to the individual also apply to the community; therefore, the recognized objective of genetic screening is informed choice, and a program designed to meet the needs of the community must be based on the observed choices of informed at-risk couples.

In carrier screening for hemoglobin disorders, carriers might use knowledge of their carrier status when choosing a partner (e.g. by avoiding marriage with another carrier), and/or in reproductive choice (by choosing to limit the number of children they have or to make use of prenatal diagnosis); alternatively, they might not use the information at all.

Thalassemia in the United Kingdom and Cyprus

β-Thalassemia is common in the Mediterranean area, particularly in the islands of Cyprus (carrier prevalence 17%) and Sardinia (carrier prevalence 12%) (WHO, 1983). Between 1957 and 1967, when the role of genetics in pediatrics was being increasingly recognized, 47,000 young Cypriots (almost 10% of the population) migrated to the United Kingdom because of civil strife and economic problems at home. Most settled in north London, and children with thalassemia major began to turn up in the pediatric wards of local hospitals. There have been close links between the United Kingdom and Cyprus in research and service development for thalassemia ever since.

Population studies in the United Kingdom determined the high prevalence of β-thalassemia carriers among Cypriots and showed that the whole community is at high genetic risk (Modell et al., 1972). At the same time, it became clear that couples who knew that they were at risk because they had an affected child tried to avoid further pregnancies, and when a pregnancy did occur most of them requested abortion despite the 75% chance that the outcome would be the birth of a wanted healthy child (Modell et al., 1980). This behavior conveyed unambiguously the parents' view of the burden of thalassemia and the need for better solutions.

The high prevalence of thalassemia in Cyprus was recognized soon afterwards because affected children were being increasingly diagnosed and referred for blood transfusion. As a result these children survived, and the total number of individuals requiring treatment began to increase very rapidly. In 1971, a WHO consultant concluded that within 20 years all available Cypriot blood donors would have to give blood once a year for thalassemia alone, and the island's drug budget would be doubled (Ashiotis et al., 1973). Thalassemia was clearly a priority health problem, and a commitment to long-term individual care could be undertaken only if the birth rate of affected children could be reduced (Angastiniotis and Hadjiminas, 1981). By 1975, carrier screening was being considered as an option both in Cyprus and in the United Kingdom.

At that time it was not known what use people would make of information on carrier status, but it seemed possible that young couples who were aware of their risk before marriage might prefer to find a different partner. Carrier screening with the objective of discouraging marriage between carriers was tested in Greece and Cyprus, but it soon became clear that the population did not feel that this was an acceptable solution (Angastiniotis and Hadjiminas, 1981).

Role of prenatal diagnosis

In the early 1970s, a combination of carrier screening and prenatal diagnosis was shown to be feasible and acceptable to communities at risk for Tay–Sachs disease, a serious recessive disorder common among European Jews (Kaback et al., 1974). Although reliable carrier testing was feasible for hemoglobin disorders, at that time prenatal diagnosis did not seem to be an option. Diagnosis would require analysis of fetal blood, and no method for obtaining blood from the fetus existed. It also seemed that prenatal diagnosis would not be possible even if fetal blood were available, because β-thalassemias and sickle cell disorders involve the β chain of hemoglobin, which is present in adult hemoglobin ($\alpha_2\beta_2$), whereas fetuses make fetal hemoglobin ($\alpha_2\gamma_2$).

Prenatal diagnosis did prove to be possible, however, because classical studies carried out in the 1950s had shown that human embryos make a small proportion of adult hemoglobin. This was confirmed in the mid-1970s (Kan et al., 1972), and it turned out that the small amount of adult hemoglobin made by the fetus could be measured reliably by combining biochemical techniques for separating the globin chains of hemoglobin with radioactive labeling methods (Alter et al., 1976). Demand from parents of affected children then soon led to the development of methods for fetal blood sampling and for prenatal diagnosis of hemoglobin disorders (Alter et al., 1976).

Fetal blood sampling can be carried out safely only after 18 weeks of gestation, and prenatal diagnosis for hemoglobin disorders shared the common limitation of all other prenatal diagnosis approaches at that time: if the fetus was affected and the parents requested termination of pregnancy, then this usually took place at or after 20 weeks of gestation. Even so, it was found that the Cypriot community wanted carrier screening and used prenatal diagnosis to ensure a healthy family (Modell et al., 1980). Prenatal diagnosis for hemoglobin disorders became part of the UK National Health Service in 1979 and also spread rapidly in Europe and the Mediterranean regions.

First trimester prenatal diagnosis

Because an approach based on mid-trimester abortion would not be acceptable on religious and social grounds in many countries where thalassemia was a public health problem, the next priority was to find a method for earlier prenatal diagnosis.

Fortunately, the hemoglobin genes were the first human genes to be studied using deoxyribonucleic acid (DNA) methods. This made it theoretically possible to use samples of tissue from the developing placenta (i.e. chorionic villi, which have the same genetic make-up as the fetus) to achieve far earlier prenatal diagnosis. An obstetric technique for chorionic villus sampling was developed by 1982, and this was used with the new DNA technology to advance prenatal diagnosis of hemoglobin disorders to the first trimester of pregnancy (Old et al., 1982). This approach has proved widely acceptable. It has permitted the global dissemination of approaches for controlling hemoglobin disorders, and it has also advanced the prenatal diagnosis of most other inherited disorders to the first trimester of pregnancy.

Role of the World Health Organization

The availability of prenatal diagnosis is a necessary, but not a sufficient, condition for a thalassemia control program. In the absence of carrier screening, at-risk couples can be identified only through diagnosis of their first affected child. Even if they then avoid having a second affected child, there is a relatively small effect on the affected birth rate. Only a carrier screening program can identify at-risk couples prospectively and allow them an informed choice in every pregnancy. Population screening can be difficult to establish, however, because it is a complex public health intervention that affects the whole community. It needs the explicit commitment of the public health authorities, involvement of the health service at many levels, careful planning, clear policies and regular service audit.

In 1980, the WHO recognized that thalassemia control programs provide a model for the control of inherited disorders in the population. A WHO Working Group on Hemoglobin Disorders was convened to work out the principles of genetic population screening. Among other methods identified, they recommended a two-step approach for measuring the effects of screening and prenatal diagnosis: first, measuring changes in the birth rate of affected children; and second, enquiring into the circumstances of affected births to assess the relative roles of informed parental choice (the objective of the program) and of service failures.

The working group also noted that it is simpler to organize a screening program in a small community (e.g. Cyprus), where thalassemia is a recognized public health problem and a single center can run the program. The challenge is greatest where screening must be delivered to diverse minority groups scattered among a large majority population that is not at risk, as in the United Kingdom. The comparison between outcomes for Cyprus and the United Kingdom summarized here confirms this prediction. It also shows that if the problem of delivering screening for hemoglobin disorders can be solved in countries such as the United Kingdom, in principle most of the problems surrounding screening for recessive disorders will be solved.

Outcomes in Cyprus

In Cyprus carrier screening was initially offered during pregnancy, as it is in the United Kingdom today. However in 1982 the Greek Orthodox Archbishop pointed out that this does not allow at-risk couples the full range of choice, including finding another partner or avoiding pregnancy in the first place. He therefore mandated, for ethical reasons, that all couples planning to marry in church should present a premarital certificate from the government laboratory, confirming that they have been tested and advised accordingly. Confidentiality is preserved and no results are given. This approach has proved acceptable and has many advantages. For example, it is equitable because almost everyone gets married in church, therefore almost all at-risk couples learn of their risk at an early stage and have the option of first trimester prenatal diagnosis in every pregnancy.

In addition, the thalassemia center records the results of all screening tests, and the choices of all at-risk couples. The data show conclusively that most people use genetic information for reproductive choice, but not in partner choice. Ninety-six per cent of intending couples who find out that they are both carriers before marriage proceed to marriage. More than 96% request prenatal diagnosis in every pregnancy, and when the fetus is affected, more than 96% decide to terminate the pregnancy. As a result, there have been very few affected births in Cyprus since 1986. The data also show that 20% of the fall in affected birth rate is due to at-risk couples settling for a smaller final family size than the population norm, whereas 75% is due to the use of prenatal diagnosis and selective abortion.

Outcomes in the United Kingdom

In the United Kingdom, the decreasing popularity of marriage rules out premarital screening. The present screening policy depends on carrier testing during pregnancy, and choices for the carrier couples identified are limited to requesting prenatal diagnosis or 'taking the risk'. Two national registers, one of individuals with thalassemia and one of prenatal diagnoses for hemoglobin disorders, identify all known affected conceptions and their outcomes. In 1985 and again in 1997, these registers showed that the birth rate of affected infants had dropped by only 50%, in contrast to a far greater drop in Italy, Greece and Cyprus. Most couples who had had a prenatal diagnosis were of Mediterranean origin, and those who had not were British Asians, mainly British Pakistanis. To many, this finding seemed to show that British Pakistanis did not want prenatal diagnosis because termination of pregnancy is unacceptable to Muslims on religious grounds.

But the second audit in 1997 told a different story. The UK National Confidential Enquiry into Genetic Counseling (Modell et al., 2000) reviewed the obstetric records of a sample of women with an affected pregnancy and found that most British Cypriots living in southeast England had access to informed choice during pregnancy, whereas most British Pakistanis living in the midlands and the north were not informed of their risk and so did not have access to choice. Their risk was identified either too late in pregnancy for the option of prenatal diagnosis, or when an affected child was diagnosed – a disappointing outcome for a service that has long been considered standard practice. Many different problems were identified, most of which reflected the lack of a clear screening policy. Perhaps this is not surprising, because in the absence of central guidance 145 separate district health authorities were expected to work out, deliver and audit their own screening programs.

The enquiry also showed that uptake of prenatal diagnosis (when it was offered) differed by ethnic group and gestation at counseling. Practically 100% of Cypriots requested prenatal diagnosis whether it was offered in the first or second trimester of pregnancy. By contrast, British Pakistanis made relatively little use of prenatal diagnosis when it was offered in the second trimester, but over 70% requested it when it was offered in the first trimester. Thus, regular service audit can elicit the views of the community, including the views of minority groups that have difficulty in articulating their needs.

The conclusion is clear. At-risk couples require both early information and an early offer of prenatal diagnosis, and antenatal screening as it is organized at present is not capable of meeting this need. Carrier screening and information should be provided by general practitioners, either before pregnancy or as part of the management of early pregnancy.

The UK National Plan for the Health Service (National Health Service, 2000) includes a commitment to a national, linked program of antenatal and neonatal screening for hemoglobin disorders. This will do much to improve service quality and equity, and it will require continuous audit using diagnosis registers.

References

Alter B P, Modell B, Fairweather D V I, et al. (1976) Prenatal diagnosis of haemoglobinopathies: a review of 15 cases. *New England Journal of Medicine* 295: 1437–1443.

Angastiniotis M A and Hadjiminas M G (1981) Prevention of thalassemia in Cyprus. *Lancet* i: 369–370.

Ashiotis T, Zachariadis Z, Sofroniadou K, Loukopoulos D and Stamatoyannopoulos G (1973) Thalassaemia in Cyprus. *British Medical Journal* ii: 38–42.

Fletcher J C, Berg K and Tranoy K E (1985) Ethical aspects of medical genetics: a proposal for guidelines in genetic counseling, prenatal diagnosis and screening. *Clinical Genetics* 27: 199–205.

Kaback M M, Zeigler R S, Reynolds L W and Sonneborn M (1974) Approaches to the control and prevention of Tay–Sachs disease. *Progress in Medical Genetics* 10: 103–134.

Kan Y W, Dozy A, Alter B P, Frigoletto F D and Nathan D G (1972) Detection of the sickle gene in the human fetus: potential for intrauterine diagnosis of sickle cell anemia. *New England Journal of Medicine* 287: 1–5.

Modell B, Benson A and Payling-Wright C R (1972) Incidence of beta thalassaemia trait among Cypriots in London. *British Medical Journal* ii: 737–738.
Modell B, Ward R H T and Fairweather D V I (1980) Effect of introducing antenatal diagnosis on the reproductive behavior of families at risk for thalassemia major. *British Medical Journal* ii: 1347–1350.
Modell B, Harris R, Lane B, et al. (2000) Informed choice in genetic screening for thalassemia during pregnancy: audit from a national confidential enquiry. *British Medical Journal* 320: 325–390.
National Health Service (2000) *A Plan for Investment, a Plan for Reform*. London: Her Majesty's Stationery Office.
Old J M, Ward R H T, Petrou M, et al. (1982) First-trimester fetal diagnosis for the haemoglobinopathies: three cases. *Lancet* ii: 1413–1416.
Weatherall D J and Clegg J B (1981) *The Thalassaemia Syndromes*, 3rd edn. Oxford: Blackwell Scientific Publications.
WHO (1983) Community control of hereditary anaemias: memorandum from a WHO meeting. *Bulletin of the World Health Organization* 61: 63–80.

Web links

Accessible Publishing of Genetic Information (APoGI) for Haemoglobin Gene Variants. [Basic information about hemoglobin disorders, designed for use by affected individuals, health workers and members of the public.] www.chime.ucl.ac.uk/APoGI/

15

Carrier Screening of Adolescents in Montreal

Charles R. Scriver and John J. Mitchell

Introduction

Communities in Montreal and Quebec province, in which there were elevated birth incidences of Tay–Sachs disease and β-thalassemia major, adopted voluntary programs of genetic screening and testing. The programs focused on carrier detection and provided opportunities for reproductive counseling and prenatal diagnosis. Participation was found to be highest among adolescents (16–18 years of age), and high schools offered an ideal setting for dispersal of information. Through the program, the incidence of these two Mendelian conditions in Quebec has declined by over 90% since the early 1980s.

Screening and testing

The screening process implies populations; testing involves individuals. Medical screening is a procedure that is used to differentiate a person with the disease from those who probably do not have it. The term assumes that signs and symptoms of the disease are already present. 'Genetic screening' is a search in a population for persons harboring susceptibility-causing or disease-causing alleles (or genetic markers) that may lead to disease in the individual or in that person's offspring. 'Genetic testing' identifies the status of an individual already expected to be at high risk for a particular disease by affiliation in a family or because of existing clinical manifestations; it may lead to investigation of other members in the extended family (so-called 'cascade screening'). Genetic screening and testing introduce the opportunity to prevent, or avoid, the associated 'genetic' disease.

The essential rationales for genetic screening/testing are the following:

- To identify persons who will benefit from early treatment.
- To introduce counseling about the inherited genetic risk and the reproductive options, where there is no effective treatment for the condition of concern.
- To gain information about the distribution of mutations and their phenotypes in populations, and about factors that may contribute to expression of the genetic phenotype.

Locales of the 'Montreal' carrier screening projects

Montreal is the largest city in the province of Quebec, Canada (Scriver, 2001). The screening programs took into account its multiethnic character. Large migrations to Montreal of Jewish, Italian and Greek citizens from abroad took place in the twentieth century. At the outset of the screening programs, the Jewish community comprised ~75,000 people, the Greek community ~100,000 and the Italian community ~250,000. These ethnic communities tended to cluster in particular regions of the city; as a result, the students attended particular schools in the school system – a feature that is helpful to the screening program.

Tay–Sachs disease occurs in the Ashkenazi Jewish community of Montreal; it also occurs, owing to a unique and different mutation, in a French Canadian community on the south shore of the lower St Lawrence river. β-Thalassemia minor (the carrier phenotype) is prevalent in the Montreal Greek and Italian communities, and also in a French Canadian community in Portneuf County on the north shore of the St Lawrence river near Quebec. Genetic screening for carrier detection took place in each of these communities.

Diseases

Tay–Sachs disease and β-thalassemia major are both autosomal recessive phenotypes at the clinical level. The classical homozygous Tay–Sachs phenotype is a fatal, untreatable neurodegenerative disorder of infancy and childhood. It is caused by various mutations in the Hexosaminidase A (HEXA) gene on chromosome 15q23–q24; the affected enzyme is lysosomal hexosaminidase (hexA).

β-Thalassemia major, in the absence of treatment, is a fatal anemia of childhood; aggressive blood transfusion, chelation therapy to remove excess iron, and early use of bone marrow transplantation provide ameliorative therapy. β-Thalassemia is caused by several mutations at the β- hemoglobin (*HBB*) locus on chromosome 11p15.5; they cause deficient synthesis of the β-globin moiety of hemoglobin.

Carrier detection tests

The 'Montreal' screening tests are designed to detect the heterozygous phenotypes reliably. The Tay–Sachs carrier detection test uses the synthetic substrate 4-methylumbelliferyl N-acetyl glucosaminidase (4MUG) to measure partial loss of hexA activity in serum (or, for example, in white blood cells) by a semiautomated heat-denaturation assay (Delvin et al., 1974). Deoxyribonucleic acid (DNA)-based tests can be used to identify HEXA mutations directly and to resolve any ambiguities in the phenotype test in both the Montreal Jewish community and the French Canadian isolate.

The β-thalassemia carrier detection test uses standard hematological methods to measure mean erythrocyte corpuscular volume and the concentration of hemoglobin A2 in a whole blood sample. Where necessary, DNA mutation analysis can be used to identify the HBB mutations causing β-thalassemia in the Montreal communities and in the Portneuf isolate. Test results in both programs are classified by Bayesian density discriminant functions (Gold et al., 1974; Zannis-Hadjopoulos et al., 1977).

Screening of adolescents

The Montreal program uses a population window (high schools) and focuses on senior high-school students. Bottom-up and top-down approaches were used to implement the programs by involving families in which there were affected members, student/parent groups, religious leaders (rabbis and priests), directors and committees of school boards, principals and teachers from whose classrooms the students would come. In this way, awareness and agreement emerged so that the programs belonged to the communities themselves; participation was voluntary and recognized collective and individual consent. Medical geneticists and others contribute technical expertise and assume responsibility for performing and interpreting the tests.

Students are assembled at their own schools for a session of 'genetics education' given by the geneticists. They receive information about the thalassemia and Tay–Sachs diseases, and also about cystic fibrosis and sickle cell disease as other examples of genetic diseases that cluster in particular populations or regions of the world. The students then have the option to participate (at some later moment) in the screening phase of the program. If they choose to be tested, they fill out data cards for program records and a self-addressed envelope to receive the test result. Parental consent was not required initially, but now it is obtained and the student is tested only when that consent is on file. The blood sample is then drawn, taken to the laboratory and analyzed; the interpreted test result is returned only to the student (except in the 'Hassidic' program, see below). The laboratory components of the program are monitored by quality control.

The fact that DNA can be analyzed in the Montreal-based programs is important in the phase of reproductive counseling. The mutations causing Tay–Sachs disease in the Montreal community are known (three mutations account for 96% of the mutant alleles) (Fernandes et al., 1992); the corresponding mutations in the French Canadian isolate are also largely known (Hechtman et al., 1990). The main mutations causing β-thalassemia in the Mediterranean-derived populations of Montreal (and also in the Asian and Oriental communities) are known (Kaplan et al., 1991), as are those in the French Canadian isolate of Portneuf (Kaplan et al., 1990).

Tay–Sachs Program

This program has been monitored several times since its formal inception in 1972 (Mitchell et al., 1996). A threefold better than average participation rate was observed early on among high-school students relative to other groups in the Jewish community. By 1977, about 40% of the participants in the screening phase were high-school students, among whom the participation rate was 75% of those eligible to be tested. The students indicated that the educational component had increased their knowledge of human genetics and Tay–Sachs disease, and they expressed positive attitudes for both the principles of genetic screening and their own experience in the program. Most individuals said that they intended to make use of the information gained from the program.

A decade later, carriers and noncarrier students, now in the age range of 21–26 years, were surveyed anonymously. Any initial anxiety about being a carrier had declined significantly, and this cohort believed that the self-knowledge had or would have little or no effect on their choice of partner. The approval rate for the principles of genetic screening, and for the high school screening project itself, was very high (~95%).

After 20 years of operation, the high school Tay–Sachs screening program was again analyzed for process and outcome variables (Table 15.1). Among the features analyzed were voluntary participation rates in the high-school cohorts, uptake rates for the screening tests, origins of carrier couples seeking the prenatal diagnosis option in the program, and change in incidence of Tay–Sachs disease in the Quebec population. Of the 14,844 students who had participated in the Tay–Sachs program, 521 were identified as carriers (a frequency of 1 in 28 in the defined population). The educational component of the program had reached 89% of the demographic cohort. The voluntary participation rate in the screening phase was 67%. The carriers identified in the high-school cohort had remembered their status and had had their partner tested if they did not already know whether they were a carrier couple. All carrier couples took up the option for reproductive counseling and prenatal diagnosis.

Table 15.1 High school carrier screening programs in Montreal (1973–1992)

	Tay–Sachs program	**β-thalassemia program**
	Screening phase	
Screened participants (n)	14,844	25,274
Carriers detected (n)	521	693
Participation rate (%) (Screening phase)	67	61
	Prenatal diagnostic phase	
Pregnancies monitored (n)	32	56
Affected fetuses (n)	8	13
Terminations (n)	8	11[a]
Unaffected live-born	24	43
Decline in birth incidence of disease (%)	90	95

[a] Two affected infants were born: one by choice and one owing to an error in fetal diagnosis (a false negative attributed to maternal contamination of a chorionic villus sample).

During the course of the screening program, the incidence of Tay–Sachs disease fell by 90% in the province, while the fertility rates of couples either at risk for Tay–Sachs disease or in the population at large remained the same.

β-thalassemia program

A preprogram survey was carried out in 1978; over 3000 citizens participated from the high-risk communities, of whom 85% were high-school students. Initiation of the program was recommended by 88% of those surveyed, although only 31% considered fetal diagnosis as an acceptable option at the time. Screening in high schools or before marriage was preferred by 56% of the participants.

The formal program began in December 1979. In a 2-year period, almost 7000 people were screened, of whom slightly more than 5000 were high-school students; participation rates by students in the screening phase was 80%. Carriers, both students and adults, shared their information with family and friends, and this led to cascade screening where appropriate. Adult-age carriers pursued testing of their partner, and there were 11 fetal diagnoses, representing 75% participation in the target population at the time (1980–1983). The findings indicated collective acceptance of the program, appropriate attitudes among carriers, general acceptance of fetal diagnosis and global cost-effectiveness.

The program was evaluated again in the 1990s (Table 15.1), by which time 25,274 students had been tested representing 67% of the Italian and Greek cohorts and 61% voluntary participation in the screening phase; 693 carriers had been detected, representing a heterozygote frequency of 1 in 36 persons. Information diffused through the communities and an estimated 100% of couples at risk were reached by the counseling phase. Thirty-two carrier couples requested prenatal diagnosis and 56 pregnancies were monitored; out of 13 affected fetuses, there were 11 terminations and the birth of two affected infants – one reflecting the choice of the couple, the other a diagnostic error. These families had 43 unaffected live-born infants. The incidence of β-thalassemia in the communities at risk declined by 95% during the 13 years monitored in the program.

Commentary

The high-school carrier screening program emerged in the Quebec Network of Genetic Medicine (Scriver et al., 1978), a government-funded structure which permitted social values, expressed as universal healthcare, to address genetic diseases in the Quebec population; the principles and practice of voluntary prenatal diagnosis were part of the Quebec Network of Genetic Medicine. A concept later called 'community genetics' has always informed the work of the Quebec Network of Genetic Medicine.

The high-school screening programs fulfilled their goals: to communicate relevant information to the communities at high risk, to identify carriers by reliable voluntary tests, to inform carriers confidentially about their test results, and to facilitate the making of informed decisions about reproductive options. We used the population window of high schools because it is an efficient one in which to reach an age-related cohort in society. It also recognizes that adolescents have mastered the ability to apply logical rules and reasoning to abstract problems: adolescents are treated as young adults here and the program validates that point of view.

We used population screening as the first approach, followed by genetic testing when necessary, and the efficiency of the programs – as measured by a reduction in incidence of the target diseases – was 90% or greater. If we had provided only testing in families identified through an affected proband, the efficiency of the programs would not have exceeded 12.5%, assuming two-child families as the norm.

Other societies will have different views about genetic screening among adolescents. For example, the committee that produced the American report Assessing Genetic Risks thought that carrier screening in high schools was not advisable (Scriver, 1995). However, each community has its individual character, and recommendations for one community may not fit another. Each community will find its own way to address its problems and, if it wishes, to implement genetic screening and testing while respecting autonomy, justice, privacy, equality of opportunity and quality of services for individuals (Scriver, 1995).

Experience from the prevention and/or avoidance of Tay–Sachs disease and β-thalassemia at the community level in Montreal has led to other studies: (1) the testing for cystic fibrosis carriers by DNA analysis (Mitchell et al., 1993), with such testing now being accepted widely in Quebec; and (2) the possibility of screening (carriers and newborns) for sickle cell genotypes, with no broad acceptance of such programs yet. The Hassidic community, which will not accept prenatal diagnosis and pregnancy termination for Tay–Sachs disease, has integrated anonymous screening into the matchmaking process, thereby retaining cultural values while adopting technology; the incidence of Tay–Sachs disease has declined markedly among the Hassidim.

References

Delvin E, Pottier A, Scriver C R and Gold R J M (1974) The application of an automated hexosaminidase assay to genetic screening. *Clinica Chimica Acta* 53: 135–142.

Fernandes M J G, Kaplan F, Clow C L, Hechtman P and Scriver C R (1992) Specificity and sensitivity of hexosaminidase assays and DNA analysis for the detection of Tay–Sachs disease gene carriers among Ashkenazi Jews. *Genetic Epidemiology* 9: 169–175.

Gold R J M, Maag U R, Neal J L and Scriver C R (1974) The use of biochemical data in screening for mutant alleles and in genetic counselling. *Annals of Human Genetics* 37: 315–326.

Hechtman P, Kaplan F, Bayleran J, et al. (1990) More than one mutant allele causes infantile Tay–Sachs disease in French Canadians. *American Journal of Human Genetics* 47: 815–822.

Kaplan F, Kokotsis G, DeBraekeleer M, Morgan K and Scriver C R (1990) β-Thalassemia genes in French Canadians: haplotype and mutation analysis of Portneuf chromosomes. *American Journal of Human Genetics* 46: 126–132.

Kaplan F, Kokotsis G, Capua A and Scriver C R (1991) Quantification of β-thalassemia in Quebec immigrants of Mediterranean Southeast Asian and Asian Indian origin. *Clinical and Investigative Medicine* 14: 325–330.

Mitchell J J, Scriver C R, Clow C L and Kaplan F (1993) What young people think and do when the option for cystic fibrosis carrier testing is available. *Journal of Medical Genetics* 30: 538–542.

Mitchell J J, Capua A, Clow C and Scriver C R (1996) Twenty-year outcome analysis of genetic screening programs for Tay–Sachs and β-thalassemia disease carriers in high schools. *American Journal of Human Genetics* 59: 793–798.

Scriver C R (1995) Book review of Andrews L B, Fullarton J E, Holtzman N A and Motulsky A G (eds.) (1994) *Assessing Genetic Risks: Implications for Health and Social Policy*. Washington, DC: National Academy Press. *American Journal of Human Genetics* 56: 814–816.

Scriver C R (2001) Human genetics: lessons from Quebec population(s). *Annual Review of Genomics and Human Genetics* 2: 69–101.

Scriver C R, Laberge C, Clow C L and Fraser F C (1978) Genetics and medicine: an evolving relationship. *Science* 200: 946–952.

Zannis-Hadjopoulos M, Gold R J M, Maag U R, Metrakos J D and Scriver C R (1977) Improved detection of β-thalassemia carriers by a two-test method. *Human Genetics* 38: 315–324.

Web links

hexosaminidase A (HEXA); LocusID: 3073. EntrezGene: http://www.ncbi.nih.gov/entrez/query.fcgi?db=gene&cmd=Retrieve&dopt=summary&list_uids=3073

β-hemoglobin (HBB); LocusID: 3043. EntrezGene: http://www.ncbi.nih.gov/entrez/query.fcgi?db=gene&cmd=Retrieve&dopt=summary&list_uids=3043

hexosaminidase A (HEXA); MIM number: 272800. OMIM: http://www3.ncbi.nlm.nih.gov/entrez/dispomim.cgi?id=272800

β-hemoglobin (HBB); MIM number: 141900. OMIM: http://www.ncbi.nlm.nih.gov/entrez/dispomim.cgi?id=141900

Part 3

Disability, Genetics and Eugenics

Introduction

Angus Clarke and Flo Ticehurst

This section addresses some of the most contentious issues around human genetics, reminding us uncomfortably of the continuities between eugenic abuses of human genetics in the past and our current social and health care practices. To what extent can these parallels be dismissed as irrelevant and outdated? To what extent should we be looking to identify contemporary practices that rest on similar assumptions to those of the eugenicists in former generations, or that denigrate our fellow citizens and treat them disrespectfully?

Weindling (chapter 16) explains the background of the Nazi concept of racial hygiene and the development of this into a policy that would rid the population of those declared 'unfit to live'. He describes how the social, political and intellectual climate after the First World War led to the development of a positive eugenics, the notion that there was a duty to promote the health of future generations. The enforced sterilization measures introduced under the Nazi regime were part of the reform of public health that targeted clinical conditions such as schizophrenia and Huntington's disease. Geneticists supported the Nazi Party in this area and, in return, were treated as a favoured professional group.

Like the policy of enforced sterilization, the program of involuntary 'euthanasia' resulted from the desire by Nazi enthusiasts to address the burden placed on society by the incurably mentally ill, and was based on the concept of 'lives no longer worth living'. Researchers in hereditary biology began to exploit euthanasia for scientific ends, but Weindling asks, 'were there biological motives for the brutal butchery of the Holocaust?'

Weindling describes the distinctive Nazi ethos that pervaded a lot of the biological and genetic research of the time, for example, the way in which medical anthropologists sought to back up their ideas about racial history with 'evidence' gathered from body parts collected from individuals who had been found to be 'unfit' in some way.

Duster (chapter 17) argues that there are echoes of eugenics in the application of genetics today, as in genetic testing and screening. He sets this claim in the context of eugenic ideas in the early twentieth century; in particular, he considers how the urbanization of the Industrial Revolution particularly affected the poor, creating the climate for the abuses carried out by the Nazis in the name of improving human civilization. He goes on to elaborate this view by drawing a parallel between the era of social change that characterized the late nineteenth/early twentieth century and the change that has been occurring more recently as we move from an industrial era to a society based on tertiary (service) industries, and from urban to suburban lifestyles. This, he argues, is leading to a resurgence of interest in biological explanations for social outcomes.

Duster then turns his attention to genetic screening programmes in the USA, as examples of measures that demonstrate that the 'the institutionalization of genetic screening programs contains a strong residue of the old image of "cleaning" or "purifying" the gene pool'. He argues that, although the screening is offered to individuals, decision-making by individuals about genetic testing and screening can still mean systematically different outcomes for different social or ethnic groups.

Sun-Wei Guo (chapter 18) describes the controversy surrounding the Maternal and Infant Healthcare Law (MIHCL) of the People's Republic of China, discussing the scope of this law and the reactions to its introduction. Concerns about the potential abuse of this law were voiced strongly in the West because of China's history of human rights abuse; Guo outlines some of the other reasons why critics thought that this law was misguided, such as 'the law is scientifically ineffective in reducing "inferior births"'.

Guo then presents a Chinese point of view, outlining how the Chinese government has improved maternal and infant health since it became a republic in 1949, and then considering the socio-economic and public health context of this law. This context has changed over the last fifty years as industrialization and urbanization have become widespread, and it is here that this article connects to the previous one. This leads to a discussion about the cultural context of this law, the author asking how far cultural differences can account for the differences in responses to the law, from both Eastern and Western perspectives. One view is that, in China, the focus is on the good of society rather than on that of the individual, as is often the case in Western countries. Guo then looks at how the meaning of the term 'eugenics' differs between countries. He is keen to point out that the many criticisms of this law have not paid enough attention to the questionable science underlying it.

Tom Shakespeare (chapter 19) discusses the standpoints of what he sees as 'two polarized paradigms' – the disability rights movement and genetics, and laments the lack of communication between these two 'opposites'. In his view the disability rights movement is largely without a voice in debates about genetics.

The author explains his exasperation with the use of language that expresses a derogatory implication, such as 'culprit genes', 'the burden of genetic disease'. He describes how opposition to genetics amongst disabled people takes a range of different forms, and how some disabled people have a more positive attitude to genetic research. He also describes 'a more nuanced perspective' that can be found within the disabled community, which supports the principle of a woman's right to choose, but questions the social context in which these choices are made. He then elaborates on fears amongst the disabled community about increased discrimination/stigmatization in the light of advances in genetics and its applications, explaining the views of some who argue that if environmental and social barriers were lowered, disability would not be perceived as such a problem.

Shakespeare closes his article by suggesting that the issue of abortion may cease to be central to these debates in the future: advances in genetic research bring the prospect of new treatments such as gene therapy, and questions about human variation and difference will mean that issues raised by disabled communities now may come to affect everyone in the future, as we will all be found to have genes that are harmful in some way.

McCarthy, Sinason and Hollins (chapter 20) explain the controversy surrounding issues of sexuality and procreation among those with intellectual impairments. They describe the history of abuse associated with past examples of controlling reproductive rights and fertility in this context – further illuminating the eugenic practices in the US and Europe during the twentieth century. They consider some of the social reasons that account for the disapproval of sexual relationships with, or among, those with intellectual impairments, such as concerns about the amount of support they might need if they have children.

The authors go on to discuss the main ethical and legal concerns in this area, especially in the English legal context. The authors stress the wide variation of genetic traits and the 'whole spectrum of social, religious and cultural correlates of mental and physical characteristics in any individual' that should be attended to in decision-making about reproductive choices for people with learning disabilities, if they are to benefit from being properly treated as citizens with rights in our society.

16
Nazi Movement and Eugenics

Paul Weindling

Background: racial hygiene

The Nazi movement of the early 1920s was strongly anti-Semitic but took little interest in eugenics. Eugenics or 'racial hygiene' has a much longer pedigree in Germany dating back to ideas of racial hygiene as chromosomal engineering. These were proposed by the physician Alfred Ploetz in 1895. For its part German racial hygiene tried to steer clear from racist mysticism in order to establish its credentials as a branch of anthropology and medicine. Ploetz was influenced by Galton, German anthropologists and the utopian schemes of the American Edward Bellamy's *Looking Backward from the Year 2000* (Weindling, 1989).

In the early 1890s the psychiatrist Wilhelm Schallmayer outlined a scheme for socialized medical services. He was critical of the physician's duty to the individual patient, as this meant that therapeutic medicine prolonged the lives of the eugenically undesirable. The physician as state official should oversee a scheme of annual genetic checks registered in health passports. There was a new concern to regulate access to marriage to prevent the spread of sexually transmitted and chronic diseases. Turn of the century studies pointed to the inheritance of a range of physical, medical and psychological traits, for example, myopia, alcoholism, schizophrenia and homosexuality. In 1904, the psychiatric researcher Ernst Rüdin proposed sterilization of alcoholics, and by 1912 developed a Mendelian scheme for the inheritance of schizophrenia.

Prior to 1914 eugenic notions pervaded schemes to eliminate chronic degenerate diseases, such as sexually transmitted diseases, tuberculosis and alcoholism. While only very few conditions were proved to be inherited on a genetic basis, contracting an infection was attributed to an inherited susceptibility. Holistic notions of genetic susceptibility created the expectation that a range of psychological and physical conditions, as well as many diseases, were the result of the inheritance of a basic genetic core.

Positive eugenics

These ideas of predisposing genetic factors gained increasing acceptance in Weimar Germany. During the First World War the German state recognized a duty to promote the well-being of future generations, and geneticists lobbied for a national institute and clinics for hereditary health. In the trauma of the loss of the war and the unstable economic conditions, biologists warned of the need to protect the 'genetic treasury' of the nation. State-sponsored eugenic committees promoted a range of positive eugenic measures. These received some influential (although far from unanimous) support from Catholic and socialist circles. The Prussian Welfare Ministry generally had a Catholic Center Party Minister. The former Jesuit biologist, Hermann Muckermann, raised funds for the Kaiser Wilhelm Institute for Anthropology, Human Heredity and Eugenics, which was established at the time of the international genetics conference of 1926/1927.

Positive eugenic measures included school health services, dietary supplements, schemes to promote exercise and fresh air, and additional welfare for the 'child rich'. Negative measures were targeted at youth psychopaths. There were proposals, which were deemed unacceptable, for sterilization, and for preventive detention. A distinctive form of eugenics lay at the core of the Weimar welfare state.

Leading geneticists such as Erwin Baur were also prominent eugenicists. Indeed, the demarcation between genetics and eugenics might have seemed to be somewhat contrived. However, there were several different forms of eugenics, each with its supporting interests. Hitler apparently read the genetics textbook of Baur, Eugen Fischer and Fritz Lenz while in Landsberg prison, and picked up on measures like sterilization. Nazi propaganda began to make more of eugenics by the early 1930s, and Lenz considered that eugenics policies were likely to have their best chance under the Nazis.

Eugenicists remained divided from Nazism over anti-Semitism, but this rapidly changed. Thus Eugen Fischer came to the Kaiser Wilhelm Institute for Anthropology as a supporter of the Catholic Center Party. He remained in office as someone prepared to work alongside the Nazis. Verschuer had deeply held religious views. Initially in 1933 the Nazi authorities looked at him suspiciously with the verdict that they liked his science, but he was naive about the ultimate aims of Nazism. Yet he too made a headlong leap into the Nazi fold. He could be characterized as serving the twin gods – of Christianity and Hitler. He was thus a fitting successor to Fischer in 1941 when the Nazi state stood on the brink of ruthless genocide.

The sterilization program

The 12 years of the Nazi era were less a unified era when the state regimented biologists than a rapidly changing sequence of biologists opportunistically reorienting their priorities and forging new alliances with factions in the Nazi power system. The rapidity with which the compulsory sterilization legislation was introduced and put into operation provides a good example. This law, already on the statute books in July 1933, proved a powerful force in the wholesale reform of public health, in that it accelerated the shift to a unified state and municipal public health system with an office for race and heredity at its core. Sterilization also achieved the widespread interest of geneticists (e.g. Fritz von Wettstein) who backed its implementation. Rüdin took a leading role in drawing up the law. This was targeted at a range of clinical conditions, notably schizophrenia, muscular dystrophy and Huntington disease, as well as epilepsy and chronic alcoholism.

There was widespread international interest in the German sterilization measures, as other countries enacted their own sterilization laws. Denmark and the Swiss canton of Vaud led the way in 1931. The German legislation, passed in July 1933, coincided with new laws in Ontario and British Columbia in 1933, and was followed by legislation in Sweden and Norway in 1934 and Finland in 1935. The centrality of incurable mental illness in all these laws makes comparison reasonable, although paths were soon to diverge. It is necessary to distinguish between the medical framework of the laws and the Nazi context of their enactment. No one – according to the law – could be sterilized for belonging to a supposedly pathogenic racial group such as the Jews or gypsies. Moreover, a heavy drinker who happened to be a Nazi might be sterilized.

That was the theory. However the practice was at times different. The black–German half-castes (offspring of the French colonial troops who occupied the Rhineland) were sterilized: although this was technically illegal, the Nazi state was permissive when it came to racial brutalities. There were variations in the actual rates of sterilization; for example, Hamburg and Erlangen were both in Protestant areas, but Hamburg saw a higher incidence of sterilization. It was mainly (but not exclusively) the working class who were sterilized.

Nazi eugenics can be seen not so much as unethical, but as deploying ethics in keeping with the Nazi values. The stress was on the priorities of the nation and race. The sick individual was seen as a burden on the fit (a category defined in physical and racial terms). The state had coercive powers in terms of intervention not only in terms of power to detain and segregate, but also to intervene in the body.

Euthanasia

Nazi Germany became notorious for coercive euthanasia. After many years of virtual neglect this area too has become a focus of major historical reinterpretation. The origins of euthanasia remain obscure: the standard story that a family petitioned the Führer about the death of a child derived from Karl Brandt's testimony at the Nuremberg Medical Trial. Benzenhöfer (1998) has shown that the child indeed existed, but was not called 'Knauer', and that the dates of birth and death mean that any petition to Hitler occurred only after the decision to unleash euthanasia had been reached. Brandt was one of a circle of medical advisers around Hitler, who convinced the Führer of the need to go for the more radical solution of euthanasia. Hitler, who in any case favored radical solutions particularly when they showed ruthless disregard for liberal and humane values, was readily convinced at a time when he foresaw that the war economy could do without the 'burdens' of the incurably mentally ill and physically incapacitated. To what extent did euthanasia have a genetic rationale? As with sterilization there was a panel of expert adjudicators but now instead of a legal procedure of notification and appeal, a single doctor made decisions on the basis of a cursory scrutiny of clinical records.

Factors precipitating the Nazi euthanasia measures included strategic concerns with clearing hospitals for war casualties, and economic concerns with the burden of the incurably mentally ill. Malformed and disabled children were targeted in separate programs. The killings occurred in special pediatric units by means of long-term starvation and lethal medication: the norms of clinical medicine were readily perverted once the concept of lives no longer worth living was accepted.

In terms of motives and organization, euthanasia shows how Nazi medicine was on the road to genocide. Anatomists exploited victims' body parts. Researchers interested in hereditary biology showed intense interest in exploiting euthanasia for scientific ends. For example, Julius Hallervorden of the Kaiser Wilhelm Institute for Brain Research obtained brain specimens from patients whose clinical records were 'of interest'. Patients were killed to order by researchers, who followed clinical studies with autopsies. The euthanasia program was expanded to concentration camps and the killing of inmates of psychiatric hospitals from occupied territories. The staff operating the carbon monoxide gas chambers in the special killing centers were transferred to extermination camps in occupied Poland. While euthanasia did not provide a direct pattern for the gas chambers of Auschwitz, it did for the smaller carbon monoxide gas chambers.

Were there biological motives for the brutal butchery of the Holocaust? The SS continued to screen for ethnic identity; for example, Wolfgang Abel of the Kaiser Wilhelm Institute for Anthropology sought to locate racially valuable elements in occupied Russia. 'Germanization' was one option: the Nazis sought to determine who among the populations under occupation were ethnically German, and the extent that their Germanic qualities had been corrupted. Gypsy children were held for scientific observation in a German children's home with a view to studying their capacities for germanization. The negative verdict by the researcher Eva Justin meant transfer to Auschwitz for these children. Eugenics thus supported the killing of a range of ethnic and supposedly asocial groups.

The overall context of the Holocaust was a vast scheme of 'clearance' of the east for resettlement. Extermination of the Jewish 'race' was considered a means of regenerating what were variously referred to as the German, Aryan or Nordic races. For example, the Nazis attempted to transfer the politically inconvenient South Tyrol 'Germans' to the east. There was also interest in a takeover of Soviet plant breeding and biological institutions to support the resettlement program.

Nazi ethics

A mass of biological and genetic activity was pervaded by a distinctively Nazi ethics. The Nazi medical anthropologists hunted for body parts to illustrate their grotesque views on racial history. Brains, eyes, skeletons and internal organs were assiduously collected. Yet it is tendentious to refer to Auschwitz as a 'genetics laboratory': the camp arose from the intersection of multiple historical factors, including economic and policing factors. Even within medicine, there were several strands operating. Joseph Mengele as camp doctor applied sanitary precepts of bacteriological hygiene ruthlessly in the control of typhus and other infections. Yet at the same time the racial killing and experimentation had the rationale of isolating and eradicating the carriers of pathogenic genes. Genetics flourished in Nazi Germany in often gruesome research programs, and as an incentive to racial policy. Many thousand of victims were killed and disabled as a result of research with its purpose in genetics.

What happened after the War

After 1945 the eugenicists experienced varying degrees of difficulty. Rüdin was interned for nearly a year, but no geneticist was brought to trial unless involved in euthanasia. A planned second medical trial at Nuremberg focusing on Verschuer's links to Auschwitz did not come to fruition. Eugenics now became human genetics. Hans Nachtsheim made a smooth transition, assisted by the Lysenkoist opponents of genetics in the Soviet zone. Lenz gained a chair in 1946, and Verschuer rather later in 1951.

References

Benzenhöfer U (1998) Der Fall 'Kind Knauer'. *Deutsches Ärzteblatt* 95: B954–B955.

Weindling P J (1989) *Health, Race and German Politics between National Unification and Nazism, 1870–1945*. Cambridge: Cambridge University Press.

17
Eugenics: Contemporary Echoes

Troy Duster

Past eugenic abuses: context

In the nineteenth century, the Industrial Revolution and rapid urbanization wreaked havoc on traditional life and traditional social roles in both Europe and the United States. Extended kinship systems that had been valued as an economic advantage on farmlands were often inverted and became economic liabilities when those families were forced off the land and moved to the teeming cities. Unemployment, homelessness, mental illness and a host of other social problems seemed, in particular, to victimize the poor, whose visibility, if not sheer numbers, dominated the public sphere of urban life.

Cholera, yellow fever, typhoid and tuberculosis were the scourge of city-dwellers and, once again, the poor were the most likely victims (Tesh, 1988). They were also the most likely to be blamed for causing the problems, as they were typically characterized as living in unclean conditions. Hygiene came first as an explanation for the better fortunes of the privileged and middle classes, and later as a challenge to the poor.

As the wealthier families began to have fewer children and to have the resources to hire the poor as servants to help them 'clean up', some observers began to notice what they thought was a disturbing pattern. The more well-to-do members of society were procreating less, while the poor were still having very large families. The dark Malthusian prediction about a population explosion took a particularly elitist turn. The upper classes theorized that 'if we could just get the poor to have fewer children, and encourage the "best families" to have more children, civilization could be saved'. This was the cornerstone of a new set of ideas that came to fruition around the time that hygienics was flowering as a strategy for staving off disease.

The term 'eugenics' was coined by Francis Galton, an English social theorist and first cousin to Charles Darwin. Galton reasoned that if animals could be improved by careful breeding, then so could humans. 'Eu' means good, and the new movement would take two forms: positive eugenics meant that the 'good families' should be encouraged to procreate more; negative eugenics meant that the poor or 'bad families' should be discouraged from procreating. These were the seeds of the methodology, if not the idea, that would later occupy the Nazi fixation with racial purification and Aryan superiority. If we are to learn anything from our past, however, it is imperative that we understand more fully that the appeal and popularity of eugenics was compelling to the full range of thinkers of all political persuasions at the beginning of the twentieth century. Very much like its sister concept 'hygiene', there was a strong association between cleanliness and order, progress and eugenics.

Just as hygiene was seen as the normal value of cleanliness to which all should aspire, eugenics was accepted widely and promoted actively by the major public figures of the period. University presidents, medical doctors, judges, academic scholars, writers, intellectuals, and political figures on both the left and right of the political spectrum all espoused the idea that the betterment of humankind would result from the practices and the techniques that would prevent the procreation of 'imbeciles' and 'mental retards' and 'criminals' and 'prostitutes' and 'homosexuals' and 'alcoholics' and 'gamblers' and .... The list goes on and on. To give just a flavor of the times, the Fabian socialist George Bernard Shaw, the anarchist Emma Goldman and the writer H. G. Wells were as ardent eugenicists as the President of Stanford, prominent chairs of psychology at Harvard, Princeton and Yale Universities, and the leading geneticists of the period (Haller, 1963; Ludmerer, 1972).

During the dawn and then heyday of the progressive era, the state of California led the nation in pioneering the involuntary sterilization of those poor and sometimes indigent women and men in state institutions, from hospitals to mental institutions to homes for wayward youth to prisons (Reilly, 1991). From 1909 to 1927, California sterilized thousands of its 'unfit' citizens.

Continuity and persistence of eugenic thought and goals

Although many are aware of the gross human rights abuses 'in the name of eugenics' (Kevles, 1985) of the early part of the century, most of the current advocates, researchers and celebrants of the putative link between genetic accounts and socially undesirable behavior (or characteristics or attributes) are either unaware of the social context of that history or trivialize the ominous shadow from the earlier period. In a time of rapid social change in which there are disruptions of the established order and the attendant challenges to authority and tradition, there is a special appeal of genetic explanations and eugenic solutions to the most privileged strata of society.

Every era is certain of its facts. At its heyday the eugenics movement was no exception, being sure that feeble-mindedness, degeneracy and criminality were inherited. In 1912, the American Breeders' Association – an organization of farmers and university-based theoreticians – created a Committee to Study and to Report on the Best Practical Means of Cutting off the Defective Germ Plasma in the American Population. At the 1913 meeting of the association, the report was delivered (Laughlin, 1914). It read in part:

> Biologists tell us that whether of wholly defective inheritance or because of an insurmountable tendency toward defect, which is innate, members of (select) classes must generally be considered as socially unfit and their supply should if possible be eliminated from the human stock if we would maintain or raise the level of quality essential to the progress of the nation and our race.

California had one of the longest-running involuntary sterilization programs in the country. In 1927, a team of prominent and respected citizens was assembled to consult on the effectiveness of this program. They extolled its virtues. In early May that same year, the Supreme Court upheld Virginia's involuntary sterilization law, opening up not only a floodgate of sterilizations in the United States, but also a model that would soon be adopted, expanded and forever made infamous by Hitler's Third Reich. In mid-July 1933, Germany enacted a eugenic sterilization law. The American eugenicists provided the intellectual and ideological underpinnings and were cited widely as the genetic authorities on behalf of this development. California was one of the leading states in the country in terms of its use of involuntary sterilization laws. From 1930 to 1944, over 11 000 Californians were sterilized under these laws. The Germans cited the California development as a model (in 1936, Heidelberg University awarded honorary degrees to several key American eugenicists), but they took it much further. In the first year of the German program, 52 000 people were placed under final order to be sterilized – a development that was in turn hailed by American eugenicists. In the period from 1933 to 1945, the best estimates indicate that the Nazis sterilized about 3 500 000 people (Reilly, 1991).

Revelations of the Holocaust after the Second World War gave eugenics a very bad name. Respectable scientific journals changed the key word in the journal's title from 'Eugenic' to 'Genetic', and the notion of human betterment through selective breeding or through sterilization of the 'unfit' fell into public disfavor (Kevles, 1985). However, this did not mean that such thinking had been eradicated. It was simply dormant or placed on the back burners for a few decades.

Parallel massive social displacements: late nineteenth and late twentieth centuries

Just as the twin shifts from agrarian to industrial and rural to urban dominated the shifting social demography of the late nineteenth century in Europe and the

United States, so the shift from industrial to service (or tertiary) and from urban to suburban dominated the shifting social demography of the late twentieth century. The United States has been in the vanguard of this development, and the massive economic displacement of African-American urban youth is the context for a renewed conception of biological thinking about social issues (Duster, 1995; Nelkin and Lindee, 1995).

As we begin the twenty-first century, the United States is heading down a subtly parallel road, entertaining the connection between genes and social outcomes. This is being played out on a stage with converging preoccupations and tangled webs that interlace youth unemployment, crime and violence, race and genetic explanations.

There is a direct link between deindustrialization, youth unemployment, and ethnic or racial or immigrant minority status. In 1954, Black and White youth unemployment in America was equal, with Blacks actually having a slightly higher rate of employment in the 16–19 age group. By 1982, the Black unemployment rate had nearly quadrupled, whereas the White unemployment rate had increased only marginally (Kasarda, 1983). Just as unemployment rates among African-American youth were skyrocketing during these three decades, so were their incarceration rates. This provides the context in which we might best review the sharply clear pattern of the historical evolution of general prison incarceration rates by race. In the second half of the twentieth century, the incarceration rate of African-Americans in relation to Whites went up in a striking manner. In 1933, Blacks were incarcerated at a rate that was roughly three times that of Whites. By 1970, it was six times; and in 1995, it was seven times that of Whites.

In the 1990s, America built more prisons and incarcerated more people than at any other moment in our history. In the decade from 1981 to 1991, the United States went from a prison population of 330 000 inmates in State and Federal prisons to 804 000 inmates. That rate constitutes substantially more than a doubling in a single decade – 'the greatest rise in a prison population in modern history' (Bureau of Justice Statistics, 1992).

So-called 'genetic studies' of criminality have a heavy dependency on incarcerated populations. Thus, for example, one of the more controversial issues in the 'genetics' of crime is whether males with an extra Y chromosome – XYY males – are more likely to be found in prisons than XY males.

The first major study to suggest a genetic link was carried out in Edinburgh in the United Kingdom. All 197 males in this account of prison hospital inmates were described as 'dangerously violent' (Jacobs et al., 1965), and seven of these had the XYY karyotype. These seven males constituted about 3.5% of the total. However, as it was estimated that only about 1.3% of all males have the XYY chromosomal make-up, the authors posited that the extra Y significantly increased one's chances of being incarcerated. Ever since, a controversy has raged as to the meaning of these findings and the methodology that produced them. The claim for a genetic link to crime is based only on studies of incarcerated populations. Yet, incarceration rates are a function of a full range of criminal justice decisions – a fact that social science research has long shown to be a function of social, economic and political factors (Skolnick, 1966; Currie, 1985; Miller, 1992; Cole, 1999; Mauer, 1999). The new forensic sciences are now attempting to use deoxyribonucleic acid (DNA) markers to identify 'ethnic affiliation estimations' of suspects in criminal investigations (Shriver et al., 1997; Lowe et al., 2001). Just as health and hygiene were the vanguard for the late-nineteenth-century screen for the unfit, so the genetic screen is first of all a health screen. The shift in use and focus to forensic science is already in the wings.

Contemporary echoes of a eugenic past: genetic screen

Genetic screening is one of the outgrowths of health screening for several public health problems, most notably tuberculosis. However, unlike tuberculosis, genetic disorders tend to cluster in populations where there have been centuries of inbreeding because of cultural endogamy rules (who can marry whom) and/or because of the long-term geographical residence of a population in which there has not been much physical mobility. In both circumstances, genes that cause diseases cluster in these populations, which makes those who are part of those populations at greater risk. Examples include: cystic fibrosis, a disease disrupting lung and digestive function that primarily affects persons of North European descent; β-thalassemia, a blood disease affecting persons living in the Mediterranean area; and sickle cell anemia, a blood disorder primarily affecting people with ancestors from West Africa and some areas of the Mediterranean.

In the last two decades of the twentieth century, many states began to offer postnatal genetic screening of all newborns; however, Massachusetts has been a leader in providing prenatal screening to pregnant women. If the screen detects changes in the amounts of certain substances in the mother's blood (α-fetoprotein, HCG, human chorionic gonadotropin [a chemical produced in pregnancy], unconjugated estriol), then the fetus is determined to have an increased risk for neural tube defects (spina bifida or anencephaly), trisomy 21 or trisomy 18. In the most literal sense, to 'screen' something means to prevent that something from getting past the screen. Thus, whether explicitly or implicitly, the institutionalization of genetic screening programs contains a strong residue of the old image of 'cleaning' or 'purifying' the gene pool. The social aspect of the eugenic implication is disguised because the screen is offered to individual women or individual families. Thus, the specter of state-sponsored screening of a particular group is diffused and obscured.

However, as noted above, because genetic diseases tend to cluster in certain ethnic and racial groupings, individual decision-making (imposed or presumed) cannot mitigate the fact of systematically different outcomes for different groups.

Getting rid of 'bad babies' with 'genetic defects' is only half of the eugenic equation. There is also the idea of a 'positive eugenics' – the active recruitment of some people to procreate and breed selectively to increase a human trait or characteristic that is considered positive. Singapore actively encourages and rewards its wealthy and middle-class citizens to have more children. This is the group approach to positive eugenics. On the individual level, contemporary residues of eugenic thinking can be seen in the emergence and increasing use of sperm banks with sperm donated by Nobel laureates, the much higher cost of ova from young women from wealthy private schools, and the exorbitant pricing of ova from the supermodels on a website. There is evidence that, given a choice, some people will try to add a bit of height to their offspring with a growth hormone. Each of these developments indicate a lingering of a eugenic past.

References

Bureau of Justice Statistics (1992) *Office of Justice Programs*, vol. 1, no. 3, NCJ-133099. Washington, DC: US Department of Justice.

Cole D (1999) *No Equal Justice: Race and Class in the American Criminal Justice System.* New York: New Press.

Currie E (1985) *Confronting Crime: An American Challenge.* New York: Pantheon.

Duster T (1995) Post-industrialism and youth unemployment. In: McFate K, Lawson R and Wilson W J (eds.) *Poverty, Inequality and the Future of Social Policy: Western States in the New World Order*, pp. 461–486. New York: Russell Sage.

Haller M (1963) *Eugenics: Hereditarian Attitudes in American Thought.* New Brunswick, NJ: Rutgers University Press.

Jacobs P, Brunton A M, Melville M, Brittain R and McClemont M (1965) Aggressive behavior, mental subnormality, and the XYY male. *Nature* 208: 1351–1352.

Kasarda J D (1983) Caught in the web of change. *Society* 2(November): 41–47.

Kevles D J (1985) *In the Name of Eugenics: Genetics and the Uses of Human Heredity.* New York: Alfred A. Knopf.

Lowe A L, Urquhart A, Foreman L A and Evett, I (2001) Inferring ethnic origin by means of an STR profile. *Forensic Science International* 119: 17–22.

Laughlin H H (1914) *The Scope of the Committee's Work, Bulletin No. 10A*, pp. 12–13. Cold Spring Harbor, NY: Eugenics Records Office.

Ludmerer K M (1972) *Genetics and American Society.* Baltimore, MD: Johns Hopkins University Press.

Mauer M (1999) *Race to Incarcerate.* New York: New Press.

Miller J G (1992) *Hobbling a Generation: Young African American Males in the Criminal Justice System of America's Cities.* Baltimore, MD: National Center on Institutions and Alternatives.

Nelkin D M and Lindee S (1995) *The DNA Mystique: The Gene as a Cultural Icon.* New York: W H Freeman.

Reilly P (1991) *The Surgical Solution: A History of Involuntary Sterilization in the United States*. Baltimore, MD: Johns Hopkins University Press.

Shriver M, Smith M W, Jin L, et al. (1997) Ethnic-affiliation estimation by means of population specific DNA markers, *American Journal of Human Genetics* 60: 957–964.

Skolnick J (1966) *Justice Without Trial: Law Enforcement in Democratic Society*. New York: John Wiley.

Tesh S N (1988) *Hidden Arguments: Political Ideology and Disease Prevention Policy*. New Brunswick, NJ: Rutgers University Press.

18

China: The Maternal and Infant Health Care Law

Sun-Wei Guo

Introduction

The Maternal and Infant Health Care Law (MIHCL) of the People's Republic of China was promulgated as Presidential Decree Number 33 on 27 October 1994 by President Jiang Ze-Min and became effective on 1 June 1995. With the aim of preventing 'new births of inferior quality', it was drafted – initially under the title of 'the Eugenics and Health Protection Law' – in response to an increasingly heavy burden of disability in the nation and some dubious local practices for regulating reproduction such as those in Gansu province, which mandated sterilization for mentally impaired people considering marriage, and those in Shanxi province, which banned anyone with an intelligence quotient lower than 40 from getting married (Pearson, 1995). Since the enactment of the law, the Ministry of Health has issued seven regulations covering the operational side of the law.

Scope of the law

As the first ever specialized law on the health and welfare of women and children since the founding of the People's Republic of China, the MIHCL codifies the need to promote maternal and infant healthcare, and aims 'to ensure maternal and infant health and to improve the quality of the birth population'. It has seven chapters and 39 articles, covering premarital, antenatal and perinatal health, and provides guidelines on technical implementation, management and legal issues surrounding maternal and infant health.

The law mandates a premarital medical examination for serious genetic diseases, some infectious diseases and certain mental disorders. If the disease is serious enough, long-term contraception or ligation will be used to enforce childlessness; otherwise, the couple will not be allowed to marry. During pregnancy, prenatal testing will also be compulsory if an abnormality in the fetus is found or suspected, followed by termination if the fetus has a serious genetic or somatic disorder.

Responses to the law

Viewed largely as a draconian eugenics law, the MIHCL sparked widespread concerns, criticisms and in some cases fierce outcries in the Western scientific community, even when it was still in an embryonic stage (Dickson, 1994; Editorial, 1994, 1995a, 1998). It also drew considerable media attention in the West.

In protest at the law, the Genetical Society in Britain, apparently apprehensive of possible abuses of genetics, boycotted the 18th International Congress of Genetics held in Beijing in 1998, and its counterparts in Argentina and Holland followed suit. Some Western observers went so far as to liken 'exporting genetic know-how to a regime that sanctions eugenics' to 'selling Semtex to countries that sanction terrorism', and to ask 'Do Western democracies need to regulate exports of genetic expertise as well as arms sales?' (Editorial, 1996).

Other scientists, well aware of the population problems facing China, were more sympathetic. The late pioneering human geneticist James Neel asked:

> If in the face of a perceived military threat to the integrity of the nation, a government can call upon its citizenry to place its lives at risk, why isn't it equally appropriate, in the face of a civil threat perceived to be of equal magnitude, for the government to pressure its citizenry to limit its reproduction?
>
> (Neel, 1997, p. 333)

John Drake, who chairs an advisory committee to the International Genetics Federation, believes that the law is intended to be advisory: 'few Westerners have an appreciation of the magnitude of the population problem China is trying to come to grips with' (Beardlsey, 1997).

Several regional and international scientific bodies, including the British Clinical Genetics Society and the Human Genetics Society of Australia, made public their concerns about certain aspects of the law. At the Ninth International Congress of Human Genetics, held in Rio de Janeiro, Brazil, in 1996, in a statement presented at the conclusion of the meeting, the Chinese government was urged to delay implementation of its controversial law, pending discussion within the world's genetics community.

The Ethics Committee of the Human Genome Organization considered the law in a meeting in 1996. Although the Committee acknowledged that 'a number of provisions of the law involve desirable statements of principle or law that are protective of the rights of mother and child and apparently conformable to international human rights standards and ethical principles' and accepted that 'different cultures and religions can affect the approach of different countries to medical practice', it nonetheless thought that 'certain provisions of the Chinese Law . . . may infringe international human rights principles' (Knoppers and Kirby, 1997). The Canadian College of Medical Geneticists made a position statement in support of 'any measures that will improve communications with Chinese geneticists, physicians, and legislators . . .' (Canadian College of Medical Geneticists, 1997). The Board of Directors of the American Society of Human Genetics also issued a statement on eugenics and the misuse of genetic information to restrict reproductive freedom.

In contrast to generally critical responses in the West, many scientists in China defended the law vehemently, saying that it would only facilitate practices that have been common 'for decades . . . in Western countries' (Editorial, 1995b), that it has caused 'few or no negative effects' (Normile, 1998), and that it was intended merely to warn of the risks of passing on hereditary diseases and to help people avoid these diseases (Wang, 1999). On the basis of polls, other scientists claimed that the controversy was due largely to cultural differences between the East and the West (Mao, 1997a, 1997b, 1998).

Focal point of the controversy

The provisions of the MIHCL that provoked most of the criticisms are the following three articles (based on the translation in Knoppers and Kirby (1997) with modifications):

- *Article 9.* Physicians shall, after performing the premarital physical examination, give medical advice to those who are under the infective period of certain designated infectious diseases, or who are in the morbid period of certain designated mental diseases; both the male and female planning to be married shall postpone their marriage for the time being.
- *Article 10.* Physicians shall, after performing the premarital physical examination, explain and provide medical advice to both the male and the female who have been diagnosed with certain genetic disease of a serious nature which is medically considered to be inappropriate for reproduction; the two may be married if both sides agree to take long-term contraceptive measures or ligation operation for sterility, with the exception of people who are not permitted to marry as dictated by the Marriage Law of the People's Republic of China.
- *Article 16.* If a physician detects or suspects that a married couple of reproductive age suffer from hereditary disease of a serious nature, the physician shall provide medical advice to them, and the couple of reproductive age shall take measures in accordance with the physician's medical advice.

Given the lack of informed consent of the mother, the vagueness in specifying what constitutes a 'hereditary disease of a serious nature' and the lack of legal recourse for the couple involved, there is a clear and real danger that the law could be abused (Knoppers and Kirby, 1997; Guo, 1999). In addition, the tainted track record of the government in protecting human rights, its coercion (albeit mostly isolated) in carrying out the one-child policy, and reported, sporadic, incidents of female infanticide and forced late-term abortions suggest that the concerns for abuse are not without any merit.

Western perspective

To the scientific community in the West, the law is misguided for several reasons.

First, the law infringes international human rights principles, conflicts with basic human reproductive freedom and is morally unacceptable (Knoppers and Kirby, 1997; Morton, 1998). Mindful of abuses of genetics in the past, ranging from compulsory sterilizations in North America, Scandinavia and other countries to ethnic 'cleansing' and genocide in Nazi Germany, many scientists opposed the law. It is perhaps no coincidence that geneticists from Britain, the birthplace of the great writer George Orwell, were among the most vocal in protesting against the law.

Second, the law is scientifically ineffective in reducing 'inferior births' (Morton, 1998; Li, 2000). With few exceptions, the hereditary nature of most mental diseases such as schizophrenia is far from clear. There is no effective way to reduce the number of deleterious genes responsible for rare recessive diseases.

Third, there is fear that the government-sanctioned legislature on eugenics, especially coupled with a totalitarian government, may be the harbinger of a Nazi-style eugenics program that would weed out unwanted individuals (Dickson, 1994).

Last, there is fear of what could happen if China's law were to become linked to advanced genetic screening techniques developed in the West, which could cause a backlash against these techniques or even against the whole of genetic research in the West (Dickson, 1994; Harper, 1995).

Chinese perspective

Much to the credit of the Chinese government, phenomenal progress has been made in improving maternal and infant health since the founding of the People's Republic of China in 1949. The government has appreciably reduced mortality and morbidity with surprisingly meager resources by controlling and, in some cases, eradicating infectious diseases, improving sanitation, establishing and expanding maternal and child health services, and improving general living standards. As a result, life expectancy doubled in two generations, increasing from 35 years in the 1940s to 71.6 years in 2001, while infant mortality fell from 200 in 1000 to 29 in 1000, and maternal mortality fell from more than 1500 in 100 000 to 53 in 100 000 during the same time period. (For comparison, current infant mortality and life expectancy are 5.54 in 1000 and 77.8 years in the United Kingdom, 6.76 in 1000 and 77.3 years in the United States, and 63.19 in 1000 and 62.86 years in India, respectively (CIA World Fact Book 2001; see Web links.)

These achievements are all the more impressive if one realizes that China, being the most populous nation in the world, contains 22% of the world's population but has only about 7% of the world's arable land. Obviously, with 1.27 billion people to feed, the population problem is acutely urgent, not only to the Chinese government but also to many other nations. China's adoption of the policy of industrialization as the way to lift its standard of living means that resources such as arable land and water have to be diverted for industrial purposes. Improving living standards also raises individual consumption, putting more pressure on the carrying capacity of the land and on water resources. In fact, the population growth has already created problems on several environmental and social fronts in China, from diminishing cropland and water availability to unemployment.

In addition to the population problem, there are many people with disabilities in China. A national sampling survey conducted in 1987 reported that the overall prevalence of disability was 4.9% (Guo and Meng, 1993). For the 2001 population of 1.27 billion, this amounts to over 62 million people with disabilities – more than the total population of either Britain (59.7 million) or France (59.6 million). For children under 14 years, the prevalence of disability is 2.66% (Chen and Simeonsson, 1993) and the prevalence of mental retardation is 1.2% (Zou et al., 1994). According to the population statistics in 2001, 25.01% of the Chinese population are aged between 0 and 14 years. Consequently, there are over 8.4 million children with disability in China, and 3.8 million of these children are mentally retarded. Any sensible government confronted with these numbers would be duly alarmed and prompted to take actions. Obviously, preventing 'new births of inferior quality' might seem to be a quick fix.

On the personal level, a disabled child will be a substantial financial toll owing to the high cost of healthcare. Because about 80% of the population reside in rural areas where there is no pension, and patients or their families have to pay for healthcare out of their own pockets (Feng et al., 1995; Liu et al., 1995), the financial toll on the families with disabled children can be devastating. Even in urban areas, people with employment must contribute financially toward the medical care that they or their offspring receive (Liu and Hsiao, 1995). In addition, because the welfare system by many Western standards is virtually nonexistent, a disabled child will probably be unable to support destitute aging parents. This is all the more serious when there is only one child in the family.

Western observers familiar with Chinese affairs also provide other explanations. Veronica Pearson, a British social scientist who has lived and worked in Hong Kong since 1981, writes:

> Account has to be taken of the fact that the Chinese do not necessarily share Western priorities. Autonomy, individuality, privacy, the right to have as many children as wanted are selfish values. What is encouraged and valued is concern for the greater good and an ability to fit into the group, rather than to stand out from it. Furthermore, they live in a harsher world. Many people alive in China today remember severe famine, civil war, the horrors of the Japanese occupation and the Cultural Revolution. These are not conditions that encourage a kinder, gentler view of the world. Sterilization and abortions are already part of their lives through the one-child policy . . . They are not inflicted on people with a mental illness or learning disability exclusively. Such a fate is part of many people's lives.
>
> (Pearson, 1995, p. 3)

The controversy: just cultural differences?

Some scientists have attributed the difference in response to the law to the cultural differences between the East and the West, claiming that 'the Chinese culture is quite different, and things are focused on the good of society, not the good of the individual' (Mao, quoted in Coghlan, 1998; Mao, 1998). Indeed, even the word 'eugenics' carries substantially different meanings inside and outside China. Whereas in Western vocabularies eugenics carries a pejorative overtone and often conjures up images of compulsory sterilization and genocide, in Chinese it is 'you shen', which literally means quality birth or giving birth to a healthy baby. Even British geneticists, after visiting China when it reopened its door to the West, were left with a strong impression that

> the word 'eugenics' is used more widely in China than currently in the West . . . Undoubtedly much of what is termed eugenics in China would be included under what we would call genetic counseling, genetic screening, or clinical genetics.
>
> (Harper and Harris, 1986, p. 387)

This view is overly simplistic, however, because it is difficult to distinguish whether the willingness to sacrifice oneself for the common good results from a deeply ingrained culture or from living in a rigid, totalitarian and sometimes oppressive country that has little regard for individualism and liberty, and that is itself a scion of an authoritarian and feudalistic regime. It is very likely that Mao's polls would be different if he stratified responses by birth cohort: for example, by those who went through incessant political turmoils, and by those who grew up in a less rigid and more tolerant political environment after the Cultural Revolution. If sociocultural attitudes should indeed dictate legislative issues, then China's one-child policy would never have been adopted because the traditional Chinese culture advocates large families. The culture also favors boys over girls, but this should not be grounds for selective abortion and female infanticide.

There is also a misplaced belief that the law would help to 'clean up the gene pool ... and to reduce the number of deleterious genes' (Mao, quoted in Coghlan, 1998). Unfortunately, this is largely wishful thinking. It is well known that discouraging carriers of the same defective gene from reproduction is not an effective way to reduce the number of deleterious genes responsible for rare recessive diseases (Li, 1978). Neither the pathogenesis nor the genetic components of many of the mental disorders, including schizophrenia and mental retardation, that were mentioned in the law is understood. The fact, as pointed out by the sponsor of the law, that births of 'inferior quality' are relatively more common among 'the old revolutionary base, ethnic minorities, the frontier and economically poor areas' (Dickson, 1994) suggests that many so-called 'inferior births' are actually caused by environmental factors and are thus preventable through improved living standards and better pre-, peri- and postnatal care, achieved, for example, by folic acid supplement, reducing perinatal trauma and eliminating iodine deficiency (Guo et al., 1998). Without teasing out the environmental (especially the preventable) components, the genetic components, and their interactions through extensive genetic epidemiological studies, it is simply too premature, or even irresponsible, to launch any intervention procedure, especially in legislative form.

During and after the heated debate over the MIHCL, it is perhaps interesting to note the attitude of Chinese geneticists toward the controversial articles in the MIHCL. Besides a lopsided defensive stance, there has been a lack of open and scientific discussion on the efficacy of the law. Given the weightiness of the issues this is puzzling. In fact, almost all defenders of the law in China have been, conspicuously, social scientists and molecular geneticists, who may have distinctive insight that reflects their vantage points but may not be fully aware of principles of genetic epidemiology and population genetics. What seems to have been disregarded completely is that the relevant issues are more than cultural or social differences. Despite the noble intention of the law, it is seriously undermined by its questionable scientific foundations.

Subsequent developments

On 20 June 2001, Premier Zhu Rong-Ji signed Decree Number 308 of the State Department of the People's Republic of China to promulgate the Implementation Methods of the Law of the People's Republic of China on Maternal and Infant Healthcare. This document, issued in the form of executive regulation of the State Department, is purported to be more authoritative and operationally more executable. Drafted by the State Department, it is based on comments, suggestions and consensus from 14 different executive branches of the government and some social bodies, and from 15 provincial governments. The document, with 8 chapters and 45 articles, covers premarital healthcare, pre- and perinatal care and infant healthcare, and provides guidelines. In many senses it is an elaboration of the MIHCL, but it also removes some vagueness in language.

Conspicuously, articles in the MIHCL that aroused controversies are now written in a manner that is more 'politically correct' by Western standards. For example, when a couple is found during the premarital examination to be in the infective period of designated infectious diseases, to be in the morbid period of certain mental diseases, to have a hereditary disease of a serious nature that is considered to be inappropriate for reproduction from a medical standpoint, or to have another illness that is considered inappropriate for marriage from a medical standpoint,

> the physician shall explain to the involved party and provide advice on prevention, treatment or other appropriate medical procedures. In accordance with the medical advice provided, the involved party can either postpone the marriage or take long-term contraceptive measures or ligation on a voluntary basis; the medical and healthcare institutions shall provide medical consultation on treatment and provide medical service.
>
> (Article 14)

Interestingly, however, Article 21 says that 'the list of hereditary diseases of a serious nature, serious birth defects, and the diseases, when contracted during pregnancy, that put the pregnant woman at risk of her life and safety if her pregnancy is continued, shall be determined by the health administrative branch of the State Department'. In contrast to the stormy reactions to the MIHCL, this document, notably, has not triggered any criticism from the Western community so far.

Time to move on?

The MIHCL was drafted in response to enormous population pressures, the burden of a huge population of people with disability, and questionable local practices that restricted the reproductive freedom of people with hereditary and mental diseases. It provoked much concern, criticism and outcry from the Western scientific community shortly after its conception. With a fresh memory of the past abuse of genetics in Nazi Germany and other countries, many Western observers' concerns were not entirely groundless, especially given the tainted track record of human rights of the Chinese government. Although cultural differences between the East and the West do exist and may indeed account for certain differences in the attitude toward the law, the differences cannot justify a scientifically questionable legislature. Without any solid data to prove that the law can prevent 'births of inferior quality' effectively, the noble intent of the law is seriously undermined.

Perhaps the most disquieting observation is the notable lack of open discussions on the efficacy of the law by genetic epidemiologists, population geneticists and public health workers in China during and after the debate. As modern societies are becoming more and more dependent for their well-being on scientifically complex technology, any law to be drafted should be based on decisions that are made within the boundaries of scientifically sound knowledge and that approximately reflect the scientific state of the art. Any other course of action will not serve the purpose well.

Ironically, the population problem confronting China today stemmed precisely from the mistake of making decisions without any regard to science in the 1950s. When the late economist Ma Yin-Chu pointed out with great courage and prophetic prescience the grave consequence of unbridled population growth, his voice was harshly quenched by an authoritarian government simply because his opinions were at odds with those higher up. To reduce the number of future births with defects and disability certainly requires hard science and solid data, and any legislation toward that goal has to be drafted with best knowledge and utmost care. As the Chinese adage says, 'If you go in haste, you'll never reach your destiny.'

References

Beardlsey T (1997) China syndrome. China's eugenics law makes trouble for science and business. *Scientific American* 276: 33–34.

Canadian College of Medical Geneticists (1997) China's eugenics law: position statement of the Canadian College of Medical Geneticists. *Journal of Medical Genetics* 34: 960.

Chen J H and Simeonsson R J (1993) Prevention of childhood disability in the People's Republic of China. *Child Care, Health, and Development* 19: 71–88.

Coghlan A (1998) Perfect People's Republic. *New Scientist* 160: 18.

Dickson D (1994) Concern grows over China's plans to reduce number of 'inferior births'. *Nature* 367: 3 (news).

Editorial (1994) China's misconception of eugenics. *Nature* 367: 1–2.

Editorial (1995a) Western eyes on China's eugenics law. *Lancet* 346: 131.

Editorial (1995b) When is prenatal diagnosis 'eugenics'? *Nature* 378: 549.

Editorial (1996) Speak, don't hold your peace. *New Scientist* 152: 3.
Editorial (1998) China's 'eugenics' law still disturbing despite relabelling. *Nature* 394: 707.
Feng X, Tang S, Bloom G, Segall M and Gu Y (1995) Cooperative medical schemes in contemporary rural China. *Social Science and Medicine* 41: 1111–1118.
Garver K L and Garver B (1991) Eugenics: past, present, and the future. *American Journal of Human Genetics* 49: 1109–1118.
Guo S W (1999) Cultural difference and the eugenics law. *American Journal of Human Genetics* 64: 1197–1199.
Guo M and Meng J X (1993) Demography of people with disabilities in China. *International Journal of Rehabilitative Research* 16: 299–301.
Guo S W, Zheng C J and Li C C (1998) Dilemma over genetics and population in China. *Nature* 394: 313–314.
Harper P S (1995) Eugenics in China. *Lancet* 346: 508 (letter).
Harper P S and Harris R (1986) Medical genetics in China: a western view. *Journal of Medical Genetics* 23: 385–388.
Knoppers B M and Kirby M (1997) Report on the law of the People's Republic of China on maternal and infant health-care. *Genome Digest* April: 6–8.
Li C C (1978) *First Course in Population Genetics*, 2nd edn. Pacific Grove, CA: Boxwood Press.
Li C C (2000) Progressing from eugenics to human genetics. celebrating the 70th birthday of Professor Newton E. Morton. *Human Heredity* 50: 22–33.
Liu Y and Hsiao W C (1995) The cost escalation of social health insurance plans in China: its implication for public policy. *Social Science and Medicine* 41: 1095–1101.
Liu Y, Hsiao W C, Li Q, et al. (1995) Transformation of China's rural healthcare financing. *Social Science and Medicine* 41: 1085–1093.
Mao X (1997a) Chinese eugenic legislation. *Lancet* 349: 139.
Mao X (1997b) Ethics and genetics in China: an inside story. *Nature Genetics* 17: 20.
Mao X (1998) Chinese geneticists' views of ethical issues in genetic testing and screening: evidence for eugenics in China. *American Journal of Human Genetics* 63: 688–695.
Morton N E (1998) Hippocratic or hypocritic: birth pangs of an ethical code. *Nature Genetics* 18: 18.
Neel J V (1997) Looking ahead: some genetic issues of the future. *Perspectives in Biology and Medicine* 40: 328–347.
Normile D (1998) Geneticists debate eugenics and China' infant health law. *Science* 281: 1118–1119.
Pearson V (1995) Population policy and eugenics in China. *British Journal of Psychiatry* 167: 1–4.
Wang Y (1999) A call for a new definition of eugenics. *Eubios Journal of Asian and International Bioethics* 9: 73–74.
Zou Q H, Lei Z W and Zhang Z X (1994) An epidemiological study on etiology of mental retardation. *Zhonghua Yi Xue Za Zhi* 74: 134–137 (in Chinese).

Web links

CIA World Fact Book 2001 http://www.cia.gov/cia/publications/factbook

19
Disability, Human Rights and Contemporary Genetics

Tom Shakespeare

The 1980s and 1990s were marked by two potentially contradictory developments. On the one hand, disabled people were organizing as a new social movement in order to challenge their exclusion from contemporary society and promote their civil rights. In many countries, antidiscrimination legislation has been passed, and barriers to the participation of disabled people are beginning to be removed. By arguing that they are disabled by society, not by their bodies, disabled people have challenged negative attitudes toward impairment, and developed the grounds for a strong political identity (Campbell and Oliver, 1996). On the other hand, advances in genetic knowledge and the development of prenatal screening programs promise improved health on the basis of selective termination of pregnancies affected by impairment, with the ultimate possibility of gene therapies to cure genetic diseases. Disability rights and genetics, at their extremes, represent two polarized paradigms.

There has been little communication between the two perspectives. The views of disabled people have been largely absent from debates on genetics; but this group of people would argue that their perspectives on impairment and quality of life are vitally important for considering the role of genetics in general and prenatal screening in particular. Many disabled people deeply resent the hyperbole surrounding discoveries in the new genetics, as well as the negative language that is often used about disability. Often, advocates of genetics use phrases such as 'culprit chromosomes', 'the burden of disability' and 'the horrors of genetic disease'. In their enthusiasm for the new science, many geneticists ignore the fact that the health improvements they promise can only be achieved, in the short term, by preventing the birth of disabled people (Shakespeare, 1999). Others celebrate this possibility, such as the pioneer of reproductive technology who announced that 'soon it will be a sin of parents to have a child that carries the heavy burden of genetic disease' (Rogers, 1999). While genetics professionals distance themselves from the human rights abuses of the eugenic past, the underlying ethos of contemporary clinical genetics may retain a similar mission (Paul, 1992). This is particularly the case in some non-Western societies such as the People's Republic of China.

Opposition to genetic research comes from those disabled people who oppose all abortion and enlist in organizations such as the UK Society for the Protection of the Unborn Child (SPUC). It also comes from members of the disability rights movement, who may support a woman's right to terminate a pregnancy for social reasons, but oppose use of prenatal diagnosis to enable selective termination of pregnancies affected by genetic or developmental conditions. It is argued that characteristics of the fetus should not be a factor in choosing abortion, otherwise patient autonomy might lead to sex selection, or abortion on grounds of trivial differences, or 'designer babies'.

The philosophy of the disability rights movement plays down the problems caused by impairment and focuses instead on the need for removal of environmental and social barriers. Genetics is perceived as a threat because it once again defines disability in medical, indeed molecular, terms and because it is predicated on the removal of disabled people from the world. Some disability radicals have used emotive rhetoric about 'Nazi eugenics' or genetics as a 'search and destroy mission directed against disabled people' (Shakespeare, 1999). Conferences, demonstrations and publications have directed criticism at the goals and methods of contemporary genetic research and clinical practice.

However, not all disabled people share this blanket opposition to genetics. Members of some genetic support groups welcome genetic research because they hope it will achieve cures via gene therapy for the conditions from which they suffer. An organization such as the UK Genetic Interest Group, which represents families affected by genetic conditions, would be generally supportive of research and works closely with the genetics profession. Similarly, many medical research charities have disabled people working for them or fund-raising for research. In particular, disabled people who have degenerative conditions, for example, Friedreich ataxia, seem more likely to be positive about genetics than those who have stable conditions which do not make them 'ill'.

A third perspective within the disability community takes a more nuanced perspective on prenatal selection than either the disabled opponents or the disabled supporters of genetics. Eschewing the extreme rhetoric of eugenics, but sceptical about the claims and promises of genetics, occupants of this middle ground support the principle of a woman's right to choose. Rather than blaming individuals for the choices which they make or condemning the decision to terminate a pregnancy affected by a genetic condition, such commentators would focus on the social context in which choices are made (Asch, 1999). Three factors directly influence choice: a lack of balanced information about disabled lives; evidence that counseling may not be nondirective and that professionals may exert implicit pressure to terminate; and the routinization of antenatal care, which undermines the possibility of consent and often provides little time for counseling or reflection (Rapp, 2000).

Broader social and cultural factors may play an indirect role, including evidence that people who choose not to have prenatal tests, or choose not to terminate, are blamed for having disabled children (Marteau and Drake, 1995). Furthermore, lack of social support for disabled people and their families means that the consequences of continuing affected pregnancies may be very negative. Fear of litigation may lead physicians to insist on tests or encourage terminations. Insurance companies or healthcare providers may be unwilling to provide support in cases of avoidable genetic disease.

Many in the disability community fear that genetics will cause problems for those disabled people who exist now, as well as reducing the numbers of disabled people born in the future. The consequences of increased genetic screening could lead to increased discrimination against disabled people, if they are seen as 'a problem which could have been avoided'. Genetic solutions may replace social policy interventions designed to support disabled people. If the number of disabled people in the population declines, disability may become more stigmatized, and services may dwindle. Disabled people who wish to have children may face pressure not to reproduce or else to take advantage of genetic screening techniques to avoid the birth of children like them. Although the rhetoric of reproductive medicine supports choice, if people with particular impairments – such as deafness – want to choose to have children sharing their condition, this may not be an option that is supported by healthcare professionals or services.

Rather than the philosophy of removing disability from the world, many disabled people argue that impairment need not make life problematic if the right social supports are in place. Ironically, medical advances during the twentieth century have revolutionized lives by increasing the life spans and reducing the morbidity of many disabled people. At the same time, new assistive technologies can mitigate communication barriers and mobility difficulties, while more accessible environments facilitate disabled people to take advantage of educational and employment opportunities. In these contexts, the disability movement has argued that disability should not be seen as a problem. Some radicals even argue that impairment itself is a good thing, and they welcome the birth of more disabled people.

Many, whether because they subscribe to disability rights or humanist or religious values, welcome the contribution of disabled people to society and celebrate the diversity of human embodiment. Difference, they argue, should be supported not reduced. Each person makes a different contribution to the world, but should be valued for themselves. The meaning of life should not be found in conventional goals such as careers or wealth, but in human relationships and spiritual development. People with learning difficulties, for instance, may be unable to get paid employment and may need social support and advocacy, but they can still be happy, successful and valued members of a community (Vanier, 1999). This approach relies on communitarian or spiritual philosophies, which contrast strongly with the individualist and competitive ethos that governs Western societies.

While the abortion debate currently dominates discussion of genetics and disability, this may change in future. First, future research may lead to gene therapies or pharmaceutical interventions that can mitigate negative aspects of genetic conditions, without avoiding the birth of people with such conditions. Second, issues raised by the debate on disability and genetics will become increasingly important for nondisabled people. Everyone has deleterious genes in their genome, and the genetic factors in many common diseases are currently being discovered. In future, genes involved in many normal variations and behaviors will be identified, and new diagnostic techniques will enable speedy identification of an individual's characteristics prenatally or after birth. Therefore questions of normality and difference, therapy and enhancement, and what constitutes 'a good life' will become salient for everyone. Deciding on the appropriate interventions for each genetic difference will be vital: these may be prenatal, postnatal, medical, psychological, environmental, social or cultural, depending on the particular condition. Abortion should no longer be the main response to the diagnosis of a genetic variation.

References

Asch A (1999) Prenatal diagnosis and selective abortion: a challenge to practice and policy. *American Journal of Public Health* 89(11): 1649–1657.

Campbell J and Oliver M (1996) *Disability Politics: Understanding Our Past, Changing Our Future*. London: Routledge.

Marteau T M and Drake H (1995) Attributions for disability: the influence of genetic screening. *Social Science and Medicine* 40: 1127–1132.

Paul D (1992) Eugenic anxieties, social realities and political choices. *Social Research* 59(3): 663–683.

Rapp R (2000) *Testing Women, Testing the Fetus: The Social Impact of Amniocentesis in America*. New York: Routledge.

Rogers A (1999) Having disabled babies will be 'sin' says scientist. *Sunday Times*, UK, July 4.

Shakespeare T (1999) 'Losing the plot'?: medical and activist discourses of contemporary genetics and disability. *Sociology of Health and Illness* 21(5): 669–688.

Vanier J (1999) *Becoming Human*. London: Darton, Longman and Todd.

20
Mentally Handicapped in Britain: Sexuality and Procreation

Jane McCarthy, Valerie Sinason and Sheila Hollins

Introduction

Genetic research does not exist in isolation. It presents complex ethical issues for policy-makers, for society as a whole and for individuals. The use of rapidly increasing genetic knowledge in medicine and related sciences may lead to further discrimination against people with intellectual disabilities.

Historical evidence shows that physically and intellectually disabled people have been prevented, in a variety of ways, from reproducing (Sinason, 1992). In much of nineteenth- and twentieth-century Western Europe, for example, those with intellectual disabilities were sexually segregated outside the community. In Britain, from the beginning of the twentieth century, people with learning disabilities (mental handicaps) were segregated into male and female facilities in large hospitals away from society. Their sexuality was denied and procreation was actively prevented. Although there was no official eugenics policy, the segregation was in part a reaction to the rise of the international eugenics movement, which was spreading rapidly at that time.

Legislation changed the situation, with the 1971 white paper *Better Services for the Mentally Handicapped*, which followed the investigation of poor care and abuse in those large hospitals. As a result, during the last 30 years such institutions have been gradually closed down. People who have been institutionalized for most of their lives are now living in smaller homes in the community, and young people with intellectual disabilities today are growing up in their families of origin. However, the controversy surrounding procreation remains. The next white paper for England, titled *Valuing People: A New Strategy for Learning Disability* (2001), took 30 years to emerge.

History of eugenics

The use or abuse of genetics in arguments advocating the control of fertility and the rights of people with intellectual disabilities to reproduce is not a modern phenomenon. In 1850, the Austrian scientist Benedict Morel, influenced by Charles Darwin, noted hereditary diseases that led to deterioration of intellectual functioning. Francis Galton, a biologist and cousin of Darwin who coined the term 'eugenics', sought an interrelationship between heredity and intelligence. He was convinced that intelligence was distributed according to 'the curve of normal distribution', which meant that there would always be small groups of both exceptionally gifted and exceptionally intellectually disabled people. Galton's wish for eugenic reform provided a corrective to the previous era's unrealistic hopes for cure. There was a new demonology of the poor and disabled, backed by influential and otherwise liberal-minded intellectuals who said that those groups would cause a national deterioration of the gene pool, and hence of civilization. Such attitudes led to a Royal Commission in England in 1908. The commission found that 'feeble-mindedness is largely inherited; that prevention of mentally defective persons from becoming parents would tend to diminish the number of such persons in the population'. Eugenics was becoming a way of expressing a socially acceptable and popular prejudice. In many countries, the applications of eugenics were taken even further. In Germany, in 1933, the Law for the Prevention of Offspring with Hereditary Diseases demanded compulsory sterilization for 'congenital feeble-mindedness', schizophrenia, manic-depressive psychosis, hereditary epilepsy, Huntington's chorea, hereditary blindness, deafness, physical deformity and severe alcoholism. Hereditary health courts consisting of doctors and lawyers were set up and more than 400 000 people were sterilized. Later, under the Nazi regime, the eugenics policy intensified, leading to the systematic 'cleansing' of people with disabilities among millions of other 'genetically inferior' men, women and children. In Sweden, some 60 000 women considered genetically inferior, including women with intellectual disabilities, were forcibly sterilized in a program that continued until the 1970s. Thirty US states and two Canadian provinces legalized sterilization in the 1920s. The procreative exclusion of people with intellectual disabilities continues in many places around the world.

In the field of learning disability, the impact of centuries of stigmatization means that genetic advances bring complex consequences. Even in countries where progress has been made, clinicians must continue to augment their skills in evaluating the scientific, emotional and psychological meaning of genetic research for those with learning disabilities and their families. At times the different disciplines can unite to aid families. For example, the discovery of fragile X syndrome which was first described in 1969 and the responsible gene, fragile X mental retardation 1 (FMR 1), was identified in 1991, brought new understanding that alleviated family distress.

Personal relationships

Concern about sexual relationships and parenthood among people with intellectual disabilities may not always be a question of true genetic concerns. Frequently it stems from the difficulty that many people in the general population have in accepting that men and women with intellectual disabilities have ordinary sexual feelings and desires, let alone that they should be allowed to act on them. People with intellectual disabilities have the same rights as other people to form sexual relationships. It is important, therefore, that they are provided with basic information about such relationships, so that they can make informed choices about what they want to do in this context. Also, since people with intellectual disabilities are at relatively high risk of being sexually exploited, there is a need to develop self-confidence and assertiveness (Hollins and Sinason, 1992, 1993).

Few adults with intellectual disabilities become parents, but low intellectual functioning does not in itself exclude the possibility of parenthood. It is likely, however, that a person with an IQ below 60 will need considerable support in order to provide adequate care for a child. Parents with intellectual disabilities often have additional circumstances that may hinder their ability to parent successfully, such as poverty, early adversity, poor social support and ill health – issues that need to be addressed just as they are in the general population. Success often depends on the support of the father or partner and grandparents of the child (Cotson et al., 2000), as well as on effective support from outside the family.

When a person with intellectual disabilities is contemplating parenthood, concern is often expressed in the form of the risk of genetic disorder. However, if one parent has an IQ below 70, for example, only one in seven of their children will have a similar level of intelligence (Reed and Reed, 1965). Four out of five children with learning disabilities have parents within the normal range of intelligence (Reed and Anderson, 1973).

Prenatal testing may be seen primarily as a public health concern, but it is also an important ethical, social justice and human rights issue. Extra care and training are needed to ensure that learning-disabled clients and their parental hopes and aspirations are treated with respect and truthfulness.

Ethical and legal issues

Within the UK (Sinason, 1992), the issue of sterilization polarizes advocates and disability rights groups on one side and parents on the other. Parents who have worked hard bringing up their learning-disabled children fear having to bring up their grandchildren too; and, in the absence of adequate support, they might support the sterilization of their children. The new British white paper *Valuing People* addresses such issues by focusing on the principles of rights, independence, choice and inclusion.

An important issue in this context is consent. Consent in healthcare is essentially a choice about whether to have a procedure or treatment. For people with intellectual disabilities, personal choice is more often than not a limited experience. The key question in any situation is whether the person is competent to make a decision. If people with intellectual disabilities are competent to make a decision, they have the right to expect to be able to make that choice (Department of Health, 2001a, 2001b, Keywood et al., 1999). For informed consent to be given, it is necessary for the individual to have an understanding of the proposed procedure or treatment, of its benefits and risks and of the consequences of not receiving the treatment or procedure. The consent must be voluntary and free of coercion.

The assessment of incapacity is devolved to the doctor who is proposing the treatment. It is therefore crucial that medical education at both undergraduate and postgraduate levels should include assessment of incapacity, a skill that doctors will need in almost every branch of medicine (Curran and Hollins, 1994).

The legal aspect of the argument is about determining who has capacity and the course of action required. For those who lack capacity to consent to treatment, doctors must act in their patients' best interest. The ethical dilemma is what is in the best interest of another person. That adds significantly to the complexity of deciding who in society can procreate.

Parents with intellectual disabilities who bring up their children with support can manage as well as other vulnerable groups for whom there is no question of genetic testing. Under the British 1989 Children Act, a child's welfare is evaluated by a range of criteria to ensure that there is no significant risk. While there are complex decisions to be made concerning the welfare of some children with learning-disabled parents (Baradon et al., 1999), it is important to acknowledge the legislative achievements in this area.

Fuller knowledge of the genetic basis of human characteristics will provide increased understanding of the risks, benefits and consequences of reproduction of individuals. However, there is a danger that decisions based on this equation of genetics and human characteristics will ignore the wide range of variation in any trait and also the whole spectrum of social, religious and cultural correlates of mental and physical characteristics in any individual. Recently gained medical knowledge, along with the new legislation, has only increased the need to view people with learning disabilities as citizens with rights that include being members of society in all its aspects.

References

Baradon T, Sinason V and Yabsley Y (1999) Assessment of parents and young children (Children Act 1989): a child psychotherapy point of view. *Child: Care, Health and Development* 25(1): 37.

Cotson D, Friend J, Hollins S and James H (2000) Implementing the framework for the assessment of children in need and their families when the parent has a learning disability. In: Horwath J (ed.) *The Child's World: Assessing Children in Need*. London: Department of Health.

Curran J and Hollins S (1994) Consent to medical treatment and people with learning disability. *Psychiatric Bulletin* 18: 691–693.

Department of Health (2001a) *Reference Guide to Consent for Examination or Treatment*. London: The Stationery Office.

Department of Health (2001b) *Valuing People: A New Strategy for Learning Disability for the Twenty-first Century*. London: The Stationery Office.

Hollins S and Sinason V (1992) *Jenny Speaks Out*. London: Books Beyond Words/St George's Mental Health Library, St George's Hospital Medical School.

Hollins S and Sinason V (1993) *Bob Tells All*. London: Books Beyond Words/St George's Mental Health Library, St George's Hospital Medical School.

Keywood K, Fovargue S and Flynn M (1999) *Best Practice? Health Care Decision Making by, with and for Adults with Learning Disabilities*. Manchester, UK: National Development Team and Institute of Medicine, Law and Bioethics.

Reed S C and Anderson V E (1973) Effects of changing sexuality on the gene pool. In: De La Cruz F F and La Veck G P (eds.) *Human Sexuality and the Mentally Retarded*, pp. 111–125. London: Butterworth.

Reed E W and Reed C S (1965) *Mental Retardation: A Family Study*. Philadelphia, PA: Saunders.

Sinason V (1992) *Mental Handicap and the Human Condition*. London: Routledge.

Part 4

Genetics and Society: Information, Interpretation and Representation

Introduction

Angus Clarke and Flo Ticehurst

This section now leads us to examine some of the broader social issues that have emerged as important in the development of human genetics. Other topics could have been selected for inclusion here, but space has permitted only some of those articles that we would like to have included.

Nelkin, now sadly deceased, discusses (chapter 21) genetic essentialism and genetic identity as employed in cultural discourse over the recent past. As a biological construct identity in reproduction lies in the passing on of genes, and Nelkin considers aspects of this as it is represented in literature and in the social value placed on a biological sense of 'roots' in family values and in giving a sense of 'true' identity – this concern being manifested in a need to know one's parentage. As the family unit is threatened in today's social and political climate, genetic knowledge is increasingly called upon to provide a secure foundation – and this is what can be reinforced by media reports.

The message conveyed in the media is often fatalistic – as 'every quality has a gene'. The implication of this is that one must accept and adapt to genetic limitations. Nelkin demonstrates how this type of media portrayal reinforces gender stereotypes and fosters gender inequalities. She then suggests that the gene has assumed a spiritual significance as the symbol of genetic essentialism, a concept encouraged by scientists who describe the genome as a Bible, or the Book of Man. Nelkin also considers the implications of these iconic images and concepts of biological destiny in relation to individual responsibility and freedom and to social policy and human rights for groups and populations.

Richards (chapter 22) considers research into the public understandings and knowledge of heredity, how this interacts with aspects of scientific and social inheritance and the implications for education about these issues in the context of advances in genetic knowledge. He discusses conflicts between lay beliefs about inheritance and biological knowledge of these processes. He presents data from a study investigating lay understandings of genetic and environmental influences on the determination of individual characteristics; this work shows that physical characteristics are seen to be influenced by heredity more than by behavioural characteristics, and that women are much more aware of family histories of disease than men and often take more responsibility for managing the information within the family – as the 'genetic housekeepers'.

Richards also describes the social and political context of the current 'public understanding of science' debates – he sees this as largely promoting the public appreciation of science – and questions the salience of current approaches to teaching about genetics in schools (i.e. the arguably inappropriate focus on Mendelian inheritance), calling for more imaginative ways to educate children about genetics that takes their beliefs into account.

Atkinson, Bharadwaj and Featherstone (chapter 23) discuss kinship and descent as human cultural processes that may be transmitted by both genetic and non-genetic means. Their perspective involves articulating some of the ideas about biological and social inheritance, focusing on the parallels and differences between the various notions of relatedness. Atkinson and his colleagues present examples of culturally different patterns of descent and systems of naming – particularly those that depart from Euro-American conceptions of kinship.

They ask, what is inherited and how? As well as personal attributes one can also inherit land, property, wealth, names, ethnicity, nationality, religion and cultural competence. They also discuss different cultural definitions of physical and moral attributes.

English and Sommerville (chapter 24) ask whether we have a duty to share genetic information and other important health data, particularly in the context of close relationships. They describe how modern ethical discourse gives prominence to the confidentiality and privacy of the individual, and discuss the implications of advances in genetic knowledge for this supremacy. Referring to social and cultural aspects of relatedness, they point out that being genetically related does not necessarily mean that familial bonds or responsibilities exist. They ask whether genetic information can be private when the genetic material is itself shared among family members; they ask whether and how it is different from other types of medical information. For them the fundamental difference is that it can have relevance to a chain of individuals, therefore producing the potential for conflict between one individual's rights and the rights of other individuals.

Consequentially, they argue that there is a duty owed by individuals to share their genetic information with relatives, and they suggest that this can depend on the strength and importance of the relationship in question. They believe that inflexible rules would not be helpful in these family contexts.

McGleenan (chapter 25) considers the implications of advances in predictive genetic medicine for the provision of insurance. He sets out some of the principles of the insurance industry and discusses how advances in our understanding of genetics may lead to the problem of adverse selection. He extends his consideration to the implications for public health and further advances in health care and medicine, leading to a discussion of the potential for genetic discrimination.

McGleenan suggests that, if insurance companies were permitted no access to genetic information, then premiums would have to rise to cope with the uncertainty, so governments must find a balance between protecting the interests of those potentially at a disadvantage and ensuring that people can purchase appropriate, affordable insurance. Current technology is still limited as to what can be predicted, so only a small number of genetic diseases are currently relevant from an insurer's point of view. In his view, until genetic testing is more widespread, the impact on the insurance industry will be minimal.

Beckwith & Alper (chapter 26) begin by describing the history of ideas in relation to social hierarchies and inherited biological differences. This takes us from the 'myth of the metals' to justify the rule of the philosopher kings in Plato's Republic, to Herbert Spencer's explanation of people's differing economic success in his interpretation of *The Origin of Species*. From Spencer we are led to Mendel's laws of inheritance and Binet's Intelligence Quotient. They describe the assertion of the craniologist Samuel George Morton that Blacks had less brain capacity and thus less intelligence than Whites and more recent explanations about a genetic basis of important behavioural traits from the field of socio-biology and evolutionary psychology.

The authors turn to research on differences between population groups, and consider whether or how such research may be (mis)used to support racist views. They assert that scientists should be socially responsible and have 'a special obligation to expose misuse of their science', and should ensure that knowledge of genetics conveyed to the public properly represents the complexities involved.

Adshead and Bostock (chapter 27) discuss the suggested existence of links between genes and criminal behaviour, considering how such links might influence accepted notions of legal and moral responsibility. Their critique of the research in this area leads them to elaborate the view that criminal or antisocial behaviour is a social construct; they also refer to some of the limitations of this research, such as its focus on delinquency rather than 'white-collar crime'. They also question the way that research into the genetic basis of criminality is represented in the media – exaggerated and overly deterministic.

The authors discuss the legal use of a 'genetic defence' giving a number of examples of cases where genetics has been appealed to as a causative factor in the crime in question. Although such genetic defences most often do fail to reach the standard for admissibility of evidence, advances in genetic research might change this and jurors would then need to understand the relationship between genetic influence on behaviour and the allocation of legal and moral responsibility. The two main schools of thought on this are assessed and their relationship to our notions of free will, as well as the potential application of such knowledge – in mitigation of a crime committed or so as to pre-empt such crimes.

Finally, the authors describe their view of the political context within which research on genes and criminal behaviour currently finds itself – a climate where situating social adversity within the individual means that a political approach may be seen as unnecessary. They suggest that the possibility of genetic manipulation as a means of dealing with criminal behaviour is feared or welcomed depending on one's social and political views.

21
Gene as a Cultural Icon

Dorothy Nelkin

Introduction

The gene in American popular culture appears in newspapers, television programs, fiction, film, advertising and childcare books as a powerful entity that determines an extraordinary range of character traits and behavioral tendencies (Nelkin and Lindee, 1995). The media convey a message of genetic essentialism – the person is defined as DNA. Genetic language has become a code, readily applied and broadly understood. To say, 'it's in the genes' renders further explanation unnecessary. What is genetic is powerful, predictable and permanent; 'hard-wired' in the human constitution.

Deoxyribonucleic acid (DNA) is a biological entity specifying the composition of the proteins carrying the information that form tissues and cells, but the popular discourse portrays the gene as less a biological entity than a cultural icon. Though the gene refers to a biological construct and derives its powers from science, its symbolic meaning is independent of its biological definition. It is rather a convenient way to define personhood, to talk about destiny, to justify social stereotypes and to shape social policies on the basis of 'natural' forces.

DNA as identity

DNA in the cultural discourse has assumed significance not only as a tool for identification, but as the essence of identity and personhood, the basis of human relationships and even a source of immortality. Genes are what make us human, what give the body life. This concept of genetic essentialism appears repeatedly in science popularizations, explicitly, for example, in stories about reproduction. Women in these stories voice despair over infertility in terms of genetic immortality: their identity lies not in accomplishments, relationships or actions, but in passing on genes.

Genetic constructs also enter debates about family values. In the popular culture media, 'real' relationships depend more on genetic connections than on emotional commitment. Numerous adoption stories portray 'true' parents as those who contribute gametes, because real connections depend on shared DNA. They show children searching for their birth parents and desperate to find their 'genetic roots'. The media message is that knowing one's genetic connections is a way not only to gain clues to possible health problems, but also to define the authenticity of family relationships. This view of the importance of genetic connections also shapes the news coverage of surrogacy and custody cases, as reporters refer to 'genetic rights'. And in real-life litigation over child custody, proven genetic connections have overridden the emotional interests of the child (Dolgin, 1993).

At a time when the family seems a troubled institution, threatened by feminism, divorce, working mothers, alternative partnerships, gay rights, and the complex arrangements enabled by new reproductive technologies, genetic ties seem to offer an apparently solid basis for human connections. Their shared DNA grounds family relationships in a stable, well-defined unit, providing the individual with indisputable roots that appear to be more reliable than the ephemeral ties of emotional love, marital vows or shared experiences. Genetic connections, after all, can never be severed, and they validate the individual as genetically placed in unambiguous relationships to others.

DNA as destiny

The cultural appeal of biological determinism – the belief that the physical and behavioral traits of individuals are caused by inborn distinctions – also pervades the popular media (Van Dijck, 1998; Conrad, 1999). Scientific advances in molecular genetics – the identification and location of disease genes – have fueled expectations of eventually identifying and predicting the genetic origins of complex behaviors and personality traits. A very wide range of behaviors have been attributed to genetics over the past few years, including alcoholism, criminality, homosexuality, mental illness, aggressive personality, exhibitionism, dyslexia, divorce, religiosity, dangerousness, job success, educational success, arson, intelligence, tendency to tease, risk taking, timidity, social potency, traditionalism and zest for life.

Beyond the curious notion that every quality has a gene, the media are conveying a fatalistic message: one must understand, accept and adapt to genetic limitations. Visual images of the double helix portray people as prisoners of their DNA. The old story about those who make it despite coming from a bad family is no longer a mystery: 'He had good genes.' Conversely, those who fail despite opportunity are written off as having bad genes. People are the way they are because they are 'born that way'.

Studies of identical twins, intended to show how their genetic make-up determines intelligence and other personality traits, have enjoyed extensive and enthusiastic media coverage. Although the scientific significance of these studies is in doubt, the media have welcomed them as evidence of genetic determinism. Most often, genetics is appropriated as a way to explain social problems. 'Evil is embedded in the coils of chromosomes that our parents pass on to us at conception', writes a *New York Times* reporter (Franklin, 1989).

Stories about genetic influences draw on research into molecular and behavioral genetics, but they ignore the complexities of genetic and environmental interactions and the distance between the molecular level of genetic systems and actual behavior. Thus, the popular culture treats the most complex behaviors as if they are passed down through the generations like brown hair or blue eyes.

Genetic stereotypes

So too, genes appear in the cultural discourse as a way to explain the social and economic differences between groups. In his book *Backdoor to Eugenics*, sociologist Troy Duster asks why scientists study the biological basis of violent crime by minority groups, but not the biological basis of white-collar crime, perceiving this as a product of opportunity structure, not predisposition. Social stereotypes, he argues, influence the focus of research, and they clearly affect how such research is interpreted (Duster, 1989).

Genetic explanations appear in stories about gender inequalities. A television news program, called 'Boys and Girls are Different', claimed that scientific evidence demonstrates important genetic differences between the sexes, for example in their intellectual and emotional skills. Women are better nurturers, men are better at mathematics, and such differences are already manifest in their behavior as toddlers. A woman's natural abilities, claimed the television host, lead her to prefer childcare to working outside the home. Repeating a narrative that has long justified the exclusion of women from the world of work, such stories suggest that the failure of women to achieve economic and professional parity with men is a consequence of genetic differences rather than gender discrimination or other historical and social forces.

The cultural appeal of genetic explanations was explicit in the press responses to the claims of sociobiologists in the late 1970s and early 1980s. The premise of sociobiology is that behavior is shaped primarily by genetic factors selected for their survival value. The media focused on the implications for race and gender differences, using the ideas to support popular stereotypes. Today, sociobiology and the related field of evolutionary psychology have become accepted and are used in debates to provide scientific justification for stereotypes.

Sacred genes

Most cultures recognize some entity – the *hun*, the soul, the spirit – that is relatively independent of the body, but gives it life and immortality. The essence of 'personhood' has often been associated with various body parts – the pineal gland, the blood, the mind, the brain, the heart; today it is the DNA. The gene has assumed a kind of spiritual significance as the location of the true self. Scientists have encouraged such sacred images, referring to the genome as a Bible, an oracle or the Book of Man. But such images also encourage a cultural resistance to genetic engineering or manipulation. Increased access to the genome has revived images of Frankenstein's monster and mad scientists (Turney, 1998). Popular novels call forth images of the callous, heartless researcher, who creates monsters while engaged in a neutral process of scientific discovery. Religious groups object to 'tampering' with genes. Fear of genetic manipulation reflects the belief that the gene is the essence of the person, and also that if such research is permitted, someone will try to create a master race.

The DNA is data without dimension, text without context, so that the equation of DNA with personhood requires a profound leap of faith. There is no fundamental reason why the DNA, rather than the heart or liver, should now be the locus of personhood. Heart surgery, even brain surgery, is not construed as tinkering with the 'soul'. Yet the gene has been identified with a sacred, God-given basis of essential humanity, the 'fundamental stuff of life'.

Genetic essentialism and social policy

Why has the gene become a cultural icon? Clearly important are the expectations invoked by the widely publicized research yielding information about the links between genes and disease. But, in part, genetic constructs hold cultural appeal because they suit the political context, providing justification for social policies and legitimization of political goals. Interpretations of this research reflect deeply embedded beliefs and prevailing public concerns. The gene is appropriated to explain current social dislocations – the threats implied by the changing role of women, the perceived decline of the family, the changes in the ethnic and racial structure of society, and the failures of social welfare programs. Much of the cultural discourse about genetics is a way to reinforce traditional family ties and the roles of the sexes, to account for deviance, and to explain economic inequalities and social hierarchies as the natural consequence of genetic differences.

As the interpreter of nature, science has long been a cultural resource, appropriated and interpreted to support social beliefs and legitimate social policies. Whatever the reasons for its cultural appeal, the surfacing of hereditarian beliefs must be taken seriously, for common and recurring images influence individual choices, public attitudes and policy agendas. If, for example, it is widely believed that biology is destiny, what will be the implications for concepts of responsibility and free will underlying the judicial system? How will beliefs in the importance of genetics influence perceptions and policies toward the disabled (Shakespeare, 1999)? Belief in genetic destiny could lend legitimacy to social practices with pernicious implications for human rights. The use of genetic testing to predict dangerousness, for example, could become acceptable, overriding values of justice or fairness (Nelkin and Tancredi, 1994), and genetic assumptions could encourage eugenic practices (Marteau and Richards, 1996).

If it is believed that some people are inherently and irretrievably incapable or dangerous, those who fail can be labeled intrinsically, genetically flawed, so why put resources into education or rehabilitation? At a time when governments, faced with cost constraints, are dismantling social welfare programs, genetic explanations can be used to avoid costly policy interventions by deflecting blame onto the individual: 'It's all in the genes.' If deterministic ideas are extended to groups, this has implications for perpetuating stereotypes, with a bearing on the rights and opportunities open to classes of people as well as to individuals.

Genetic research is yielding increased access to the human genome at a remarkable rate. The information from this provocative science will contribute to the understanding and, eventually, management of devastating genetic diseases. However, the information will also be appropriated to support prevailing beliefs, for assumptions about the transcendent power of the gene translate easily into moral guidelines about normal or natural behavior. Embedded in the futurism of contemporary genomics and conveyed through popular culture are fundamentally conservative assumptions about social and moral order that implicitly legitimize existing arrangements and social hierarchies. These cultural and symbolic meanings of the gene will ultimately shape the uses of a powerful science, one that offers prospects for promising applications, but also opens possibilities for pernicious abuse.

References

Conrad P (1999) Media images, genetics and culture. In: Hoffmaster B (ed.) *Bioethics in Context*. Cambridge: Cambridge University Press.

Dolgin J (1993) Just a gene: judicial assumptions about parenthood. *UCLA Law Review* 40: 637–694.

Duster T (1989) *Backdoor to Eugenics*. New York: Routledge.

Franklin D (1989) What a child is given. *New York Times* September 3: 36.

Marteau T and Richards M (eds) (1996) *The Troubled Helix*. Cambridge: Cambridge University Press.

Nelkin D and Lindee M S (1995) *The DNA Mystique: The Gene as Cultural Icon*. New York: WH Freeman.

Nelkin D and Tancredi L (1994) *Dangerous Diagnostics*, 2nd edn. Chicago, IL: University of Chicago Press.
Shakespeare T (1999) Losing the plot. In: Conrad P and Gabe J (eds.) *Sociological Perspectives on the New Genetics*. Oxford: Blackwell.
Turney J (1998) *Frankenstein's Footsteps: Science, Genetics and Popular Culture*. New Haven, CT: Yale University Press.
Van Dijck J (1998) *Imagenation: The Public Image of Genetics*. Basingstoke: Macmillan Press.

22
Heredity: Lay Understanding

Martin P. M. Richards

Introduction

The lay member of the public has wide knowledge of inheritance and an interest in it, particularly as it is relevant to their own family. This knowledge often relates to features that may run in the family and may be derived from knowledge of social relationships. There is little general understanding of Mendelian genetics. Indeed, notions of inheritance based on segregating particulate genes conflict with the widely held concepts of inheritance and this may be why Mendelian explanations offered in school or the genetics clinic are often poorly assimilated. Lay accounts may be more compatible with epigenetic perspectives of development.

While scientists sometimes believe that failures of the public to appreciate their work is the result of either ignorance or irrationality, many people see the bulk of scientific knowledge as simply irrelevant to their needs or interests. However, when members of the public become involved in issues about the use or misuse of scientific information or the deployment of technology, interest in scientific knowledge and its assimilation may rapidly increase. Research on the public understanding of science, or what is more appropriately called the public appreciation of science, would suggest that we need to review both the approach and the nature of the scientific knowledge related to genetics and inheritance that is taught to nonspecialists in school and in adult education.

General knowledge of inheritance

Much discussion within families is about particular characteristics that family members may share, from whom they may have been inherited and which children they are passed on to. Such discussions concern both physical and behavioral characteristics, as well as disorders that may run in the family

(Ponder et al., 1996; Richards, 1996a, 1996b). Given that photographs and pictures are often passed down through families, there is often a concentration on resemblance of physical appearance. Lay knowledge holds that characteristics that run in the family are determined by what we inherited, with varying contributions from the environment and our lifestyle. In general, the two sides of the family are seen to provide equal contributions to an individual's inheritance. However, characteristics particularly associated with one or other sex may be seen to be inherited exclusively from the same-sex parent. Thus, for example, many female clinical patients find it hard to understand that breast cancer can be inherited via a father (Green et al., 1997). A practical consequence of this is that the majority of women attending genetics clinics because of a family history of breast or ovarian cancer have a maternal and not a paternal history of the disease.

Members of families who carry inherited disorders are often well aware of their family history and may have particular explanations for the way in which cases are distributed in the family; for example, women in families with inherited breast/ovarian cancer variously talk of the cancer 'skipping a generation' or always affecting firstborn daughters, while others see the inheritance remaining latent until it is provoked by an event such as a blow to the breast. Others believe that those who in physical or behavioral characteristics resemble forebears that had cancer are more likely to contract the disease, while those who 'do not take after' such relatives are safe.

It has been suggested that these concepts of inheritance are derived from knowledge of social kinship relationships. In support of this hypothesis it has been shown that (preschool) children expect relatives (physically) to resemble one another (Springer, 1992) and genetic links are seen to be closer for parents and children than for siblings, with whom kin ties are weaker (Richards and Ponder, 1996).

There are obvious conflicts between these widely held ideas about inheritance and Mendelian explanations of the nature of the hereditary substance and the segregation of phenotypic characters. While the terms gene, deoxyribonucleic acid (DNA) and chromosomes are widely used in discussion of inheritance in many contexts, these may be used simply to refer to a process of biological inheritance rather than carrying the meaning they would in the context of a scientific discussion of genetics.

Inherited and environmental factors in development

Table 22.1 shows some of the results of an interview study in which respondents were asked about the relative effect of inherited and environmental factors in the determination of individual characteristics. Typically in such

Table 22.1 Inherited and environmental factors in the development of various characteristics

Characteristic/ illness	**Totally inherited (nature)**				**Totally environmental (nurture)**	**Don't know**	**Mean**
	1	2	3	4	5		
Antisocial behavior	3	7	26	34	28	2	3.7
Intelligence	22	29	33	12	4	1	2.4
Eye color	74	13	6	2	1	4	1.3
Body height and weight	30	35	28	4	2	–	2.1
Flu	3	6	18	15	54	4	4.1
Measles	4	6	20	16	45	8	3.9
Heart disease	19	22	41	12	2	3	2.9
Cancer	13	23	40	11	7	5	2.7
Huntington's disease	34	16	15	2	3	30	1.9
Cystic fibrosis	35	19	16	5	4	21	2.0

Notes: Respondents' views of whether particular characteristics may come about because they are inherited (nature) or because of environmental factors (nurture) or a combination of these. These data come from face-to-face interviews with members of the People's Panel carried out by the Market and Opinion Research Institute (MORI) in 2000. Data are weighted to the demographic profile of the UK population. Respondents were read a list of 'characteristics that may come about because they are inherited (nature) or because of environmental factors (nurture) such as lifestyle, upbringing, etc. or because of a combination of these'. For each characteristic respondents were asked to indicate 'how each characteristic comes about from 1, if you think it is totally inherited, to 5 if you think its development is entirely dependent upon environmental factors'. All figures are percentages of responses for each characteristic (n = 938).

studies, physical characteristics are seen to be more a matter of inheritance than behavioral characteristics. However, intelligence is seen to be more strongly inherited than other behavioral characteristics, perhaps because of the long-standing discussion of the heritability of intelligence quotient (Sternberg and Grigorenko, 1997). While the Mendelian diseases cited in the study are seen by more respondents to be totally inherited than cancer and heart disease, it is noticeable that the former attracted far more 'don't knows', indicating the lack of familiarity of even the best known of the Mendelian diseases.

A number of lines of evidence suggest that women are much more aware of family histories of disease than men (Richards, 1996b) and may take more family responsibility for such conditions by, for example, informing other family members, clarifying their own or others' risks by collecting family history information and undergoing predictive and other genetic testing and involvement in risk-reducing activities (Hallowell, 1999). The term 'genetic housekeeper' has been used to describe this role.

Public understanding of science

While there may be general agreement about the desirability of increasing public understanding of science for a number of reasons, including improving the quality of both public and private decisions and enriching our lives and culture, the concept itself is now seen to be more complex and contentious than in earlier debates (Turner, 1996). The term itself is increasingly being dropped in favor of others such as 'public appreciation of science'. Part of the reason for this is a move away from the over-simple deficit model, which held that members of the public merely lacked information held by scientists. In rejecting the notion that science is revealed truth, there is a growing understanding that attitudes to expertise are inevitably bound up with questions of trust. While at least in the genetics field, trust in clinicians and 'independent' medical research remains strong, the same is not true for scientists in general and, in particular, those involved in the biotech industries (see Office of Science and Technology and the Wellcome Trust, 2000). As, for example, Turney (1996) argues (p. 1088):

> we have to pay close attention to the character and interests of a range of different publics, rather than discussing a single monolithic 'public'; that understanding may not mean the same, theoretically or practically, for various publics as it does for scientists; and that science is to be seen neither as certain knowledge nor as the only source of authentic expertise ... The implication is that improving public understanding of science requires a willingness to work toward a scientists' better understanding of science.

As others have done, Turney goes on to advocate a bottom-up approach, which begins with the interests and concerns of specific groups. While it is often suggested that a general scientific literacy would allow engagement in debates about science policy and technological development, evidence suggests the opposite that if people see an opportunity to participate in debates, understanding will follow. An example of this was a consensus conference on plant biotechnology involving a panel of lay people. After cross-examining specialists, and much reading, a well-informed position paper was produced by the panel members who had little prior knowledge of the issues (Joss and Durant, 1995). Increasingly, these lessons are being acted on by public bodies concerned with developing policy related to human genetics (e.g., in the UK, The Human Genetic Commission (2000)).

As yet, little attempt has been made to consider the explanations of inheritance that are offered in schools and clinics in the light of the research into lay knowledge. School teaching of inheritance, at least in Britain, remains firmly rooted in the classic early twentieth century model of genes and chromosomes illustrated by Mendelian disease or the few obvious human characteristics such as eye or hair color that can be interpreted in similar ways. While such teaching may have some salience for the few children whose families carry a Mendelian disorder, it is unlikely to relate to the lay knowledge of inheritance of the many that is not based on highly penetrant single-gene effects. Lay concepts of blending inheritance are much closer to notions of developmental systems, which display high degrees of redundancy and achieve similar ends by multiple means. We need imaginative ways of working with what children (and indeed adults) understand of inheritance, and working from these toward whatever biology may be salient. Mendelian diseases need to be seen as the exceptions for a developmental system that produces broadly similar (and functional) phenotypes from varying genotypes.

We need to follow the same approach in the clinic. What is it that patients want to know? Again, the research gives strong hints. While few of those from families with Mendelian disorders may understand Mendelian ratios, after counseling most do understand the information that is most salient to them; for example, their chance of producing an affected child (Snowdon and Green, 1997).

References

Green J, Richards M, Murton F, Statham H and Hallowell N (1997) Family communication and genetic counseling: the case of hereditary breast and ovarian cancer. *Journal of Genetic Counseling* 6: 45–60.

Hallowell N (1999) Doing the right thing: genetic risk and responsibility. *Sociology of Health and Illness* 21: 597–621.

Human Genetic Commission (2000) Whose Hands on Your Genes? A Discussion Document on the Storage, Protection and Use of Personal Genetic Information. http://www.hgc.gov.uk/Client/document.asp?DocId=38&CAtegoryId=8.

Joss S and Durant J (1995) The UK National Consensus Conference on Plant Biotechnology. *Public Understanding of Science* 4: 195–206.

Office of Science and Technology and the Wellcome Trust (2000) A Review of Science Communication and Public Attitudes to Science in Britain. http://www.wellcome.ac.uk/doc%5Fwtd003420.html.

Ponder M, Lee J, Green J and Richards M (1996) Family history and perceived vulnerability to some common diseases: a study of young people and their parents. *Journal of Medical Genetics* 33: 485–492.

Richards M (1996a) Lay and professional knowledge of genetics and inheritance. *Public Understanding of Science* 5: 217–230.

Richards M P M (1996b) Families, kinship and genetics. In: Marteau T and Richards M (eds.) *The Troubled Helix: Social and Psychological Implications of the New Human Genetics*, chap. 12, pp. 249–273. Cambridge: Cambridge University Press.

Richards M P M and Ponder M (1996) Lay understanding of genetics: a test of an hypothesis. *Journal of Medical Genetics* 33: 1032–1036.

Snowdon C and Green J M (1997) Preimplantation diagnosis and other reproductive options: attitudes of male and female carriers of recessive disorders. *Human Reproduction* 10: 101–110.

Springer K (1992) Children's awareness of the biological implications of kinship. *Child Development* 67: 151–171.
Sternberg R J and Grigorenko E (eds.) (1997) *Intelligence, Heredity and Environment.* Cambridge: Cambridge University Press.
Turner J (1996) The public understanding of genetics – where next? *European Journal of Genetics and Society* 1: 5–20.
Turney J (1996) Public understanding of science. *Lancet* 347: 1087–1090.

23
Inheritance and Society

Paul Atkinson, Aditya Bharadwaj and Katie Featherstone

Cultural variation

There are numerous ways in which inheritance is managed socially and culturally. Insofar as cultural forms are learned and are passed on from generation to generation, the array of human cultures, with all their extraordinary variety, is inherited. Although there may be some contentious areas of disagreement between the sociologists and social biologists that pertain to the genetic and cultural components of such transmission, most cultural phenomena are maintained and transmitted by cultural means. Equally, there are specific social mechanisms to ensure the intergenerational transmission of particular attributes or goods such as property, nationality, religious faith, group membership, and so on. These vary greatly between different cultures and present a subject matter too vast to be examined in one essay. We therefore concentrate on the key topics of kinship and descent – an area of human cultural process that lies at the cusp of genetic and nongenetic inheritance.

The relationship between kinship and nongenetic inheritance is not transparent. Contrary to the assumptions of Euro-American kinship – that is, the biological fact of having sex and transmitting genes resulting in conception and eventual birth is succeeded by the social fact of kin relations – the shared beliefs about inheritance in their cross-cultural complexity deeply affect ideas and practices of transmission. This is significant because kinship systems and theories of procreation across the range of human cultures predate modern molecular biology.

We first focus on the concept of kinship and descent in order to explain the complexities that are inherent in geneticizing cultural models of kinship and connectivity. We then argue that to talk about the cultural context of nongenetic inheritance within the kinship domain is to talk about 'what' is inherited and 'how' – a problem that we broadly subsume under two generic terms: 'substance' and 'routes'.

Kinship, descent and inheritance

We begin by discussing the cultural practices of kinship because the parallels and differences between biological and social definitions of relatedness are especially important and raise interesting issues concerning the relationships between biology and culture. If genetic relatedness is a matter of biology and is governed by universal natural processes, then kinship is a cultural phenomenon that significantly varies from culture to culture. Moreover, cultural patterns of descent differ significantly from biomedical understandings of inheritance. This is a longstanding research interest for social scientists. Anthropological excursions into the cultural domain from the very beginning became preoccupied with the study of kinship, but by the 1960s it had become a less central concern. Since the early 1990s, however, reproductive technologies and the implications of genomic science have revitalized this field of study. These developments have focused attention on the Euro-American reproductive model and its assumptions about the connections between natural facts and social relations. Social scientists recognize that it is a modern and twentieth-century view that the social recognition of parenthood must follow the biological fact of having sex, transmitting genes, and setting into motion the biological development of the embryo.

Modern biomedical science itself has made the relationships between biological and social phenomena problematic. For example, technologically assisted conception creates different possibilities of configuring biological kinship. The twenty-first-century 'natural parent' – one who embodies the genetic and social kin credentials – is dispersed either by enabling fertilization outside the body (as in *in vitro* fertilization) or by involving donated third-party gametes (eggs or sperm). The anthropologist Sarah Franklin writes of the dilemma of contemporary kinship: 'how to make sense of new forms of assisted conception which create more flexible and uncertain relation'. In contemporary Euro-American societies, therefore, the relationships between the 'natural' facts of procreation and the social arrangements of relatedness have become increasingly problematic. The boundaries between nature and culture, and their implications for inheritance, have become more fluid. The new biological possibilities of reproductive cloning and the social possibilities that include two lesbian partners contributing genetic material to a shared child have the potential to transform further the cultural links between natural facts and social arrangements.

However, it would be quite wrong to assume that innovations in biomedical science disrupt natural relationships that are grounded in the biological principles of inheritance. If we begin from the context in which most genomic research has been done and in which most cosmopolitan ('Western') biomedicine has been generated, then the relationship between biology and kinship seems transparently straightforward. The simplicity is deceptive, however, as a consideration of other kinship systems helps to reveal. Like all cultural systems, kinship depends on the recognition of similarity and difference. Between individuals it provides systems of coding relations, which are based on relations between categories of person. Whether or not one ultimately attributes kinship to the natural categories of sex and generation, there are key differences between kinship systems.

Bilateral relationships

The Euro-American system of kinship is based on the recognition of bilateral relations. It makes no classificatory distinctions between one's maternal and paternal kin. For example, one's father's sister and one's mother's sister are both classified as 'aunts', and the equivalent male kin are classified equally as 'uncles'. Reciprocally, one can have nieces and nephews irrespective of whether they are the offspring of one's brother or one's sister. The offspring of aunts and uncles are all equally cousins. Paternal and maternal kin are indistinguishable, in a classificatory sense, in preceding generations: the terms 'grandfather' and 'grandmother' are not normally separated into mother's father and mother and father's father and mother.

Now it might be thought that such a kinship pattern is an obvious one. Tracing relatedness equally through maternal and paternal relations of consanguinity seems to mirror the natural facts of biology in that genetic inheritance is also bilateral. It must be emphasized, however, that the kinship system predates modern molecular biology and that modern molecular biology mirrors cultural practice rather than the other way round. In other words, genealogical relations are not relations of biological or genetic connection but are rather relations that derive from the engendering and bearing of children as the processes of human reproduction are understood in any given society. All human cultures have their own procreation theories about how children are brought forth into the world. Regardless of the differences that those societies may ascribe to men and women in the procreation of children, such notions become the basis of kinship differences from one society to another.

'Western' systems of descent obviously predate even nineteenth-century biological theories. Some of these are visible in significant representations of Euro-American kinship. They include various manifestations of the family tree or pedigree. The family tree is a major, iconic, visual symbol of kinship as a cultural system. We are familiar with elaborate family trees from the European nobility and royal families, and with Christian religious iconography. In the latter, the ancestry of Jesus Christ may be traced, establishing that he descended from the House of David in the Tree of Jesse. More simple family trees are to be found elsewhere – such as in the family Bible. These representations again remind us that 'pedigrees' predate any contemporary sense of genetics and biological mechanisms of inheritance. Pedigrees as artifacts are not confined to the portrayal of human family trees: animal husbandry and the rearing of prize farm animals or race-winning horses is also a major source for common images of inheritance and descent. Such practical theories of inheritance also predate modern biological understandings.

Naming practices

A bilateral kinship system, such as that encountered in Euro-American culture, does not generate unequivocal lines of descent that result in lineages or 'houses' without the invocation of other principles. The general issue can be illustrated with reference to 'naming'. Part of one's identity and sense of 'family' membership is conveyed by naming processes. In traditional 'Anglo' naming practices, the family name is normally inherited through only the male line, whereas the personal name is a unique one. Naming practices in this cultural system thus work simultaneously to differentiate and to establish intergenerational patterns of identity. The fact that this aspect of identity is inherited through the male line introduces an asymmetrical element within a bilateral system. This pattern of identity transmission has been reflected in the convention (no more than that – it has not been a legal requirement) of a wife adopting her husband's family name.

However, conventions have been superimposed – notably by the custom of women adding their husband's or partner's family name to their own. This naming practice remains asymmetrical, because most male partners in such relationships do not 'hyphenate' their family name with their partner's. In addition, the female partner's name is normally her father's. Even when the mother's family name is self-consciously adopted, in effect it is usually the mother's father's name. Furthermore, this practice is in itself virtually impossible to sustain within a bilateral kinship system. Generation one may combine two family (father's) names; generation two would have to combine four family names; generation three, eight names; and so on. The impossibility of maintaining a cumulatively symmetrical naming system illustrates the cultural imperative for imposing asymmetrical preferences rather than the equal recognition of all possible lines of ancestry. Moreover, naming practices are far from universal, even in European cultures. For example, the convention of the patronymic (as in Russian usage) that names a person as the son or daughter of their father establishes a form of social inheritance in the male line. In contrast, Spanish naming practices attach the name of the father and the mother.

Fatherhood is a good example. In the Euro-American model of kinship the genitor (the genetic father) and the pater (the social father) is ideally the same man, and kinship is traced through that man. However, there are numerous cultural contexts where the relationship is not that straightforward. The most startling difference between Euro-American systems of kinship and some other systems is the principle of matrilineal descent. This has been documented in many cultures. It is not as simple as implying that descent through women alone is what counts, and it certainly does not equate with matriarchy. But matrilineal descent means that the father is not one's main male relative. That role is reserved for one's mother's brother. Indeed, one's father is significant as an affine, one's mother's spouse, rather than as kin.

Unilineal descent

In one of the most famous accounts of matrilineal kinship – Malinowski's description of the Melanesian Trobriand Islanders – it was claimed that the islanders' practices of kinship and reproduction had no role for the father (i.e. the mother's husband), whom the Trobrianders referred to as *tomakava* meaning 'stranger' or 'outsider'. This is an example of the cultural belief in parthenogenesis, the so-called 'virgin birth', in which some role for sex is granted in shaping a child, for example, by opening the way for the ancestral spirit, baloma, to enter but not for its original creation. The nature of beliefs in parthenogenesis have been subject to considerable debate among anthropologists.

There are other important cultural variations that similarly depart from Euro-American kinship conceptions. For example, in the Nayar castes of Kerala in southern India, after the ceremony of tying the gold ornament *tali* around a woman's neck by a chosen bridegroom, the woman was traditionally free to have sexual intercourse with any number of men. In turn, such 'visiting husbands' were enjoined to acknowledge paternity and make appropriate gift payments at birth. Similarly, the Canela of Brazil (studied by Crocker and Crocker) believe that once a woman is pregnant, any semen added to her womb during her pregnancy becomes a biological part of the fetus. Thus, in Canela culture children have one mother but several 'contributing fathers' or cofathers. Once the baby is born, the mother-in-law asks her daughter-in-law to designate the baby's contributing fathers; on being named by the woman, these cofathers are identified by a child who is sent to walk around the village to announce their names at the appropriate doors.

Whatever the niceties of interpretation may be, one point is clear: the existence of a matrilineal system of descent and a reproductive system in which the social father has no role in procreation together creates a cultural system of kinship and affinity that are certainly not grounded in the 'natural' phenomena of biological relatedness as defined by the biomedical science of genetics. In other words, a deoxyribonucleic acid (DNA) test for paternity, as practiced in most Euro-American contexts, will be deeply problematic in such cultures.

By contrast, in certain patrilineal cultures, in which descent is counted exclusively through the male line, it is women who are bypassed completely in the procreation theories. In the north African countries of Morocco and Egypt, for example, there are numerous 'monogenetic' procreation theories in which men are entirely responsible for the creation of a child by ejaculating preformed fetuses into women, who in turn bring them to fruition. In New Guinea the child is conceptualized as being solely created by the semen, and in other Melanesian societies, such as the Madak or Gimi, a child's body is thought to be made solely by the father's substance.

There are many cultures that employ idioms of procreation and inheritance that allow for some influence on the part of the mother and the father, but attribute different contributions to the two sexes. For example, Carol Delaney (1991) reports that among Anatolian Turks men are granted a primary, active role in fashioning children and women are accorded only a secondary and passive one. Men are described as 'planting the seed', and women are described as 'the field'. This asymmetry in turn reflects broader cultural differences between the sexes. From elsewhere in Turkey, a similar model has been described by Ilakó Bellé-Hann (1999) that identifies women as 'baskets', with men solely responsible for filling the basket. A parallel example has been described by Leela Dube (1986) from India, where the 'seed and the soil' belief is encountered and also reflects the broader asymmetrical relations between men and women. Matrilineal descent involves one kind of asymmetry. Patrilineal descent creates a different asymmetry – one that is less 'strange' to Euro-American observers but is equally divergent from the model of bilateral kinship systems. Both matrilineal and patrilineal systems can create unilineal descent groups. Under conditions of unilineal descent, a person's identity is constructed and inherited through descent groups ('lineages') that endure over time and are defined through relations traced through one sex.

Routes and substance: what is inherited and how

If kinship and descent are cultural systems that partly define the transmission of personhood and identity from generation to generation, then we also need to look more closely at systems of inheritance. What is inherited and how it is inherited are matters of cultural definition and are sometimes enshrined in legal codes. What is inherited can include land, property, wealth, ethnicity, nationality, religion and cultural competence. Here we concentrate on the inheritance of personal characteristics in order to illustrate the diversity of cultural practices.

There are many ways in which cultures define the transmission of physical and moral qualities. How one 'takes after' one's parents and forebears is not a matter of chance in most societies. In some cultural systems, physical and nonphysical attributes are inherited differentially. In many parts of India, one's caste purity is thought of as being situated in one's blood. This notion of blood goes far beyond an understanding of human biology. For example, Fruzzetti (and Östör, 1976) has studied south Indian Tamils who endow blood with several meanings, including beliefs about procreation and inheritance. South Indians say that condensed blood can become semen, a repository of purity and power. It can also accumulate at the base of the brain through sexual abstinence. In a woman, condensed blood can become breast milk. The child is formed from these aspects of condensed blood, as well as from the mother's blood transferred directly in the womb. The child, therefore, is formed from the parents' blood and inherits the purity contained in that blood. Blood purity itself is divided into two parts: *utampu*, the body or matter; and *uyir*, the spirit or motion. *Utampu* is the male aspect of blood and *uyir* is the female.

In the same study, Fruzzetti and Östör (1976) compared this conception of blood with that found in east Indian Bengali culture. There, the term *kul* is used to refer to blood purity, quality, highness and nobility, which must be preserved as it is handed down from one generation to the next. The wife of a man is a vehicle through which he establishes his line, transmitting his blood to the children through his wife. Madan's study of the Kashmiri Pandits from the extreme north of India found that in the Pandit worldview it is generally maintained that conception occurs when husband and wife reach orgasm simultaneously. Female orgasm is believed to result in the discharge of vital fluids into the womb, where the male seed is received. The male seed is believed to contain all the requirements for making the complete human being: bones, flesh, blood, all internal and external organs, hair, nails, intellect, knowledge, ignorance, health and disease. Similarly, among the Canela of Brazil one finds that pregnant women usually seek good hunters, or good providers of certain types of food, to be the contributing fathers of their fetuses because they believe such characteristics are inherited.

Such cultural beliefs about inheritance are, however, not an exclusive preserve of 'other' non-Euro-American cultures. Cultures everywhere have complex beliefs about people, which serve as cultural maps that help trace the routes that substances take in the creation of social and moral personhood. Our own research (Featherstone et al. 2006) in south Wales in the United Kingdom among individuals seeking genetic services and their wider kindred amply demonstrates such cross-cultural complexity and variability. People in Wales refer to blood in accounting for the transmission of personal characteristics. 'Bad blood' is an especially persistent way of accounting for personal traits. When talking about inheritance, for example, many respondents describe having 'bad blood within the family' or discuss the purity of their blood.

Bad blood is a general way of attributing undesirable personal characteristics to descent from one or the other side of the family; it links personal traits with biological processes of transmission. Consider, for example, the two sisters who describe how they believe that one of them may be free of a particular neurological degenerative condition that the other has inherited; the unaffected sister had a blood transfusion shortly after birth and this was linked to her identity within the family, marking her as different – in other words, 'not one of us'. There was also an expectation that the male policemen's blood that she received had additional potency and protective abilities that would provide her with strength and good health. (The source of the blood was known as the hospital needed to contact a police college to secure blood of her type when she needed it.) We do not have to assume that there is a literal belief in the capacity of blood to transmit personal qualities but to recognize this as a well-established idiom that expresses principles of transmission that are not grounded in biomedical theories of genetic inheritance.

The south Wales family members that we have studied identify other traces and mechanisms of inheritance. In addition to the specific reference to blood, they associate the inheritance of personal and physical traits with the different 'sides' of the family – tracing characteristics through the 'mother's' or 'father's' side. Such theories are also associated with naming conventions. One may be thought to inherit a particular character, physical appearance or predisposition to illness by virtue of being a member of a named familial line ('being an Atkinson', say). Such genealogical links can also be attributed to taking after a particular named relative, thus establishing individually personalized intergenerational links.

Beliefs about blood and inheritance are deeply entrenched in European culture, especially in relation to notions of racial and other 'purity'. Membership of the European aristocratic elite, as judged by being listed in the Almanac of Gotha, depended on unbroken generations of noble-born ancestors. So too did membership of one of the national *langues* of the Knights of Malta. The inheritance of social status and ethnicity is therefore subject to variable constructions. Cultural definitions of purity can be found among social elites, and their counterpart can be found under conditions of racist categorization. Just as one ancestor might be enough to spoil one's noble purity, so one ancestor might be enough to spoil one's racial purity. This was certainly so for the racist laws of South Africa under the apartheid system. Racial identity is something that is inherited culturally, and its patterns of inheritance depend on the imposition of cultural boundaries.

In some systems there may be elaborate proliferations of racial categories that are held to reflect that different kinds of racial ancestry had different racial 'mixes' (octaroons, mulattos, etc.). The elaborate racial laws of Germany's National Socialist period were based not only on generalized antisemitism, but also on a calculus of the purity of German blood and degrees of pollution from Jewish blood. The systems of racial types and subcategories reflect the construction of cultural categories of continuous natural phenomena. The inheritance of race is the outcome of cultural creations of natural facts. So too is the inheritance of other social and personal attributes. Different legal codes enshrine assumptions about how one can inherit a national identity and can offer different constructions of citizenship and its transmission, such as whether one can inherit through both parents or through the father only. Equivalent asymmetries may apply to the social inheritance of similar characteristics. For example, Jewish identity is bestowed through the female line: the consequences of 'mixed' marriages are thus different and depend on whether the Jewish partner is the male or the female.

Conclusion

We have done no more than illustrate the culturally variable views that exist about kinship, relatedness and inheritance. To speak of kinship is to view how relatedness is mapped and traced in social contexts independently of the biomedical domain of genetic inheritance. This essay shows how cross-culturally kin relations are not about reflecting and re-using 'biological truths' but are instead a feature of cultural patterning of biological processes into the social tapestry. It remains to be seen how far and how quickly the idioms of genomic science will be joined with other cultural models of inheritance. One should not assume, however, that social inheritance will ever be determined exclusively by the theories of biomedical science any more in the future than they are in the present and recent past.

References

Bellé-Hann I (1999) Women, work and procreation beliefs in two muslim communities. In: Loizos P and Heady P (eds.) *Conceiving Persons: Ethnographies of Procreation, Fertility and Growth*. London: Athlone Press.

Delaney C (1991) *The Seed and the Soil: Gender and Cosmology in Turkish Village Society*. Berkeley CA: University of California Press.

Dube L (1986) Seed and earth: the symbolism of biological reproduction and sexual relations of production. In: Leacock E and Ardner S (eds.) *Visibility and Power. Essays on Women in Society and Development*. Delhi: Oxford University Press.

Featherstone K, Atkinson P, Bharadwaj A, and Clarke A (2006) *Risky Relations: Family, Kinship and the New Genetics*. Oxford: Berg.

Fruzetti L and Östör Á (1976) The seed and the earth: a cultural analysis of kinship in a Bengali town. *Contributions to Indian Sociolog* 10 (1): 97–133.

Madan T N (1989) *Family and Kinship: a Study of the Pandits of Rural Kashmir*. Delhi and New York: Oxford University Press.

Malinowski B (2005 [1929]) *The Sexual Life of Savages in North-Western Melanesia: An Ethnographic Account of Courtship, Marriage and Family Life Among the Natives of the Trobriand Islands, British New Guinea*. Montana, USA: Kessinger Publishing.

24
Privacy and Genetic Information

Veronica English and Ann Sommerville

Introduction

The right of individuals to make free choices is highly valued. No opprobrium is attached to selfish decisions. Competent people can refuse medical treatment, even if this causes their death or the loss of a viable fetus. Making such decisions, people are not obliged to consider the impact on others, even on those closest to them. Notions of familial duty or responsibility may seem outmoded in a world where many families have little contact and often consist of different partners, step-children and half-siblings. At the same time, increasing public awareness of genetics is leading to recognition of the inescapable interconnectedness of individuals, at least at the biological level. Information obtained by one individual about a genetic disorder means that others sharing the same genetic heritage are also at risk. Questions arise about the moral duty to share that information. They fundamentally challenge the expectation that health data are a purely personal matter. It is not only societal values and expectations that need to be examined; the basic terminology of this debate also needs to be reviewed.

'Privacy' describes a general human right to control and limit information about oneself. In reality, however, it lacks substance. Detailed information about individuals' personal affairs, including where they live, what they earn, their credit-worthiness, criminal convictions and purchasing patterns, are held by governmental and commercial agencies. Public records such as registers of births, deaths and marriages, and the electoral rolls disclose information about family connections: who lives with whom and where.

'Medical confidentiality' describes the right to control identifiable personal health information. Confidentiality also describes an ethical duty owed to all patients by health professionals. Medical information has been accorded more protection than other data, but neither the right nor the duty of confidentiality is absolute. Health professionals are legally and ethically empowered to breach confidentiality in order to avoid a significant, foreseeable harm. Indeed, failure to disclose information about a person who is a risk to others may constitute negligence.

'Public interest' is the common justification for breaching privacy and medical confidentiality. Individuals have little control over any aspect of their social, personal, financial or health information that might seriously affect other people.

'Personal obligation' summarizes the expectations, rights and duties of individuals who have special emotional, social or genetic ties with each other. There may not be a 'public interest' to justify an enforced disclosure of information, but patients may be seen as having a personal obligation to share information, such as information about sexually transmitted disease or relevant genetic data, with those with whom they have close relationships. It is debatable whether health professionals should ever take on the role of 'enforcer' if individuals fail to disclose important data to people close to them.

The connection between privacy and 'family' is reflected in the UK Human Rights Act, where privacy is not a purely individual right but the 'right to respect for private and family life' (Article 8). This suggests that privacy can be a shared possession of a group of individuals in an intimate relationship. In modern ethics discourse, however, confidentiality is almost invariably seen as an individual, rather than a family or group, attribute. Yet any consideration of genetic privacy raises questions about how to define obligations within families, although 'families' involve more individuals than just those who are genetically related. Furthermore, the mere fact of being genetically related does not necessarily mean that familial bonds or responsibilities exist, as is shown by the example of sperm or egg donors. Whether such donors have duties to their biological offspring and, if so, how strong such duties might be, are issues addressed in the section 'To whom are duties owed?' (page 194).

Can genetic information be private?

Opinions vary about whether genetic information should be seen as different from other medical information. A fundamental difference is that genetic information has direct relevance for a chain of related individuals who may not agree about what should be done with it. A patient obtaining a positive result from a presymptomatic test for Huntington disease, when her maternal grandfather has the condition, also knows that her mother carries that gene. The mother may have decided against predictive testing to avoid knowing her own status, but the information becomes available as an incidental finding of her daughter's test. One individual's right to know about her health denies another person's right to ignorance. Women who are carriers of an X-linked disorder know that the information is also important for siblings making their own reproductive decisions. One person's right to confidentiality conflicts with the rights of others to access knowledge that would enable them to make informed choices. The aim of medical ethics is to balance such conflicting rights, but part of our argument is that individual and family duties (rather than doctors' duties) must be at the heart of this debate.

One solution is to classify confidentiality in genetics as a duty owed to families rather than to individuals. The Royal College of Physicians, for example, has suggested that:

> genetic information about any individual should not be regarded as personal to that individual, but as the common property of other people who may share those genes, and who need the information in order to find out their own genetic constitution. If so, an individual's prima facie right to confidentiality and privacy might be regarded as overridden by the rights of others to have access to information about themselves.
>
> (Royal College of Physicians, 1991)

This was rejected, however, by the House of Commons Science and Technology Committee, which argued that 'the individual's decision to withhold information should be paramount' (House of Commons Science and Technology Committee, 1995). A middle course was adopted by the British Medical Association (BMA), emphasizing doctors' duty of confidentiality to individuals, but arguing that individuals should recognize their own moral obligations. While contrary to the trend for isolated individual decision-making, this has particular relevance in genetics. It may also eventually impact on more general notions of personal rights and duties, as the BMA stresses that genetics should not be seen as raising completely different issues, but rather that 'all patients' decisions, not only those in the genetic context, should take account of serious implications for others' (British Medical Association, 1998).

To whom are duties owed?

Whether articulated or not, most people accept that they owe some moral obligations to individuals with whom they have a significant relationship and that these obligations are over and above the more general civic duty to avoid harming other citizens. There is an expectation that we will positively help people close to us rather than merely refraining from harming them, and that the degree of assistance owed to them mirrors the strength and importance of the relationship. These connections with other people are of a social, emotional or genetic type. A widely accepted alternative to talking about patients' 'relatives' is to refer to 'people close to the patient'. This tacitly acknowledges that biological relatives may not be the most significant people in patients' lives, but that 'closeness' or 'proximity' of other kinds can also be important, whether it be geographical proximity (neighborliness) or emotional closeness.

Traditionally, blood has been seen as thicker than water, but individuals are not necessarily perceived as owing the same duty to all of those with whom they share a genetic heritage. The case of gamete donors exemplifies this. By donating sperm or eggs to other potential parents, donors relinquish their 'rights' over the future individuals who may be born as a result. They cannot expect to be supported in their old age by their biological descendants, but nor do they owe those future people the same special duties that they owe to their other children (biological or adopted) for whom they retain parental responsibility in moral and legal terms. The act of donation is intended to be a gift, not a loan. Nevertheless, we might still expect that an egg donor who later discovers some important genetic information would make reasonable effort to ensure that it is passed on to her offspring. The moral connection is significantly diluted, but the biological connection can be still seen as relevant. The degree of effort required would probably not be as onerous as parents would make for their own child, because the moral obligations are of a weaker order but more than the minimal harm avoidance owed to strangers.

Rhodes argues that blood ties are not necessarily the basis of people's ethical responsibilities:

> If genetic similarity were the source of our moral relationships, genetic maps could identify our most similar sibling, or even some distant DNA-matching stranger, as the one to whom we owed the most. But if anyone maintained that we had different degrees of responsibilities to different siblings, it is not likely that they would attribute that distinction to our degrees of genetic matching. More likely reasons would be related to the intimacy and dependency of our previous relationship, or the strength of our feelings, or the history of our interactions, or something about our relative wherewithal and neediness.
>
> (Rhodes, 1998)

Genetically unrelated individuals can also have a shared personal interest in knowing genetic information. Potential parents, for example, need to be honest with each other in identifying any genetic risk to the child. So, the extent of such duties is determined by a mixture of blood ties, social bonds and individual circumstances.*

* Since this paper was written the law has changed with regard to anonymity of gamete donors. On the 31st March 2005, the Human Fertilisation and Embryology Authority (Disclosure of Donor Information) Regulations 2004 removed the protection of anonymity that donors were previously afforded. After the transitional period that ends on 31st March 2006 only gametes or embryos from identifiable donors may be used (see http://www.hfea.gov.uk/ForDonors/FAQsaroundliftingofdonoranonymity for further information and exemptions)

A duty to share information or a duty to know?

Is sharing existing genetic information enough or is there a duty to seek out genetic information? Individuals aware of their risks of being carriers of a serious X-linked disorder arguably owe it to partners and future children to have a test. It might be argued that one also has a duty to oneself. Meyer (1992) discusses individuals' 'duty to engage in responsible self-care' and to make responsible, informed decisions. Nevertheless, such decisions also depend on the implications of testing and the harm for individuals burdened with unwanted knowledge. Denial is a common defense mechanism, the enforced removal of which can be damaging. Again, therefore, the extent of the moral obligation will depend on the circumstances. Inflexible rules should be avoided.

Can moral obligations be enforced?

Accepting that individuals have some moral obligations to share information, should those obligations be enforced? The duty of confidentiality is not absolute. Situations arise where the public interest justifies disclosure, such as where failure to disclose 'may expose the patient or others to risk of death or serious harm' (General Medical Council, 2000). Arguably, there is a difference between disclosure to avoid serious harm and disclosure to provide a benefit, such as enabling a relative to make informed choices. In both cases, the certainty of the risk, the degree of harm and the possibility of taking some avoiding action would be relevant considerations. Yet it seems contradictory to speak of an enforceable 'moral' duty, and most people, when aware of the implications for others, willingly share genetic information. There is general agreement that this is an issue on which the usual rule of nondirective counseling should be flexible, so doctors can help patients develop their own ethics. Buchanan points out:

> where the prevention of serious harm is at stake, one should not assume that clinicians must abstain from communicating to the patient any ethical judgment whatsoever. There are instances in which a patient's ethical obligation to inform relatives is clear and uncontroverted, in the light of the widely accepted principle that we ought to prevent serious harm. Informing the patient that this is so need not involve 'imposing values'.
>
> (Buchanan, 1998)

Research about public attitudes to this question is limited. One survey of 200 Jewish women in the USA found that the vast majority supported the voluntary sharing of genetic information, but less than a quarter thought that doctors should breach confidentiality contrary to the patient's wishes, even where a serious and preventable condition was involved (Lehmann et al., 2000). In the absence of further studies, however, it is impossible to predict whether this reflects a general view or one that might not be susceptible to change with greater public awareness of genetics. Nevertheless, it echoes the reluctance felt by most health professionals to try and force patients to act ethically.

Conclusion

We have argued that individuals have moral obligations to share information with people close to them, but that there are many factors to be considered and inflexible rules are unhelpful. Genetic privacy implies that individuals can make decisions in isolation, without having to consider the interests of others. Just as the more general notion of privacy is a fiction rather than a real right, so the notion of genetic privacy is impractical. With growing public understanding of the benefits of sharing genetic information within families, the concept of genetic privacy becomes even less relevant.

References

British Medical Association (1998) *Human Genetics: Choice and Responsibility*. Oxford: Oxford University Press.

Buchanan A (1998) Ethical responsibilities of patients and clinical geneticists. *Journal of Health Care Law and Policy* 1(2): 391–420.

General Medical Council (2000) *Confidentiality: Protecting and Providing Information*, para 36. London: General Medical Council.

House of Commons Science and Technology Committee (1995) *Human Genetics: The Science and its Consequences*, para 228. London: HMSO.

Lehmann L S, Weeks J C, Klar N, Biener L and Garber J E (2000) Disclosure of familial genetic information: perceptions of the duty to inform. *American Journal of Medicine* 109(9): 705–711.

Meyer M (1992) Patients' duties. *Journal of Medicine and Philosophy* 17: 541–555.

Rhodes R (1998) Genetic links, family ties and social bonds: rights and responsibilities in the face of genetic knowledge. *Journal of Medicine and Philosophy* 3: 10–30.

Royal College of Physicians (1991) *Ethical Issues in Clinical Genetics*, para 4.10. London: Royal College of Physicians.

25
Insurance and Genetic Information

Tony McGleenan

Introduction

Advances in predictive genetic medicine have raised the prospect of individual patients being able to determine their future healthcare status by taking genetic tests. The healthcare benefits that would flow from widely available genetic tests are readily apparent. People found to be at risk for particular genetic diseases would have new options open to them. Those with such a heightened susceptibility could take appropriate action by electing to have prophylactic medical or surgical interventions. For those who are concerned that they may be carriers of an inherited genetic disease, the availability of testing might facilitate the making of informed choices about reproduction. However, aside from the positive healthcare implications, there may be other, negative social consequences that could flow from having an insight into future genetic health (MacDonald, 1999). In particular, insurance companies are keen to have access to any information that predicts future risk. In the private insurance market, the amount that an individual policyholder pays for insurance cover is determined by assessing their level of risk. Genetic tests can, in some cases, provide results that indicate the risk of future illness.

Insurance principles

The insurance industry operates according to the principle of actuarial fairness. This means that underwriters, who assess insurance risk, classify individuals according to their level of risk. They are then charged a premium, which is equivalent to that paid by those who have a similar level of risk. This is also sometimes described as fair discrimination (Beckwith and Alper, 1998). In most societies insurers have been exempt from the usual provisions of discrimination laws, because this practice is seen as both fair and socially beneficial. Some insurers argue that, because their business depends on the concept of actuarial fairness, they must have access to the results of any genetic tests that people have taken. The insurance industry relies on symmetry of information, which means that the person seeking an insurance policy and the company offering insurance cover must have access to the same information. Therefore insurance contracts are described as *uberrimae fides* – contracts made in the utmost good faith. If it subsequently emerges that the person seeking insurance has not made a full and frank disclosure of all material risks, then the insurance company may seek to declare the contract void. It is because of these traditional principles that insurers argue that they must have access to genetic test results. However, many people are concerned about the potential implications of revealing sensitive genetic information to commercial organizations. As genetic testing becomes more prevalent, there are concerns that sections of the population will be denied insurance because of their genetic profile. The question of whether this is a real concern or not and, if so, what governments should do about it, is one which has been debated in many countries (Hall and Rich, 2000).

Adverse selection

At the center of insurance concerns about genetic information is the concept of adverse selection (Berry, 1996). If insurance companies are not allowed to seek the results of genetic tests, then they are concerned that people with an unfavorable genetic test result, which indicates that they have an increased risk of becoming ill, will be more likely to purchase life insurance cover, health insurance or critical illness insurance. If this happens it will mean that the overall claims experience of the insurance fund will be worse than expected. Insurance companies argue that they will therefore be compelled to raise insurance premiums. This will have a further adverse effect in that people who have discovered that they have a favorable genetic profile by taking a test will be less likely to buy insurance and will be easily deterred by any increase in insurance premiums. As people with unfavorable test results purchase larger amounts of insurance, the situation may deteriorate further. Consequently, it is argued that the insurance market may enter an adverse selection spiral. This occurs when those who are at low risk leave the market and those at high risk remain, making insurance claims at an earlier age than expected or for higher sums than normal. There have been very few actual recorded instances of adverse selection occurring in the insurance industry, but most experts agree that it is a theoretical possibility when the principles of actuarial fairness and symmetrical information are not applied.

Genetic testing and insurance

Not all genetic tests are relevant from an insurance perspective (MacDonald, 1999). Life insurers are keen to know predictive healthcare information about illnesses that are likely to affect policyholders during the term of the insurance policy. Most people only take out life insurance to cover the term of a mortgage for their home. Therefore the most relevant genetic diseases for the life insurer are those that affect people in the period between their mid-twenties and mid-sixties. There are, however, actually only a small number of diseases for which reliable genetic tests are available, which affect people in this time frame. An obvious example would be Huntington disease (Gin, 1997). This is a debilitating and ultimately fatal disorder, which usually does not affect people until they are in their early forties. However, it is a rare disorder and most of those affected by it have already been identified through analyses of their family history. Diseases such as Huntington are caused by a defect in a single gene and are called monogenic disorders. These diseases are rare, and insurance companies already screen carefully for sufferers by considering family history information. Multifactorial genetic disease is considerably more common. These disorders occur when a combination of environmental and genetic factors combine to make a person ill; examples include many of the common cancers. There are very few reliable genetic tests available for multifactorial genetic disorders.

Genetic discrimination and insurance

Patient groups are worried that insurance companies will discriminate against those who already suffer disadvantage because of genetic disease by denying them access to essential insurance products (Low et al., 1998). In the United States, where there is no system of universal health insurance, there are reports of people being denied access to healthcare because of their genetic profile. These concerns are not so relevant in Europe, where most countries have advanced systems of social insurance that pay for healthcare regardless of an individual patient's level of risk. There is only very limited evidence available which suggests that insurers are involved in genetic discrimination. Perhaps a more pressing concern is that patients will be deterred from taking genetic tests because of fears that an unfavorable result will lead to difficulties with insurance. If people refuse to participate in genetic research programs because of concerns about insurance, it could have a negative impact on public health and advances in medicine. However, studies suggest that, while people are concerned about the social implications of genetics, they are more willing to take genetic tests when there is a treatment option available. Advances in genetic medicine may, therefore, remove some of the concerns raised by insurance.

Genetics, insurance and public policy

The difficult relationship that is emerging between advances in genetics and our reliance on insurance has led some governments to take radical action.

In Belgium, for example, insurance companies are banned from using the results of genetic tests when setting premium levels. In The Netherlands, insurers are only permitted to seek genetic information when the insurance policy is for an unusually large sum (Sandberg, 1995). In the United Kingdom, the government and the insurance industry have agreed to observe a moratorium on using genetic tests for 5 years* to enable all the implications of this difficult issue to be explored. If insurance companies are completely prohibited from using genetic test results, as the number of reliable tests increases, the potential costs of such an approach will inevitably rise. This increased cost will, in most countries, be passed on to the individual insurance policyholders. If the price becomes too high, some people will opt not to buy certain insurance products at all. Governments must, therefore, strike a fair balance between protecting the interests of the genetically disadvantaged and ensuring that people continue to protect themselves by purchasing appropriate insurance products.

Conclusion

Genetic information may allow us to predict our future healthcare status. However, the technology is currently limited both in terms of the accuracy of the predictions and in the number of diseases for which tests are available. The relevance of genetic testing for insurance companies is limited for similar reasons. Only a small number of genetic diseases are relevant from an actuarial point of view. Until accurate genetic testing becomes more widespread for common multifactorial diseases, the impact of genetics on insurance is likely to be minimal.

References

Beckwith J and Alper J S (1998) Reconsidering genetic discrimination legislation. *Journal of Law, Medicine and Ethics* 26: 205–223.

Berry R M (1996) The Human Genome Project and the end of insurance. *University of Florida Journal of Law and Public Policy* 7: 206–243.

Gin B R (1997) Genetic discrimination: Huntington disease and the Americans with Disabilities Act. *Columbia Law Review* 97: 1406–1452.

Hall M A and Rich S S (2000) Laws restricting health insurers' use of genetic information: impact on genetic discrimination. *American Journal of Human Genetics* 66: 293–307.

Low L, King S and Wilkie T (1998) Genetic discrimination in life insurance: empirical evidence from a cross-sectional survey of genetic support groups in the United Kingdom. *British Medical Journal* 317: 1632–1635.

---

* Since this paper was written it has been agreed that this moratorium should be extended until 2011.

MacDonald A S (1999) Modeling the impact of genetics on insurance. *North American Actuarial Journal* 3: 83–105.
Sandberg P (1995) Genetic information and life insurance: a proposal for an ethical European policy. *Social Science and Medicine* 40: 1549–1553.

Web links

Human Genetics Commission (HGC). Access to Consultation Paper: Whose Hands on Your Genes? London (2000) http://www.hgc.gov.uk

26

'Race', IQ and Genes

Jon Beckwith and Joseph S. Alper

Throughout recorded history, privileged individuals have justified their social position by deeming the less advantaged members of society inherently inferior. Intellectuals often provided support for social hierarchies by explaining differences in status as facts of life. In Greece in the fourth century BC, Plato's *Republic* offered the 'myth of the metals' to justify to the populace the rule of the philosopher kings. In the nineteenth century, Herbert Spencer (1864) explained the different degrees of people's economic success according to his interpretation of Charles Darwin's *The Origin of Species*. Throughout the twentieth century, academics from leading universities in the United States have argued that differences in intelligence, and thus presumably success in society, are rooted in inherited biological differences.

The successes of the Scientific Revolution convinced many intellectuals that science was the best means by which to study the origins of social arrangements. They used biology not only to explain variations in social status among individuals but also to account for group differences. During the nineteenth century, scholars in the United States who rationalized the continuing existence of slavery were at the forefront of this work. Much of their research was designed to prove that African slaves were inherently inferior to their White masters. On the basis of his flawed measurement of skull sizes, the influential craniologist Samuel George Morton concluded that Blacks had less brain capacity and thus less intelligence than Whites (Gould, 1981).

At the dawn of the twentieth century, two important scientific developments provided new tools for scientists interested in the biological basis of group differences. The rediscovery in 1900 of Mendel's laws of inheritance allowed the genetic analysis of all human traits. Contemporaneously, the test developed by French psychologist Alfred Binet to identify school children who needed additional instruction was seen by some as a test that measured innate intelligence. Psychologists in the United States used the results of these intelligence quotient (IQ) tests performed on immigrants entering the country to claim that

83% of Jews, 80% of Hungarians, 79% of Italians and 86% of Russians were 'feeble-minded' (Kamin, 1974).

As a result of such findings suggesting a genetic basis of important behavioral traits, many prestigious geneticists became supporters of the eugenics movement. Eugenicists believed in encouraging procreation among people with 'good' genes and discouraging it (or even preventing it) among those with 'bad' genes. In the United States, genetic arguments were used to justify the passage of eugenics laws that mandated the sterilization of criminals and mentally retarded people (Chase, 1977).

In Nazi Germany, the 'Racial Hygiene' movement, which was supported by leading geneticists (Müller-Hill, 1988), espoused an even stronger version of the eugenics program: a return to the original (genetic) purity of the 'Aryan race'. In reaction to the Holocaust, the ultimate consequence of this movement, eugenics and the genetic studies used to support it fell into disrepute. However, discussions of genetic inequality among races began to reappear within a generation of the Holocaust. In the late 1960s and early 1970s, United States psychologists Arthur Jensen and Richard Herrnstein attracted widespread attention to their conclusions that racial and class differences in achievement were attributable to genetic differences and were, therefore, largely immutable. They criticized social policies such as the 'head start' early education program, which was designed to reduce societal inequalities. Although the overwhelming majority of commentators reject Jensen's and Herrnstein's research, some scholars continue to espouse their ideas (Block and Dworkin, 1976; Herrnstein and Murray, 1994).

Jensen and Herrnstein supported their arguments by citing studies of families, identical twins and adopted children. These studies purported to uncover the genetic basis of social behaviors and aptitudes, but were not designed to identify specific genes. However, during the last quarter of the twentieth century, remarkable advances in molecular genetics, which culminated with the establishment of the Human Genome Project in 1989, suggested to some researchers that specific genes might now be connected with specific human personality traits. It seemed likely that this approach would supersede the indirect familial studies hitherto utilized by behavior geneticists. In promoting the new molecular approach, geneticists referred to the deoxyribonucleic acid (DNA) sequence of the human genome with such grandiose phrases as the 'book of life', the 'Holy Grail' and the 'blueprint for life' (Beckwith, 1999).

At first, this enthusiasm appeared to be justified. Geneticists reported the existence of human behavioral genes that specify many different 'traits'. A partial list includes novelty-seeking, risk-taking, homosexuality, happiness, criminality, manic depression and schizophrenia. Yet, nearly all of these studies were either retracted or subsequently failed the test of reproducibility. Unfortunately, only the 'cognoscenti' became aware of these setbacks to the genetic program. Whereas the initial scientific papers received widespread publicity, the retractions were usually ignored by the media (Alper and Beckwith, 2000).

In March 2001, the two rival groups (one public, one private) that were working on the Human Genome Project announced success: the human genome was essentially sequenced. The completion of the project triggered a series of questions. Would the new genetics have similar social consequences to those of the old genetics of the early 1900s? Would genetic information be used to bolster racist views, or would it eradicate biological explanations of group differences?

At press conferences announcing the completion of the Human Genome Project, leaders of the two groups emphasized their belief that the DNA sequence information showed that racist views of human group differences were untenable. Craig Venter, head of the Celera Corporation effort, noted that the genetic differences among the people of different ethnicity whose genomes his group had sequenced were insignificant. Moreover, both groups found about 30 000 genes rather than 100 000 as previously thought. This smaller number of genes seemed to leave less room for genetic variation among groups. These findings added support to previous work that had indicated that about 85l of the total variation in human genes occurs within groups; the variation between one group and another is much smaller. According to Venter and others, it is impossible to distinguish races on the basis of any genetic criteria; the biological concept of race is meaningless.

One might then wonder why it is so easy to distinguish the typical Swede from the typical Tanzanian if genetic differences are so unimportant. Evolutionary biologists suggest that visible distinctions between groups that have been used to label races are due to relatively few genetic variations. These differences, it is thought, resulted from genetic adaptations to different environments, for example, subarctic Sweden versus tropical Tanzania. However, the number of genes involved in these obvious genetic differences between groups is clearly small when compared with the overwhelming degree to which different groups share the same array of genetic variations.

Despite the unanimity among genome investigators that the concept of race is biologically meaningless, some observers remain concerned that research into the genetics of group differences will provide fodder for racist views. Much of this concern has focused on the Human Genome Diversity Project, which will analyze the genomes of several genetically isolated groups. The project aims to determine the relationships among these groups, and consequently to learn more about human migrations and the development of human languages. Because the project will search for genetic differences among groups to achieve these goals, it is reasonable to worry about the use of this information by others who are looking to support their racist ideas (Alper and Beckwith, 1999).

Because small homogeneous populations that have experienced little interbreeding with other groups carry distinctive alleles, the Human Genome Diversity Project will surely find genetic differences among groups. These differences have already been noted in susceptibilities to genetic diseases. In the United States, Whites of northern European ancestry are at greater risk for cystic fibrosis, Ashkenazic Jews for Tay–Sachs disease, and African Americans whose ancestors lived in malarial belts for sickle cell disease. Other so-called polymorphisms that are unrelated to disease will almost certainly be found.

However, finding genetic differences that would provide a test to classify individuals as belonging to the 'white', 'black', 'yellow' or 'red' race is another matter. Races are broad categories that contain large numbers of subpopulations. African subgroups, for example, differ widely in body type and blood type, and in their susceptibility to various diseases. In addition, a classification based on one of these characteristics will probably differ from one based on another characteristic.

Even though the concept of race is scientifically bankrupt, it exists as a very real social construct. Classifying people on the basis of skin color tells a great deal about, for example, their health, wealth and access to corporate executive positions. In view of these real differences, and especially the correlation between skin color and performance in IQ tests, scientific and genetic arguments against the concept of race have been largely ineffectual. Furthermore, academics such as Jensen and Herrnstein have argued that IQ tests do measure innate intelligence, that science shows that intelligence as measured by IQ tests is an inherited trait, and thus that the economic and social differences between Blacks and Whites are largely due to these innate differences in intelligence.

Will the arguments against the concept of racial differences of Venter and others prevail? It is not difficult to foresee the response of the hereditarians. They will dismiss as irrelevant the finding that there is only about a third of the number of human genes that was estimated previously. The coding portions of single genes – the exons – can combine in various ways such that each gene can effectively function as many genes. In addition, scholars determined to justify racist beliefs will discount the finding that genetic variation among groups is extraordinarily small. As efforts such as the Human Genome Diversity Project discover new differences in polymorphisms between groups, advocates of a racialist point of view can propose that, despite their small number, the ones that are found are significant. They may argue that these polymorphisms are responsible for the variations in the small number of behavioral traits that are of crucial importance, in particular, intelligence. As mistaken claims for genes for several other behavioral traits have appeared and continue to be published, it would not be surprising if reports of variation in 'intelligence' genes were to follow the discovery of new genetic variants between groups.

As a result of the evidence suggesting that the human genome contains only 30 000 rather than 100 000 genes, genome scientists began to soften their 'book of life' characterizations of the human genome sequence. They suggested that, with the smaller number of genes, a one-to-one correspondence between genes and behaviors seems far less likely than thought previously. The difficulty in finding genes for complex diseases and behavioral traits was a reflection of a new 'fact of life' – genetic complexity. There is increasing evidence that interactions among genes and between genes and the environment result in a 'book of life' that is much more difficult to read than was imagined previously.

The increasing recognition of genetic complexity has dampened the optimism of behavioral geneticists that they will quickly find genes of significant effect on human behaviors and aptitudes. The expectation that family studies will be superseded by gene-mapping approaches is no longer a certainty. In the near future, family studies may remain the only approach to issues concerning the inheritance of behavior.

However, the traditional studies used by behavioral geneticists remain highly problematic. To quantify the degree of inheritance of a trait, researchers estimate its heritability coefficient. This coefficient is a quantitative measure of the degree of variance (variation) in a trait that is due to genetic variance rather than to environmental variance. Even if heritabilities could be estimated accurately (they cannot because of the difficulty in disentangling environmental and genetic influences), they would be of limited utility. Heritabilities depend on the particular population studied and on the range of environments experienced by that population. Consequently, the heritability coefficient provides no information about whether a trait or behavior can be changed or by how much it can be changed when environmental factors, such as economic support, new educational programs, cultural attitudes and so on, are altered. In addition, a heritability coefficient calculated for a White population in the United States cannot be applied to Blacks or other groups in the same country, nor to identified groups in other countries. Because the environments experienced by Blacks is so different from those of Whites in the United States, any calculations for heritability of IQ among African Americans must come from studies of African Americans (Sober, 2000).

To sum up the state of the science, we conclude that familial studies have not provided convincing evidence that strong genetic factors are involved in most human behaviors and aptitudes. The molecular genetic studies in this area, rather than leading to the identification of genes that specify different behaviors, have led to increasing appreciation of genetic–environmental complexity. Even so, some scientists and others continue to point to genetics as an explanation for racial and class differences. As projects such as the Human Genome Diversity Project begin to report genetic markers that distinguish groups, it seems likely that hereditarians and racists will seize on these findings to support their views.

Racist attitudes will not be eliminated by scientific discovery. Those with such attitudes may even claim support from science by misrepresenting the findings of genomic studies on populations. Geneticists who seek genes associated with human behaviors have often exaggerated the implications of their findings when presenting their results in scientific journals. An example of the dangers of overextending findings occurred in the recent reports of an extremely rare mutation that seemed to cause antisocial behavior in one single family. The authors suggested in their scientific paper that such mutations might help explain 'aggressive behavior' in a population more generally (Brunner et al., 1993). Within a week, media headlines proclaimed the discovery of a 'criminal gene' and described individuals carrying it as 'born bad'. There is no doubt that such distortions will continue to occur.

Because of the continuing and probable future use of genetic arguments to support racist theories, geneticists have a special obligation to expose the misuse of their science and to explain concepts that are often misrepresented to the public. Those who study genes and human behavior should take special care to avoid even a hint of exaggerating the implications of their results. Geneticists can emphasize the complexity that exists in the path from a gene to a phenotype – a complexity revealed by current studies on human genes and their manifestations. They should insist that neither the finding of a gene correlated with a behavior nor the reports of the heritability of a behavior tell us anything about how much that trait can be changed. Genetic does not mean fated, in contrast to the impression the public has gained (Gannon, 2000; Beckwith, 1999). When racists seize inappropriately on genetic findings to support their arguments, geneticists must be at the forefront of those criticizing this misuse of genetics. The Human Genome Project has elevated genetics to the status of 'big science'. It is now time that geneticists follow the lead of their colleagues in nuclear physics – the first big science – and take very seriously their social responsibility.

References

Alper J S and Beckwith J (1999) Racism: a central problem for the Human Genome Diversity Project. *Politics and the Life Sciences* 18: 285–288.

Alper J S and Beckwith J (2000) Genes found and lost: misperceptions of behavior genetics. *Medical Humanities Review* 142: 35–39.

Beckwith J (1999) Genes and human behavior: scientific and ethical implications of the human genome project. In: Crusio W and Gerlai R (eds.) *Handbook of Molecular-Genetic Techniques for Brain and Behavior Research*, pp. 917–926. Amsterdam: Elsevier.

Block N J and Dworkin G (1976) *The IQ Controversy: Critical Readings*. New York: Pantheon.

Brunner H G, Nelen M, Breakefield X O, Ropers H H and van Oost B A (1993) Abnormal behavior associated with a point mutation in the structural gene for monoamine oxidase A. *Science* 262: 578–583.
Chase A (1977) *The Legacy of Malthus: The Social Costs of Scientific Racism*. New York: Knopf.
Gannon F (2000) Genocentric promises. *EMBO Reports* 1: 91.
Gould S J (1981) *The Mismeasure of Man*. New York: W W Norton & Co.
Herrnstein R J and Murray C (1994) *The Bell Curve*. New York: Free Press.
Kamin L (1974) *The Science and Politics of I.Q.* Potomac, MD: Earlbaum Associates.
Müller-Hill B (1988) *Murderous Science: Elimination by Scientific Selection of Jews, Gypsies and Others, Germany 1933–1945*. Oxford: Oxford University Press.
Sober E (2000) The meaning of genetic causation. In: Buchanan A, Brock D W, Daniels N and Wikler D (eds.) *From Chance to Choice: Genetics and Justice*, pp. 347–370. Cambridge: Cambridge University Press.
Spencer H (1864) *Social Statics*. New York: D. Appleton & Co.

27
Criminal Responsibility and Genetics

Jennifer Bostock and Gwen Adshead

Introduction – the genetic defense

> 'I honestly believe crime is a case for the doctor, not the policeman or the parson. In the future perhaps, there won't be any such thing.'
> 'You'll have cured it?'
> 'We'll have cured it.'
>
> (Agatha Christie, 1930/1977)

On 17 February 1991, Stephen Mobley robbed a pizza store and shot dead the manager. At trial the defense lawyers asked the judge to allow Mobley to be tested for a possible genetic abnormality, which they believed might help to explain his violent acts and save him from the death sentence (for specific case details see Table 27.1). They cited Dutch research by Brunner et al. (1993) that claimed that violent behavior was associated with a genetically determined neurochemical deficiency and argued that this might apply to Mr Mobley. The judge did not permit the tests, and in November 2002, Mobley was still on death row awaiting execution.* This case indicates how modern research into the association between genes and criminal behavior is already having an influence on accepted notions of legal and moral responsibility. The idea that crime might be a disorder caused by our genetic make-up is a powerful one in Western culture, as the opening quote shows.

Definitional problems

Research into the genetics of criminal, violent and antisocial behavior is highly complex and controversial. The first problem is that crime, antisocial behavior and violence are not synonymous terms, but rather social constructs that may

---

*Since this paper was written Stephen Mobley has been executed by lethal injection in the US state of Georgia, 1 March 2005.

Table 27.1 Summary of legal cases

Case	Year	Source
Mobley v. State	1993	426 S.E.2d 150
Mobley v. State	1995	265 Ga. 292, 455 S.E.2d 61
Mobley v. Georgia	1995	516 U.S. 942, 116 S.Ct. 377, 133 L.Ed.2d 301
Mobley v. Georgia	1998	269 Ga. 635, 502 S.E.2d 458
State of Georgia v. Glenda Sue Caldwell	1987	257 Ga. 10, 354 SE 2d 124

be altered by time, culture and history. Behavior may be deemed criminal in one place and time, but not another – homosexuality being an obvious example. A wide range of activities may be defined as violent, criminal or antisocial, making it difficult to generalize for research purposes; for example, the notion of 'antisocial behavior' would include illegal parking, insurance fraud, inner-city rioting, some types of political protest and serial killing, each of which would be perceived as qualitatively different from the other (not least by any victims of these activities).

A further problem relates to selection bias. Studies of heritability seem to be selective in the kinds of behavior examined. Most studies focus on delinquency or violence rather than white-collar crime, which is more common than violence and just as socially expensive. Interpersonal violence, although perhaps more serious in terms of outcome, is much less common than other types of antisocial behavior and constitutes only 6% of recorded crime. Even allowing for underreporting, an emphasis on the rarest type of crime may make it difficult to generalize about more common crimes. For the most part, geneticists do not consider the problems of poor definition, heterogeneity and social construction as major obstacles to their work; rather they regard them as challenges that they have confidence in overcoming (Wasserman, 1996).

Most responsible researchers agree that there is no 'gene for crime', but that there may be genetic predispositions that may increase the risk of certain types of behaviors, e.g. rule-breaking or impulsivity (Reznek, 1997). Furthermore, criminal behavior is the multifactorial product of an interaction between our genes and our environment and, although genetic influences may be relevant in the genesis of some types of crime, it is unlikely that this influence is any greater than that made by the social and cultural environment. However, referring to predisposing factors rather than 'genes for behavior' will not solve the moral or legal problem; and not all researchers are as cautious in their claims as others, nor is the media, where reports of such research are often exaggerated and deterministic (Nelkin and Lindee, 1995). The concern must be that it is these reports that the 'public', who are potential jurors, tend to read.

Legal uses of the genetic defense 'determined to kill'

How will the law deal with a generation of Justified Sinners? (Jones, 1997)

The first case to raise the question of the possible role of genetic factors in homicide was the trial of Richard Loeb and Nathan Leopold in 1924. The famed criminal defense lawyer Clarence Darrow appealed to the court:

> Is Dickie Loeb to blame because out of the infinite forces that conspired to form him, the infinite forces that were at work producing him ages before he was born, that because out of this infinite combination he was born without it [the capacity to feel emotion]? If he is, then there should be a new definition for justice.
>
> (Botkin et al., 1999)

Although Darrow made no direct claim about his clients' genetic heritage, he implied that their lack of empathy should be understood as a constitutional or inherited condition, which they could not choose or affect; and that in some sense the defendants were 'determined to kill'. This argument was successful, and Loeb and Leopold were given life sentences plus 99 years, instead of the death penalty. Such a decision clearly raised the legal possibility that inherited or constitutional factors might be used in future cases to avoid or diminish responsibility and to reduce punishment.

The first direct claims that genetic constitution might affect responsibility were raised in 1968 in relation to the XYY condition. It was argued that those men with the XYY condition (i.e. an extra Y chromosome) could not help acting criminally. However, there were many XYY men who were not criminal, and this led to the abandonment of the defense in the 1970s, although it had already been attempted in criminal trials in Australia, West Germany, the USA and the UK.

Since then, there have been other legal attempts to base a defense to crime on genetic characteristics. In 1985, Glenda Sue Caldwell killed her son and unsuccessfully pleaded insanity; she was sentenced to life imprisonment (see Table 27.1). In prison, she developed symptoms of Huntington disease (a genetic condition that is associated with episodic violence) and she was then released following a retrial. The influential factor in the retrial seems to have been the perception that a genetic condition could have made a significant contribution to the probability that the crime had been carried out and, further, that this genetic condition was more determining than other factors thought to contribute to crime, e.g. biological or social factors such as a history of abuse (Ellis, 1994; Summer, 1999).

This raises the important question of whether there is anything special or uniquely deterministic in the nature of genetic defenses compared with those relying upon environmental evidence. Psychiatric defenses based on depression or bipolar disorder have also used genetic evidence (Summer, 1999).

However, defenses to criminal behavior based on genetic research typically fail, because the current evidence base does not reach the standard required for admissibility. This position may change; the question then for jurors will be how to understand the relationships between genetic influence on behavior and the allocation of legal and moral responsibility.

Crime and moral responsibility

Reductionist molecular science harbors a hopeful vision of a world in which the genetic bases of many diseases and their behavioral manifestations will be discovered and therapeutically prevented or treated. It is a notable dream, though not unproblematic for our moral self-understanding. The tension between these two tendencies – our sense of our lives as fated, and the desire for mastery of what ails us – tugs at our understanding of ourselves as actors in the morality play of life (Carson and Rothstein, 1999).

If, in the future, it could be shown that genes significantly influence behavioral controls, there are two broad schools of thought concerning responsibility. First, criminals might be seen as unable to change their behavior and therefore would not be held responsible or as responsible for it. This argument rests on Aristotelian notions of compulsion or ignorance as a basis for excuse; people who cannot choose their behavior are in some sense 'compelled' and therefore should be shown some leniency. Alternatively, it might be argued that those who are genetically at more risk of committing crime (i.e. 'born bad') and who fail to make use of the available tests, interventions and preventative measures should be considered more, not less, responsible for their actions. It might even be argued that people with such 'faulty genes' should not become parents, but if they do then they might be held responsible for the actions of subsequent criminal offspring.

The issue of punishment needs to be distinguished from the issue of public safety. Even if one argued that the offender was in some sense less responsible, an individual who cannot control their behavior is likely to require detention on the grounds of public safety. This argument has a predecessor in the old English law on automatism. If a person is not conscious of their behavior, they are not responsible for it, e.g., sleepwalking into another person's property. If, however, a person's behavior when asleep is violent, then the law argues that, although not responsible, that person must be detained indefinitely, because their sleepwalking made them dangerous. Thus, it is possible that in the future all criminals who are known to possess a genetic defect associated with criminal behavior will be kept in a secure, secluded environment. This may not be prison, but it certainly seems like punishment irrespective of responsibility.

Even if science is never able to establish a causal connection between genes and criminal behavior, any research into this area might undermine our belief that humans are responsible agents. Claims such as the following may result in an increasing public doubt that any of us possess free will or can realistically see ourselves as autonomous agents:

> Evil is embedded in the coils of chromosomes that our parents pass on to us at conception.
>
> (Franklin, 1989)

This does not concern only those involved in the criminal law, but all of us and our societies, because our belief in free will and responsibility is so fundamental to our identity that to shake it might result in a collapse of our understanding of what it is to be human. However, it could also be argued that similarly materialist and reductionist accounts of human action and behavior in the past have not yet managed to undermine the concept of free will. For example, it is clear that all mental events are to some extent 'caused' by the brain, insofar as we need the material neural substance to have mental activity. However, our conscious experience of having and making choices is not negated by this knowledge. It might also be argued that ethical and legal discourses may utilize information from the domain of science, but cannot be determined by such information, insofar as their moral and social purposes are so qualitatively different.

Conclusion: risk assessment, and preventing and 'curing' crime

Genetic research holds out the prospect of identifying individuals who may be predisposed to certain kinds of criminal conduct, of isolating environmental features which trigger those predispositions, and of treating some predispositions with drugs or unobtrusive therapies. (US Human Genome Initiative brochure, quoted in Spallone, 1998)

If research into the genetics of criminal behavior progresses, the kinds of genetic tests sought by Stephen Mobley's lawyers may become widely available to determine whether some people are at increased risk of behaving criminally. Advocates of such testing believe that this will be good for the individuals concerned, by providing a destigmatizing explanation for their criminal behavior, ensuring a just outcome and provision of the appropriate help. Such tests will also be good for society by helping to reduce and, in some cases, prevent violent crime from occurring by providing 'treatment' (Fishbein, 1996). 'Treatments' or 'cures for crime' might include preventative measures such as sterilization of affected parents, abortion of affected fetuses, genetic manipulation of some kind, or even to 'treat crime with pills' (McGuffin, quoted in Recer, 2001).

Others have argued that such methods of 'risk assessment' will be at best worthless, because the causes of and solutions to crime are not to be found in our genes, but in our environments. At worst, such approaches promote an inhumane attitude to offenders which will lead to increased stigmatization, inequality, discrimination and prejudice. Furthermore, they argue that, by concentrating on the alleged genetic explanations for crime, this approach detracts attention and resources away from environmental problems that might help to explain crime and social programs that might help to reduce it (Zimring, 1996). Ultimately, genetic explanation for crime might be understood as a political strategy that locates social adversity in the individual and thus obviates the need for any political approach to rule-breaking and inequality.

It might be argued that these problems are not new and that new developments in genetic research merely reframe the debate in slightly different scientific language. For example, the British Government's review of mental health legislation includes a proposal to preventively detain individuals with a 'severe personality disorder' who may pose a danger to others (Department of Health, 2000). These individuals will be assessed for risk and then detained and offered 'treatment' intended to alter their behavior rather than any underlying condition. All the familiar arguments have been raised: the uncertain validity of risk assessment tools, the justice of preventive detention and the 'medicalization' of a social and political problem.

One thing is certain: people who behave criminally arouse fear in others, and, when there is fear, good-quality legal and moral reasoning may come under threat. A future where genetic research into criminal behavior is accepted, genetic defenses are widespread and it is possible to reduce crime via some form of genetic manipulation is welcomed by some and feared by others, dependent upon political, social and professional perspectives.

References

Botkin J R, McMahon W M and Pickering F L (eds.) (1999) *Genetics and Criminality*, p. 182. Washington DC: American Psychological Association.

Brunner H G, Nelen M, Breakefield X O, Ropers H H and van Oost B A (1993) Abnormal behavior associated with a point mutation in the structural gene for monoamine oxidase A. *Science* 262: 578–580.

Carson R and Rothstein M (1999) *Behavioral Genetics: The Clash of Culture and Biology*, p. 189. Baltimore, MD: Johns Hopkins University Press.

Christie A (1930/1977) *A Murder at the Vicarage*. London: HarperCollins Publishers Limited.

Department of Health (2000) *The Review of the Mental Health Act 1983*. London: Department of Health.

Ellis R (1994) She's not a cold-blooded killer: unique defense frees mom convicted of killing son. *Atlanta Journal – Constitution* 127: 1.

Fishbein D (1996) Prospects for the application of genetic findings to crime and violence prevention. *Politics and the Life Sciences: the Journal of the Association for Politics and the Life Sciences* 15(1), March: 91–94.

Franklin D (1989) What a child is given. *New York Times Magazine* 3 September: 36.

Jones S (1997) *In the Blood – God, Genes and Destiny*, p. 236. London: HarperCollins Publishers Limited.

Nelkin D and Lindee M S (1995) *The DNA Mystique*. New York: W H Freeman and Company.

Recer P (2001) *Peter McGuffin on the Genome and Psychiatry: Genome Map May Revolutionize Some Health Care*. Washington DC: Associated Press, 11 February [http://taxa.psyc.missouri.edu/bgnews/2001/msg00037.html].

Reznek L (1997) *Evil or Ill? Justifying the Insanity Defense*, p. 136. London: Routledge.

Spallone P (1998) The new biology of violence. *Body and Society* 4(4): 47–65.

Summer D (1999) The use of human genome research in criminal defense and mitigation of punishment. In: Botkin J R, McMahon W M and Pickering F L (eds.) *Genetics and Criminality*, pp. 182–191. Washington DC: American Psychological Association.

Wasserman D (1996) Research into Genetics and Crime: Consensus and Controversy. *Politics and the Life Sciences: the Journal of the Association for Politics and the Life Sciences* 15(1): 107–109.

Zimring F (1996) The genetics of crime: a sceptic's vision of the future. *Politics and the Life Sciences: the Journal of the Association for Politics and the Life Sciences* 15(1), March: 105–106.

Part 5

Genetic Explanations: Understanding Origins and Outcomes

Introduction

Angus Clarke and Flo Ticehurst

This section examines the scope for the possible contribution of genetics to explanatory accounts, especially of behaviour and of evolution. Although this topic may appear rather abstract, and some of these articles present challenging discussions of complex topics, causal accounts can have thoroughly practical consequences and so must be taken seriously.

Cohen and Atlan (chapter 28) discuss common metaphors for genes, genomes and evolution and whether these are compatible with our scientific understanding of how these actually work. They present a number of examples of why many computer-based metaphors, such as the 'gene-as-program' metaphor, are inappropriate. Instead, they put forward the view that the connection is a recursive loop – 'the meaning is the process itself' – describing the relation between the idea of a program and the self-organization evident in living systems.

This precedes a discussion of the concepts of complexity, reductionism and emergence in biological systems – 'Our point is that reduction to component parts is only the beginning of wisdom. The essence of biology, like that of other complex systems, is the emergence of high-level complexity created by the interactions of component parts.'

Further detail about emergent properties and entities suggest that 'present models of emergence will need upgrading to deal with the complexity of actual biological systems' and that the development of a shared language for biology and informatics offers the potential to achieve a more successful modelling of complexity and emergent processes.

Dover (chapter 29) presents us with his view of the origins and development of the concept of natural selection, particularly with regard to human behaviour, from Darwin (and Wallace) to modern-day genetic understanding. He describes how the insights of Darwin have managed to maintain their authority within evolutionary theory but have become unstuck in the face of Dawkins' assertion that the gene is the primary driver in natural selection, rather than the individual organism.

> In order to understand the origins of the confusion between the roles of the individual and of the gene in determining the course of natural selection, it is necessary to recognize the biological distinction between the individual phenotype and the individual genotype (or genome).

Through his examination of phenotype and genotype relationship and their role in evolution he gives a convincing account of why it is inappropriate to consider the gene as the unit of selection.

Bateson (chapter 30) also addresses the extent to which genetics can usefully contribute to causal accounts of evolution, but he focuses in particular on developmental processes and on behaviour. Bateson employs a historical perspective to view the rise of sociobiology and evolutionary psychology and to illustrate the problematic associations between gene-centred explanations and their application in support of social injustice.

Ruse (chapter 31) describes the origins of creationism and how this way of thinking has consistently opposed evolutionary theory. He charts the history of creationist ideas in the US over the last 150 years or so, giving accounts of the famous Scopes trial in Tennessee, the publication of *The Genesis Flood* (a book he describes as 'the bible of the antievolutionist movement') and the 'showdown' of the Arkansas creationism trial in 1981. Turning his attention to the present, he explains how the 'new generation of creationists' propose the 'intelligent design' theory as the alternative to the theory of evolution.

He ends with a discussion of why 'intelligent design' arguments are proposed by these 'evangelists': one reason may be so that they can evade the US constitution's ban on teaching religion in schools. In addition to this, he suggests that the use of these arguments also relates to questions of morality, in particular the disapproval of the religious right towards many aspects of the current social and political climate.

Rose (chapter 32) introduces the origins of reductionist thinking as the discovery of the 'physical laws of nature' in the mid-nineteenth century, stating that the task of natural scientists has been seen as being to reduce all observations to these laws. In relation to the HGP, Rose argues that its scientific claims have been characterized by a reductionist enthusiasm.

In giving an example of what 'gene for' means, Rose laments that the concept of development is being lost, and discusses different meanings of the word '*gene*'. He claims that 'gene talk' is confusing – because of the mismatch between these multiple meanings and how genes *really* work. He says that the way genes work is better described by the term *autopoiesis*: 'It is more helpful to understand developmental processes as ones of autopoiesis, in which organisms, and perhaps especially humans, actively construct themselves out of the raw materials of their genes and environment.'

Rose goes on to ask why genetic metaphors (particularly deterministic ones) persist in the way they do. Part of the reason for this way of thinking may be due to the claims that genetic knowledge will lead to cures for disease and will enhance the lives of humans. In addition, it could be that we see that many important problems are global in scope and cannot be solved by political or social means; in other words, we might have to live in an unjust and unequal world, because it is 'written in our genes'. Rose counters this view, however, as a profound misunderstanding of what a gene is.

Chadwick (chapter 33) draws our attention to a necessary distinction between personal and collective identity. She describes both physical and psychological accounts of where the essence of a person lies.

The account in question here is the idea of DNA as essence – she explains that this could be an attractive proposition as DNA can be seen to transcend both the physical or psychological attributes of a person and to take on a soul-like quality; DNA may be seen as more than physical because it also carries information. Chadwick discusses potentially important consequences of our views of genetic identity, such as the construction of identity in how we think about ourselves, for example through the establishment of ideas of community among members of an ethnic group; and our considerations about whether identity changes over time and how this may impact on methods of identification within the criminal justice system.

If the genome is thought to be the essence of a person, she asks, then how much of the genome might we be able to change before bringing about a change in identity? On the other hand, if we take the view that identity does not consist of someone's genome then a significant change in the genome would not constitute an identity change.

Bradby (chapter 34) discusses difficulties in using the terms race and ethnicity when considering biological differences between individuals and groups. She notes the contrast between self-assigned ethnicity and observer-assigned racial categories: 'The slippage between the vocabulary of race and that of ethnicity has impeded the development of means of conceptualizing human diversity without recourse to essentialism.' Bradby describes how, although there is a significant amount of agreement that there is no genetic basis for categorization of people into discrete races, there is a continued use of racial categorization; one reason is the strength of racism as a social reality.

As a background to this discussion, Bradby presents an overview of the relationship between eugenics and racism from past to present. This includes a consideration of how eugenics permeates even today's scientific thinking through the use of such outmoded terms as 'Caucasian' to describe White people. She further illustrates racialized thinking in science in her discussion of sickle cell anemia and rickets.

28
Genetics as Explanation: Limits to the Human Genome Project

Irun R. Cohen and Henri Atlan

Definitions

Genetics refers to the structure and function of genes in living organisms. Genes can be defined in various ways and at various scales of interest: consider evolution, populations, species, organisms, cells and molecules, or hereditary transmission, embryonic development and life management. These are quite diverse subjects, and the people who study them would seem to use the term 'gene' in distinctly different ways. But genetics as a whole is organized by a single unifying principle, the deoxyribonucleic acid (DNA) code; all would agree that the information borne by a gene is linked to particular sequences of DNA. At the chemical level, we can define a gene as a sequence (or combination of sequences) of DNA that ultimately encodes a protein. The genome refers to the germ-line DNA that an organism has inherited from its progenitors. The genome includes DNA genes along with DNA sequences that do not appear to encode proteins.

Now we can define the Human Genome Project: the genome project is a translation project. Its objective is to translate the chemical sequence information borne by the genome into the verbal information of human language and thought; the aim is to translate DNA sequences into words and ideas that can develop and spread among human minds. What we can manage to do with this information depends on how well we understand the functions of genomic DNA within the organism.

Metaphors and programs

Most minds use metaphors to understand and explain; we grasp the essence of the unfamiliar (or the complex) by seeing its likeness to the familiar (or the simple). Metaphors are not merely literary devices; metaphors, which also include mathematical models, can aid precise thinking. What metaphor is suitable for explaining the function of the genome?

Metaphorically, the genome is often likened to a computer program; just as the computer reads and executes the instructions of its program, the body is proposed to read and execute the instructions borne by the genome. The body, from this point of view, is mere hardware. The genome is the boss.

The computer program metaphor is often extended to explain evolution: evolution is thought to improve DNA programs. Diversification of genomes by random mutation combined with the selection of the most successful variants (survival of the fittest) leads, it is claimed, to the continuous upgrading of existing DNA programs. The evolution of genomic DNA is automatic but costly – the death of the less fit drives the process.

Metaphors and expectations

The computer program metaphor fosters high expectations of the Human Genome Project. Theoretically, if you know all the information borne by a computer program, you can expect to know how a computer using that program will operate; you can understand the computer's present behavior and can predict its future behavior with a high degree of accuracy. You would even be able to repair mistakes in the program, if that program were simple enough.

Metaphorically then, if the genome is really like a computer program, the genome project will empower us to understand the organism, predict its response to the changing environment and provide a key to the cure of its maladies. Or so many would have wished to believe.

Here we shall discuss what a program means to most people and then test whether the genome actually fits the bill. We shall see that the program metaphor is a misleading way to describe the genome; knowing the genome will not explain the organism. We might do well to consider other metaphors for the genome.

Genome is not a simple program

The *Oxford English Dictionary* (second edition, 1989) defines a computer program as 'A series of coded instructions which when fed into a computer will automatically direct its operation in carrying out a specific task.' A computer program is usually written intentionally by a computer programmer; the DNA program, by contrast, is written by evolution, without intention. But irrespective of who or what writes a program, at the very least, a program is a plan for a sequence of events. So most people would like a program to be unambiguous, coherent and definite. The program's task should be inherent in the program itself; the information in a program should be sufficient for the job. A program, like a blueprint, is a type of representation. But the genome, as every working biologist knows, is ambiguous, incoherent and indefinite. Most debilitating to the genetic program metaphor, the genome is not autonomous or complete. Consider the following examples:

- A DNA sequence that encodes a protein in a multicellular organism is usually discontinuous and is interrupted by chains of meaningless DNA (introns). The gene transcript (messenger ribonucleic acid (mRNA)) has to be spliced together by proteins that cut out the introns. Thus most DNA sequences are not intrinsically coherent.
- Many DNA coding sequences, perhaps as many as a third, can undergo alternative splicing to produce different proteins. In other words, a single DNA sequence can give rise to more than one species of protein. Moreover, the way the DNA actually gets spliced is not governed by the DNA sequence itself; proteins actually determine the gene – the spliced DNA sequence that is expressed in particular circumstances. Thus, the information encoded in many DNA sequences is intrinsically ambiguous until realized by the action of proteins.
- The protein products encoded by a gene may also be indefinite: a single protein may assume several functionally different conformations, and so the gene that gives rise to the protein may be said to function in more than one way. Moreover, the protein encoded by the gene can (and does) undergo chemical modifications (enzymatic cleavage, aggregation with other molecules, phosphorylation, glycosylation and so forth) to carry out further functions independent of the gene. The protein glyceraldehyde-3-phosphate dehydrogenase first discovered as an enzyme, for example, is now known to have a role in membrane fusion, microtubule bundling, RNA export, DNA replication and repair, apoptosis, cancer, viral infection and neural degeneration. The protein's gene is obviously not the program of the protein.
- The sets of genes expressed at a particular time are determined by molecules external to the genome; the previous history of the DNA can be overridden. For example, the sheep Dolly was cloned by transplanting a nucleus from an udder cell into an ovum. The molecular environment of the udder cell activated the milk genes of the nucleus; after the nucleus was transplanted to the ovum, the genes needed for making a new sheep became activated. The cellular environment 'reprograms' the genome, epigenetically.
- A single protein can function in very different ways during prenatal development and later in life after development is completed. Thus the gene encoding the protein can be seen to perform different functions at different times; the meaning of the gene varies with the stage of development.
- Some genes can be removed from the genomes of experimental animals (knocked-out genes) without producing an overt change in the form or

behavior of the animal. Knocking out other genes, in contrast, can lead to severe and unexpected effects. Scientists who knock out genes are not infrequently surprised by the resulting phenotype of the animal. In other words, the impact of a gene on an organism is not readily deducible from knowledge of its DNA sequence.

- The immune system exploits the genome to create novel genes. Each clone of lymphocytes in the immune system constructs its unique antigen receptor by recombining otherwise unexpressed minigene elements inherited in the germ line. The immune system thus manufactures millions of different genes that are not encoded as such in the genome. The immune system functions to heal the organism and protect it from foreign invaders, and is also a key factor in causing autoimmune diseases. Yet the ability of the immune system to recognize antigens, a major determinant of health or disease, is a property removed from the germ line.

In their summation, these and other facts well known to biologists lead to the conclusion that the meaning of the information encoded in the genome is variable and conditional; the meaning of a DNA sequence cannot be derived from the sequence itself. Thus the genome does not encode a coherent plan for a sequence of events.

One may argue that the genome, despite its lack of intrinsic meaning, is still a set of instructions, albeit with many possible branching points. Even so, the extragenomic environment and the history of the organism determine the path through which the genome is executed. Since the given state of an actual person is not determined by the person's genome, the genome is not a representation of the person. For this reason, the master-program metaphor does not clarify the role of the genome, but rather obscures it. The mere encoding of amino acid sequences within DNA nucleotide sequences is not programming. On the contrary, the organism uses, manipulates and, in the case of the immune system, creates genes. The genome acts as the organism's servant, not as its master. Why then have knowledgeable people likened the genome to a master program?

Meaning: line or loop?

The concept of the genome-as-program is associated with the idea that the connection between a gene and its meaning is linear: DNA → messenger RNA → protein → meaning.

A specific DNA sequence was seen as the plan for making, through the agency of messenger RNA molecules, a particular protein. The protein (for example, an enzyme that builds or degrades molecules, or a transcription factor that activates genes) is the agent that carries out a defined activity. Since the DNA

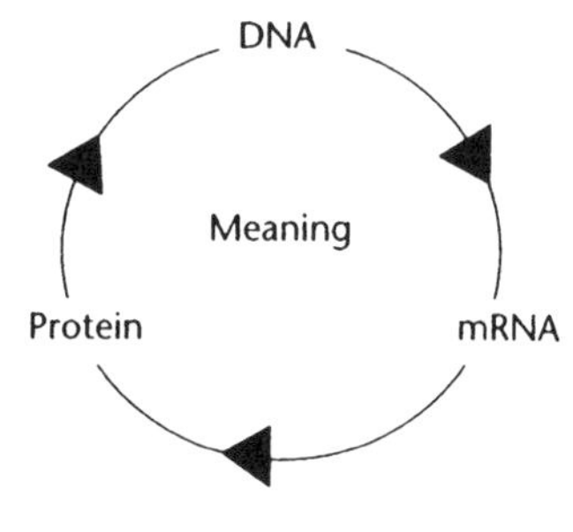

Figure 1

encodes the protein, the meaning of the information borne by the DNA is transformed ultimately into the precise action performed by the protein as an enzyme, transcription factor or other agent. A one-to-one relationship was envisioned: one gene for each protein, and one protein for each function. Thus the activities of the proteins – the meaning – were held to be inherent in the gene – the information.

But, in reality, the living system is not a linear progression from DNA information to protein function; the system is a recursive loop. Proteins, as we have discussed above, are required to make sense out of the DNA sequence; the proteins are required to activate and even to manufacture the very genes that encode the proteins. This way of drawing the connection is closer to reality (Figure 1).

A circle has no beginning and no end: the information actually expressed by DNA is formatted by proteins recursively generated in the process. There is no linear transformation of information (DNA) into meaning (protein action). Genetic information itself is one of the products of protein action; meaning (proteins), as it were, can be said to generate information (legible DNA) in a loop. We might even add the environment to the loop; the influence of proteins on genes is modulated by intracellular and extracellular factors. There is no fixed hierarchy, no one-to-one relationship. The living system is not transformational. The living system is an ongoing process. The meaning of the process that connects DNA and protein is not an outcome of the process; the meaning of the process is the process itself.

Self-organization and program

Scientists had hoped that the genome might function as a simple program because people, especially scientists, think programmatically. Planning is a characteristic of the human mind. We have intentions and goals; we scheme and we plot. We implement programs, so we take programs for granted; every building has to have an architect; a watch implies a watchmaker.

But we also know of many complex natural phenomena that organize themselves without recourse to a master plan; the world is filled with them. A colony of ants or a hive of bees seem wonderfully organized, yet no single ant or bee, not even the queen, has an idea in mind of what a colony or hive should look like. (Queens are just egg-laying machines.) Each ant and each bee only responds mindlessly to what it senses. What seems to us to be a master plan actualized by each insect colony or hive emerges from the combined actions of the insects themselves, each insect autonomous and entirely ignorant of a world beyond its own sensations.

Similarly, an organism is built and operates with the help of its genome; but the genome is only one element in a recursive process. The iterating cycle of genes that form proteins that form genes is the self-organizing process from which the organism emerges. If there be a genetic program, then such a program writes itself collectively. The action, as it were, precedes the plan. But how can that be?

Complexity, reduction and emergence

Physics is the paradigm of sciences; the others try to emulate physics. Physicists explain the behavior of matter by reducing material phenomena to the basic laws of matter and energy. Underlying the physical world are fundamental laws that account for what we see; the material world is explainable by these laws, and so is reducible to these laws. Reduction is done by analyzing the data of sense and experiment to uncover the underlying elements (laws or component parts) that give rise to or 'cause' the data.

Biologists, noting the success of reduction in physics, have attempted to reduce the phenomena of living organisms to the DNA code. Unfortunately, it does not work; life is far too complex to be explained entirely by genomes.

We do not mean to say that reduction should not be done in biology. On the contrary, scientific reduction has been the key to the identification and characterization of the elements – the cells and molecules – that constitute living organisms. The power of modern biology must be credited to reductive analysis. Our point is that reduction to component parts is only the beginning of wisdom. The essence of biology, like that of other complex systems, is the emergence of high-level complexity created by the interactions of component parts.

Emergence is not a mystical concept. A physical basis for the emergence of self-organization has been established in studies of nonequilibrium thermodynamics: open systems that exchange matter and energy with their surroundings can maintain themselves in steady states far from equilibrium. The decrease in internal entropy in such systems can be offset by increased entropy in the surroundings; this makes it possible for macroscopic organization to emerge from the coupling of multiple microscopic reactions. Certain coupled chemical reactions exemplify such processes experimentally. Computer simulations of networks of automata have also provided examples of the emergence of high-level nonprogrammed functions created by the interactions of component parts. But these simple examples only illustrate the bare principle; present models of emergence will need upgrading to deal with the complexity of actual biological systems.

Emergence in biology is difficult to study because we have not yet devised a mathematical language suitable for modeling and simulating the generation of high-level complexity out of simple parts. Fortunately, the Human Genome Project, with its need for advanced bioinformatic technology, has invigorated collaborations between biologists, mathematicians, physicists and computer scientists. New ways to model and study the emergence of complexity are already emerging from these activities. But until biology and the informatic sciences develop a common language, we shall have to make do with examples; fortunately examples of emergence abound. Think of your mind. The mind emerges not from neurons but from the interactions of neurons; all the neurons may be intact and alive, but there will be no mind unless the neurons interact. The mind is not reducible to neurons; the mind emerges from the ongoing interactions of neurons. The interactions create a new entity: the mind. Emergent entities, like your mind, are not mere abstractions; they work.

Genetic program metaphor revisited

The discovery that DNA sequences encode proteins led to the hope that the inherited DNA, the genome, would embody the programs needed to generate and operate the organism; the genome was seen as a representation of the organism. But now biologists have learned that the genome, standing alone, is not an independent program. The organism develops as an emergent process, and not by way of a preexisting plan. The genome is not a representation. What looks like a program is a process. Representation emerges from process.

So we have two alternatives: we can drop the term program and come up with another word for the emergent process we call the program. Or we can continue to use the term program, but understand it to be a metaphor. In either case, the genome by itself is not the master program either in fact or in metaphor.

Evolution

Evolution, we should note, does write genomes, but does not improve genomes, even metaphorically. Genomes, at the level of the species, develop from the processes that adapt an organism to its world. Now a bacterium is no less adapted to its environment than is a human being to its environment. A bacterium as a form of life might, in fact, enjoy a more robust future than the fragile and pugnacious human species. The life and survival of a bacterium would not be improved by making the bacterium more like a human. Self-consciousness would not help a bacterium. Improvement is relative to one's point of view; people like to see themselves as superior to bacteria.

So what does evolution accomplish, if improvement is spurious? Evolution leads to accumulating complexity; humans are objectively more complex than are bacteria. Evolution is a process that, rather than generating improvement, generates new information. But that issue is beyond the scope of this discussion.

Genetic causality: the case of sickle cell disease

Detailed knowledge of the DNA sequence, the outcome of the genome project, will not suffice to explain health or disease. We shall have to look to the activation of genes and the dynamic functions of proteins and other molecules involved in the processes of life. Biological and cultural evolution, and the environment too, have their place in the action. Take, for example, the case of sickle cell disease.

Sickle cell disease is a deficiency of red blood cells (anemia) characterized by an abnormality in the hemoglobin molecule, such that the affected red blood cells assume an elongated shape (like a sickle) at low oxygen tension. The sickle-shaped red blood cells stick together and are destroyed, producing the anemia. Small blood vessels get clogged by the clumps of sickled red cells and tissues suffer from the lack of blood flow. The disease, untreated, results in an early death. What is the cause of sickle cell disease?

The answer is deceptively simple. Sickle cell disease is a genetic disease; the hemoglobin molecule is abnormal because of a mutation in the gene encoding the β chain of the molecule: a single glutamic acid in the protein is replaced by the amino acid valine – this abnormal hemoglobin is called hemoglobin S. Persons who have inherited the gene for hemoglobin S from both parents can make only hemoglobin S and so manifest the disease. Persons who have inherited one hemoglobin S gene and one normal gene (heterozygotes) are essentially free of the disease. Thus we could define the disease as caused by having inherited two copies of the hemoglobin S gene. But this is not the whole story.

The hemoglobin S gene is present mostly in populations of people who have originated in equatorial Africa, and the incidence of the gene is much higher than would be expected from the spontaneous mutation rate of standard hemoglobin to hemoglobin S. The relatively high frequency of a potentially lethal mutation suggests that the mutated gene must have some selective advantage, must contribute to fitness. Well, it turns out that children infected with a certain type of malaria (and who are untreated) will die of the infection if their genome contains only the standard hemoglobin gene. The children who carry one gene for hemoglobin S (heterozyotes), however, are relatively resistant to malaria, and so do not die of that disease. The heterozygous children also do not die of sickle cell disease. Of course, children whose genomes include two of the hemoglobin S genes (homozygotes) die of sickle cell disease. Thus, we might say that hemoglobin S is an advantageous adaptation to malaria, and the gene is maintained in the population, despite the loss of homozyzgous children, at a rate that reflects the selective pressure exerted by the rate and severity of malaria infection. The death of homozygous individuals is the price paid by the population for heterozygous resistance to malaria. So we could say that the high frequency of hemoglobin S (and sickle cell disease) is caused by malaria. By this reasoning, we might say that sickle cell disease has value as a trade-off in exchange for malaria. In environments free of malaria, hemoglobin S provides no advantage.

Should we now conclude that one of the causes of sickle cell disease as a disease (rather than as a trade-off) is the absence of malaria? Sickle cell disease is a serious health problem in the African Americans whose ancestors were taken to America as slaves from West Africa. Should we include the slave trade among the causes of sickle cell disease in an African-American child? Or has the disease been caused by two heterozygotes falling in love?

Persons who have inherited two hemoglobin S genes do much better clinically if they continue to produce fetal hemoglobin after birth; the fetal hemoglobin makes the affected red cell more resistant to the deleterious effects of hemoglobin S. Indeed, homozygous persons are now treated with a drug that induces the production of fetal hemoglobin after birth. Are we to conclude that the normal termination of fetal hemoglobin production is a causal factor in sickle cell disease?

In short, a 'simple' genetic disease like sickle cell disease, which is associated with a defined mutation resulting in a defined molecular abnormality, presents us with a complex causal chain of events. How much more complex are the possible genetic explanations for diseases such as type 1 diabetes or multiple sclerosis, diseases that have been associated with many different susceptibility genes. Indeed, inheriting susceptibility genes does not make the disease inevitable. Take for example identical twins, who bear identical genomic DNA; if one twin develops type 1 diabetes or multiple sclerosis, the other twin will develop the disease in only about a third of the pairs. Having susceptible DNA does not suffice to explain the disease. Indeed, prevalent genes that are associated with susceptibility to complex diseases are probably advantageous trade-offs.

The environment

We can summarize the limitations of the genome most easily by repeating what has already been said many, many times: one's genes are only an incomplete explanation for one's being; the present environment and its history, at the scales of the person, the group and the biosphere, interact with the genome to determine its expressions and effects.

Genome metaphors

What metaphor might be generally useful for appreciating the function of the genome? Genomic DNA is a reservoir of raw information that, suitably processed, can be translated into the amino acid sequence of functional proteins. Perhaps we could think of the genome as akin to a list of words, a vocabulary book, that can be used to build sentences. One may argue about the meaning or range of potential meanings inherent in any word standing alone; but it is clear that a word gains its fullest, most particular meaning when the word is used as part of a sentence. DNA sequences, like words in natural language, are essentially passive; they may be spliced into different genes that give rise to proteins that function in different ways under different circumstances. Fragments of genomic DNA, like words, acquire different meanings in different contexts. They can be used artfully to tell different stories. The genome, like a vocabulary, is information, transmissible from generation to generation, that is available for processing into meaning. The process itself, as we have discussed, is the story.

We might extend the language metaphor and consider that particular three-member sets of bases – the codons – form the alphabet sequences of the metaphorical DNA words; 'stop', start, 'splice' and other codon signals may be said to encode syntax.

If you would prefer a computer metaphor, consider this one: the genome may be viewed as a compacted substrate for storing information; protein synthesis involves the processing and amplification of genomic DNA into a multitude of RNA molecules and then into an amino acid sequence expressed in the millions of proteins produced using that DNA. From this point of view, genomic DNA is like a database stored in computer memory. Genomic data is special because it is transmitted from generation to generation; such transmission can take place because the genome serves as a template for its own replication between generations. Genomic data is used by the individual organism for development and for maintenance. The organism, of course, also uses data obtained from its evolving self and from the environment. The master program for using this data, as we discussed above, emerges from the living process itself.

Conclusions

The Human Genome Project, like putting a man on the moon, is a costly undertaking of great technical virtuosity. It is good that the project has been done for the daring of it and because it has already provided much important information about genetics and the organism. No less important, the genome project has spawned powerful technologies and has opened biology to the age of informatics. Biology has learned that it is an informatic science. Finally, the very success of the genome project has dispelled the simplistic illusion of the genetic program; biology is now aware of its true complexity. The genome project, wittingly or not, has built the foundation for deeper probes into the complexity of life. The limitations of the project are only the limitations of the genome itself.

Further Reading

Atlan H (1983) Information theory: basic elements and recent developments. In: Trappl R (ed.) *Cybernetics: Theory and Applications*, pp. 9–41. New York: Hemisphere Publications Springer.

Atlan H (1987) Self-creation of meaning. *Physica Scripta* 36: 563–576.

Atlan H and Cohen I R (1998) Immune information, self-organization and meaning. *International Immunology* 10: 711–717.

Atlan H and Koppel M (1990) The cellular computer DNA: program or data? *Bulletin of Mathematical Biology* 52: 3335–3348.

Cohen I R (2000a) *Tending Adam's Garden: Evolving the Cognitive Immune Self*. San Diego, CA: Academic Press.

Cohen I R (2000b) Discrimination and dialogue in the immune system. *Seminars in Immunology* 12: 215–219, 269–271, 321–323.

Chong L and Ray L B (2002) Whole-istic biology. *Science* 295: 1661.

Fox Keller E (1995) *Refiguring Life. Metaphors of Twentieth-century Biology*. New York: Columbia University Press.

Kono T (1997) Nuclear transfer and reprogramming. *Reviews of Reproduction* 2: 74–80.

Louzoun Y, Solomon S, Atlan H and Cohen I R (2001) Modeling complexity in biology. *Physica A* 297: 242–252.

Nicolis G and Prigogine I (1977) *Self-organization in Nonequilibrium Systems: From Dissipative Structures to Order through Fluctuations*. New York: Wiley.

Science (1999) Complex systems [special issue]. *Science* 284: 79–109.

Strohman R C (1997) Epigenesis and complexity: the coming Kuhnian revolution in biology. *Nature Biotechnology* 15: 194–200.

29
Darwin and the Idea of Natural Selection

Gabriel Dover

The original concept of natural selection

Darwin's (and Wallace's) idea of natural selection is simple to comprehend, yet since the mid-1800s it has yielded as many definitions as there are textbooks and monographs on evolution; and arguments abound over the levels, units, rates, circularity and testability of natural selection. Bizarrely, the word 'adaptation' is taken to mean both the mechanism of natural selection and its product. This problem of definition is not some arcane irritation among academics who love a good quarrel; it is of major concern to the understanding of our own human behavior, given the claim that Darwin has replaced Marx and Freud, at the start of the twenty-first century, as the intellectual father of us all.

If genetics is at the heart of natural selection, then the range of definitions of natural selection might reflect a wide disparity in the understanding and experience of genetics among biologists, philosophers, psychologists, sociologists, linguists and novelists – all of which have attempted to explain, by one device or another, the naturally selected basis of human nature in recent times. Yet, neither Darwin nor Wallace had any idea of the material basis of heritable variation or of the processes that govern the distribution of variation from one generation to the next. However, such ignorance did not prevent either man from formulating, independently, a means of evolution by natural selection, which has stood the test of time. If Darwin and Wallace could get it right, why do so many contemporary thinkers get it wrong when seeking evolutionary explanations based on technically incorrect notions of the roles of genes in evolution and development?

Descent with modification

Darwin (and Wallace) can lay claim to two significant insights: first, that any proposed mechanism for evolutionary change needs to ensure 'descent with modification' and, second, that natural selection is just such a mechanism. Natural selection is not the only means of achieving 'descent with modification', but was the first, scientifically plausible mechanism by which a population of individuals of one type of organism could be modified and become another type of organism with the gradual passing of the generations. Both men recognized, along with the world's population of parents, that children are more likely to resemble their biological parents than resemble unrelated adults. And what was true for humans was also true for all other forms of life, observable to discerning eyes. Hence, some of the characteristics in form and behavior, peculiar and unique to one individual, are passed on vertically to the direct descendants in the next generation, and not horizontally to surrounding, unrelated individuals. Accordingly, an explanation of evolutionary change demands a mechanism that modifies the heritable material as it descends from one generation to another. Hence, the central idea of natural selection is that some individuals reproduce more frequently than others and so influence, disproportionately, the representation of their heritable material at the next generation. Differential reproduction would result, as both Darwin and Wallace recognized, from Malthus' principle of biological populations outstripping their resources. Accordingly, Darwin and Wallace reasoned that minor differences in the development, form and behavior of individual organisms (now known to be due in part to genetic mutations and the sexual mixing of genes) would influence the outcome of each individual's interactive struggle with its local biotic and abiotic environment.

Natural selection is not a process

With such original formulations of the idea of natural selection (Darwin, 1859), it is clear that 'descent with modification' of the heritable material is a passive outcome (or, in today's parlance, an inevitable sorting of genetic variants) arising from distinct differences in individual interactions with their ecology, as they affect reproductive success. The interactive, individual organism is at the heart of natural selection as it goes about its business of surviving and reproducing at a fleeting moment in evolutionary time. The natural sorting of genes is a one-off outcome as a consequence of many unrepeatable, fleeting moments of differential interactions by ephemeral, uniquely shaped individuals at a given generation. On this basis, natural selection is not a process as such in that the degree of heritable modification in one generation, resulting from unique interactions between unique individuals with unique surroundings, does not influence the degree of heritable modification resulting from differential interactions among the new set of uniquely formed individuals in the next generation.

Notwithstanding such passivity and uniqueness, Darwin and Wallace recognized that, because selection at each generation is a consequence of differences in reproductive success, those phenotypic features, and their underlying genetic influences that confer reproductive success will accumulate and evolve. Such features are called adaptations – often reflected as a close functional match between a phenotype and some component of the environment (niche).

Darwin's and Wallace's insight into the local, natural causes of selection has not been disturbed with the advent of Mendelian genetics, although it has been turned inside-out with the advocacy of the gene (Dawkins, 1976), rather than the individual, as the determining unit of selection. Significantly, Mendel's rules of inheritance, based on the random shuffling of the genetic material during sex, cannot in themselves lead to evolution ('descent with modification').

In order to understand the origins of the confusion between the roles of the individual and of the gene in determining the course of natural selection, it is necessary to recognize the biological distinction between the individual phenotype and the individual genotype (or genome).

The unbearable mystery of being a phenotype

The phenotype is the hardwired organism, composed of biological macromolecules that keep the phenotype alive. The genotype of an individual phenotype is the sum total of deoxyribonucleic acid (DNA) that is distributed to the next generation. The functional part of the genotype consists largely of: (i) a 'genetic code' in nucleotide sequences (genes) that specify the amino acid sequences of proteins, plus sequences that specify functional ribonucleic acid (RNA); and (ii) sequences that control the time, place and quantity of gene activity during the life cycle of an individual phenotype.

There is no one-to-one correspondence between genotype and phenotype, and no direct transmission of an intact individual genotype from one generation to another in sexually reproducing organisms, for the following reasons. Firstly, the primary effect of sex is the shuffling of inherited sets of parental chromosomes (and hence genes) in an individual such that each new gamete (sperm or egg) produced by that individual contains a mixture of paternal and maternal chromosomes. Hence, genotypically, every individual is unique: the particular sets of genes and controlling elements inherited by an individual would never before have coexisted in an individual in all preceding evolution. Secondly, the development of an individual multicellular phenotype from a unique, single-cell zygote (fertilized egg) is influenced by complex networks of interactions between multipurpose genes in that one gene can be involved in many different biological functions and a given function can be influenced synergistically by many diverse genes. By and large, there are no genes for any specific part of the phenotype: there is a many-on-many relationship between genotype and phenotype. Thirdly, the networks of gene interactions are influenced, to a variety of extents, by the internal and external environments, locally unique to a developing individual phenotype.

The unpredictability of phenotype from genotype

Taking these three features of the relationship between genotype and phenotype together, it is not possible to predict the unique phenotypic outcome of an individual from its inherited, unique combinations of genes undergoing complex interactions and subject to the vagaries of the environment. Every individual phenotype is an island unto itself, including identical twins to a lesser extent. The sheer, illogical complexity of genetic interactions that contribute to the development, form and behavior of an individual is such that the influence of any given gene on phenotype is contingent on the particular combination of genetic variants inherited by that individual. For example, the clinical manifestation of a thalassemic blood disorder in humans can range from zero to full expression, depending on the totality of an individual's genetic background in which the relevant mutant globin gene finds itself (Weatherall, 2001). As the unique, interactive combination of genetic variants unfolds in the production of a unique phenotype, there are no 'rules' by which the outcome could be predicted, even were we to ignore the influence of the environment on the unique set of genetic interactions. Essentially, genes influence everything but control nothing, notwithstanding arguments to the contrary regarding genetic predictions and predilections, for example with regard to the emergence of individual human nature (Wilson, 1978). Generalized predictions can be made (e.g. men are, on average, taller than women), but no useful predictive statement can be made about the height of any specific man or woman. The same operational difficulties arise concerning individual levels of intelligence, spatial perception, humor, sexuality, ability to read maps, aggression and so on. It is naïve to assume that there are independent genes for each and every characteristic that have accumulated through past episodes of natural selection. The nature of biology is such that the basis of individuality is largely uncapturable, making all talk of the evolutionary origins of the unknown premature at best and vacuous at worst.

The misunderstood 'selfish gene'

The ephemeral, unpredictable nature of individual phenotypes at each and every generation undermines, rather than supports, the notion that the 'selfish', eternal, 'self-replicating' gene is the unit of selection. Selection is not a process as such with predictable outcomes based on fixed, selective 'powers' of individual genes controlling aspects of phenotype. Selection involves whole phenotypes, which are in part influenced by their unique combination of genetic interactions; hence, evolution involves descent with modification of genetic interactions. Genetic interactions are phenotypic, not genotypic. Advocacy of the gene as the unit of selection is operationally incoherent and genetically misconceived.

Other levels of selection

The phenotype can be as small as the simplest, independently reproducing unit of biology, the cell, or as large as groups of individuals collectively behaving in ways that affect the differential reproductive success of groups (Sober and Wilson, 1998). Hence, selection can take place among cells (e.g. among the different antibody-producing cells in response to invasions of specific pathogens) in multicellular organisms, and among groups (potentially giving rise to altruistic, cooperative behavior). There is no essential difference in operation of selection at the levels of cell, individual, sex, kin and group, as there is between the level of the gene and all the rest. The gene is the unit of inheritance, but it cannot be the focus of natural selection without turning biology inside-out. The arena for natural selection is not down among the genes, it is between uniquely formed, genetically influenced phenotypes with reproductive capabilities and finite life spans. It is for these reasons that evolution by means of natural selection of phenotypes cannot be comprehensively modeled unless phenotypes are reduced simplistically to mathematical symbols, or to individual nucleotides in a gene or in its regulatory sequences. While such reduction makes life easy for modelers, it does not mean that an intellectually satisfying understanding of the forces that govern the evolution of life by natural selection is at hand. Such reducing devices are indirectly responsible for the faux pas of the gene as the unit of selection and the misplaced emphasis on 'genetic determinism' and 'genetic predictability' in some academic and social pursuits.

Other means of 'descent with modification'

Finally, there are two other mechanisms capable of inducing 'descent with modification'. These are 'neutral drift' (Kimura, 1983) and 'molecular drive' (Dover, 1982, 2000). The former is based on the vagaries of stochastic diffusion in fluctuating populations of phenotypes, and the latter is based on ubiquitous processes of non-Mendelian turnover (transposition, gene conversion, etc.) that promote or demote the representation of genes and/or control elements with the passing of the generations in sexual populations. Evolution of life on earth, including the evolution of genomes, is an interactive mix of natural selection, neutral drift and molecular drive that influences the representation of genetic variants and their phenotypic consequences generation after generation.

References

Darwin C (1859) *On the Origin of Species by Means of Natural Selection.* London: John Murray.

Dawkins R (1976) *The Selfish Gene.* Oxford: Oxford University Press.

Dover G A (1982) Molecular drive: a cohesive mode of species evolution. *Nature* 299: 111–117.

Dover G A (2000) *Dear Mr Darwin: Letters on the Evolution of Life and Human Nature.* London: Weidenfeld Nicolson/Berkeley, CA: University of California Press.

Kimura M (1983) *The Neutral Theory of Molecular Evolution.* Cambridge: Cambridge University Press.

Sober E and Wilson D S (1998) *Unto Others: The Evolution and Psychology of Unselfish Behaviour.* Cambridge, MA: Harvard University Press.

Weatherall D J (2001) Phenotype–genotype relationships in monogenic disease: lessons from the thalassemias. *Nature Reviews* 2: 245–255.

Wilson E O (1978) *On Human Nature.* London: Penguin Books.

30
Sociobiology, Evolutionary Psychology, and Genetics

Patrick Bateson

Rise of sociobiology

In the 1960s, field studies relating behavior patterns of animals to the social and ecological conditions in which they normally occur led to the enormous popularity and success of behavioral ecology (Krebs and Davies, 1981). A new subject called 'sociobiology' brought to the study of behavior important concepts and methods from population biology, together with some all-embracing claims of its own. The pivotal moment for the growth of sociobiology was the publication of an important book by E. O. Wilson (1975). Imaginations were captured by the way the ideas from evolutionary biology were used. The appeal of evolutionary theory, in which sociobiology was embedded, was that it seemed once again to make a complicated subject manageable. This was particularly true of the gene-centered writings of Richard Dawkins (e.g. Dawkins, 1976), who provided a crutch for understanding the complex dynamics of evolution by attributing intentions to genes.

Individual animals interact with others, have relationships and collectively form societies. Social behavior often seems to involve cooperation and from Charles Darwin onwards this aspect of behavior had continued to tease the theorists. If evolution depended on competition, how could cooperation have evolved? Three types of explanation were offered. The first benefited enormously from the thinking of Bill Hamilton (e.g. Hamilton, 1996). He pointed out that if two parties are related then the benefits of helping a cousin, say, are logically the same as helping a child – although quantitatively less effective in terms of gene propagation. This idea fostered the Dawkins 'selfish gene' approach.

The second explanation was that both parties would benefit from the act of cooperation. Robert Trivers (1985) coined the term 'reciprocal altruism'. An example of mutualism between species is when a large predator fish opens its mouth to cleaner-fish which pick food remnants from between the teeth of the predator without danger to themselves. The third explanation for cooperation is that the individual is part of a group of unrelated individuals that may survive better as an entity than another group as a result of actions taken by the individuals within it. In the face of gene-centered theories, this explanation was widely thought to be implausible, but it may occur if individuals die out more rapidly than groups and immigration between groups is difficult – a not implausible set of circumstances in human tribal societies. Group-selected behavior may also evolve when the actions of the group cannot be subverted by individual free riders (Bateson, 1988).

The gene-centered approach to behavioral biology brought a new look to studies of communication where well-known instances of supposed transfer of information were reinterpreted in terms of selfish manipulation (Krebs and Dawkins, 1984). It also led to re-examination of apparent shared interests of both parents caring for young or the mutually beneficial interactions of parent and offspring (Trivers, 1985). Most adults of most species are able to produce more than one offspring during their lifetime. The characteristics of those offspring that are most successful in leaving descendants will tend to predominate in subsequent generations. Parents who sacrifice too much for one offspring will have fewer descendants.

By the same token, offspring that do less to ensure their own survival than others will have fewer descendants. Such are the broad rules that shape any life span, but just how they look in detail will depend on the species and, within a species, on local conditions. In some species, parental care for the tiny progeny consists of nothing more than providing a small amount of yolk, enough to sustain the offspring until they can feed for themselves. Marine fish such as herring, for instance, produce vast numbers of eggs and sperm, which fuse in the sea. Neither parent provides any care for their progeny. Most fertilized herring eggs consequently die at an early stage in their lives. Other fish produce far fewer fertilized eggs and care for them in a variety of different ways, some keeping them like a mammal in their bodies until the young are born, others gathering the fertilized eggs into their mouths where they are protected.

Of all animal groups, birds and mammals produce the smallest number of young and take the greatest care of those that they do produce. Birds encase their fertilized eggs in a hard shell and eject them from their bodies, whereupon both sexes usually take turns in keeping the developing egg at body temperature until the chick hatches. While it is developing inside the egg the embryonic chick is fed from the yolk, which at the outset is enormous relative to the embryo. After the egg has hatched, both parents usually protect and bring food to the developing young. This has important consequences for the amount of care that is given by the parent and the time taken to become adult. The long period of development that is particularly characteristic of humans, but also true to a lesser extent of most other mammals, is made possible by the protection and the provisioning by the parent. As they prepare for their own eventual reproductive life, children meanwhile have to survive.

The traditional image of parenthood had been one of complete harmony between the mother and her unborn child. However, evolutionary theory in the hands of the sociobiologists cast doubt on this blissful picture. In sexually reproducing species, parents are not genetically identical to their offspring. Consequently, offspring may require more from parents than parents are prepared to give, creating the possibility of a conflict of long-term interests. Trivers (1985) called this 'parent – offspring conflict', a term that refers strictly to a conflict of reproductive interests, not conflict in the sense of overt squabbling (Bateson, 1994). The parent may sacrifice some of the needs of its current offspring for others that it has yet to produce; the offspring maximizes its own chances of survival. Parent and offspring 'disagree' about how much the offspring should receive. The result of such evolutionary conflicts of interest is sometimes portrayed as a form of arms race, with escalating fetal manipulation of the mother being opposed by ever more sophisticated maternal countermeasures.

However, limits must be encountered in the course of evolution. If the offspring is too aggressive in its demands it will kill its maternal host and, of course, itself. Likewise, if the mother is too mean, her parasitic offspring will not thrive and she might as well have not bred. Moreover, mutually beneficial communication often occurs between parent and offspring so that independence comes at a time that is beneficial to both parties (Bateson, 1994).

Despite the invigorating debates about the function and evolution of social behavior, grand claims were made for the relevance of sociobiology to human behavior. In the final chapter of his seminal book, Wilson (1975) offered biological explanations for much-fought-over areas such as the differences between the sexes, homosexuality, xenophobia and religion. Even more provocatively he predicted that before long the social sciences would be incorporated into biology. The zeal of the hard-selling proponents of sociobiology made them unpopular with other academics who felt threatened or insulted by the attempted takeover. Moreover, some sociobiological views have been known to appeal to people with a strong interest in maintaining their power and their customary privileges. These less savory aspects of sociobiology led fierce debates (Segerstråle, 2000; Laland and Brown, 2002).

The subject was deemed by many to have over-reached itself, particularly in its claims about the relevance of evolutionary biology to human social behavior. Nonetheless, interest in the links from evolutionary biology to human behavior persisted and flowered in a number of subdisciplines such as human ethology and human behavioral ecology. Many of these studies focused on observables. A view grew up that many of the most interesting legacies of human evolution lay not so much in the surface features of behavior but in the rules underlying their development. The most striking example of this would be language, which differs dramatically from culture to culture but, it is argued, develops according to rules that acquired their present form in the course of human evolution (Pinker, 1994).

Rise of evolutionary psychology

Two of the major figures in this development were Tooby and Cosmides (e.g. Tooby and Cosmides, 1992). Taking a lead from philosophical discussions of the modularity of mind (Fodor, 1983) and from advances in the design of computer architecture, evolutionary psychologists suggested that neural modules in the brain were adapted for specific functions. Famously, the workings of the brain were likened to a Swiss Army knife in which each tool serves a particular job. Each module was thought to generate a unitary group of instinctive behaviour patterns. While some functions such as speech are well described by this style of thought, the concept is loosely used and its generality has been much criticized (Heyes and Huber, 2000).

The idea of a neural module generating behavior suffers from much the same defects as the concept of instinct itself. Apart from its colloquial use, the term 'instinct' has at least eight scientific meanings: present at birth (or at a particular stage of development), not learned, develops before it can be used, unchanged once developed, shared by all members of the species (or at least of the same sex and age), organized into a distinct behavioral system (such as foraging), served by a distinct neural module, adapted during evolution, and differences between individuals are due to genetic differences (Bateson, 2000).

The problem is that one use does not necessarily imply another even though it is often assumed, without evidence, that it does. Behavior that has been probably shaped by Darwinian evolution and appears ready formed without opportunities for learning may be changed in form and the circumstances of expression by subsequent experience. The smile of the human behavior is a good example (Bateson and Martin, 2000).

Despite the weakness of some of its concepts, evolutionary psychology provided a research strategy based on asking the following six questions:

- What problems did humans have to solve in the ancestral environment?
- What were the likely adaptations to that environment?
- What information processing problems had to be solved in evolving such adaptations?
- What are the design features of such cognitive processes likely to be?
- How do the predictions of such models compare with those of other models of cognitive neuroscience?
- What do the models predict about the behavior of humans living in modern conditions?

This general approach has raised doubts about how the characteristics of the so-called environment of evolutionary adaptedness could be ever ascertained. Even so it has led to some interesting studies. For example, sex differences in behavior could be linked to the different evolutionary pressures on the reproduction of human males and human females. Women were expected to rate highly the characteristics of men that would facilitate the care of their offspring. In one major study, the desirable characteristics of individuals of the opposite sex were investigated in more than 10 000 men and women from 37 cultures around the world. Both men and women rated the capacity for love, dependability and emotional stability highly. But big differences were found between men and women in what characteristics were valued highly in a member of the opposite sex. While men rated youth and physical attractiveness more highly than did women, women rated highly the resources held by a man and all the characteristics associated with acquiring such resources, such as health and intelligence. These sex differences were consistent across all cultures (Buss, 1994). Universality does not imply, however, that if conditions changed, the same sex differences would be found. Darwinian mechanisms of evolution could have operated in the past to produce consistent sex differences so long as the rearing and social enviroments on which the differences might depend were stable from one generation to the next.

Role of genes

The great majority of biologists would agree that the evolution of life has generally involved changes in the genetic composition of the evolving organisms. A second point of consensus is that, if the change over time involves an adaptation to the environment and the postulated process of Darwinian selection has worked at all, then the genes are likely to influence the characteristics of the organism. This is not to exclude the importance of constant features of the environment. The disagreements start over the particular ways in which genes affect the outcome of an individual's development.

Some sociobiologists thought that a straightforward correspondence could be found between the genes and behavior. Wilson (1976) was quite candid about it when replying to reviewers' criticisms that he was naïve about behavioral development in his great book. He argued that most phenotypic traits of parents and offspring are correlated to some extent by virtue of a higher than average possession of the same genes. Therefore, he went on, the exact structure of the nervous system, the developmental pathways and endocrine – behavioral interactions were modules that could be decoupled from the explanatory scheme. The implication was that the development of the individual is merely a complex process by which genes are decoded.

If genes code for structures or behavior patterns, they must bear a straightforward relationship to them. Usually they do not and the correct way to describe what is observed is to state that a genetic difference between two individuals gives rise to a difference in behavior without anything being said about the mode of development. When details are known about the mode of development, the point of this intellectual discipline becomes obvious. For example, people with Kallmann syndrome associated with a lack of sexual interest may differ from others in only one gene. The Kallmann syndrome is caused by damage at one of several specific genetic loci. Cells that are specialized to produce a chemical messenger called gonadotrophin-releasing hormone (GnRH) are formed initially in the nose region of the fetus. Normally the hormone-producing cells would migrate into the brain. As a result of the genetic defect, however, their surface properties are changed and the cells remain dammed up in the nose subsequently causing anosmia. The activated GnRH cells, not being in the right place, do not deliver their hormone to the pituitary gland at the base of the brain. Without this hormonal stimulation, the pituitary gland does not produce the normal levels of two other chemical messengers, luteinizing hormone and follicle-stimulating hormone. Without these hormones in men, the testes do not produce normal levels of the male hormone testosterone. Without normal levels of testosterone, the man shows little sign of normal adult male sexual behavior. As a result, men with Kallmann syndrome who are not given hormone replacement therapy have a reduced libido and are not attracted to either sex. The pathway from gene to behavior is long, complicated and indirect. Each step along the causal pathway requires the products of many genes and has ramifying effects, some of which may be apparent and some not (Pfaff, 1997).

If a gene coding for altruism is proposed, the implication is that the gene represents the behavior and that a one-to-one correspondence will be found between the gene and altruistic behavior. If this were true, as it is for the link between gene and protein, the gene could be properly treated as an absolute unit coding for altruism and insistence on properly referring to differences in genes giving rise to differences in behavior would be mere pedantry. However, in most cases the slippage in meaning is unjustified.

The modern view about development is that the processes involve systems influenced by many different things with properties that are not easily anticipated, even when all the influences are known. Like many artificial systems, developmental processes may be strikingly conditional in character, particularly those in complex organisms. In one set of conditions they proceed in a particular and appropriate direction, in another set they do something different but equally appropriate. In yet others entirely new forms of behavior may be generated. One human example of alternative modes of development, which has been much discussed, is that in poor conditions individuals develop a small body type and a metabolism that is well adapted to those conditions – the so-called thrifty phenotype; in affluent conditions, individuals with identical genotypes develop larger bodies and have biochemical pathways well able to cope with ample food supplies (Bateson, 2001). This conditional dependence on the state of the environment means that the expression of genes and the characteristics that they influence are not inevitable.

An appropriate view of an individual's development is clarified by a culinary metaphor, which copes with the fact that in most cases many factors have been responsible for the detailed specification of behavior. In the baking of a cake, the flour, the eggs, the butter and all the rest react together to form a product that is different from the sum of the parts. The actions of adding ingredients, preparing the mixture and baking all contribute to the final effect. Nobody expects to recognize each of the ingredients and each of the actions involved in cooking as separate components in the finished cake. The biological equivalent of raisins will be found from time to time, but a simple relationship between the developmental determinants and the behavior will be exceptional. The issue then becomes squarely one of understanding the developmental processes.

Conclusion

Applying the biologists' knowledge of social behavior to humans has obvious and sometimes damaging political implications. However, the impact of sociobiology and evolutionary psychology is by no means all on the negative side. Indeed, some of the apparent support for social injustice was based on a muddle about how genes actually work. As this muddle was straightened out and genetic determinism fell away as a serious issue, biological knowledge has served to help the understanding of social issues by showing precisely how human potential is expressed in some conditions and not seen in others. The promise of fruitfully combining the insights of different disciplines is real. Sex differences in humans in terms of biological function bind together and make sense of what is observed without implying that the differences are inevitable, unchangeable or even desirable in a modern context (Hinde, 1984).

It is not necessary to appeal to biology at all, of course, and many would continue to argue that to do so in many cases remains utterly misleading and dangerous. They may well be right so long as people continue to suppose that 'natural' means 'desirable'. However, 'natural' by no means always means 'nasty and selfish' and, in as much as biological arguments are brought into debates about social issues, it is appropriate that the biological value of cooperation, for instance, should be fully appreciated.

References

Bateson P (1988) The biological evolution of cooperation and trust. In: Gambetta D (ed.) *Trust: Making and Breaking Cooperative Relations*, pp. 14–30. Oxford: Blackwell.
Bateson P (1994) The dynamics of parent – offspring relationships in mammals. *Trends in Ecology and Evolution* 9: 399–403.
Bateson P (2000) Taking the stink out of instinct. In: Rose HRS (ed.) *Alas, Poor Darwin*, pp. 189–207. New York: Harmony.
Bateson P (2001) Fetal experience and good adult design. *International Journal of Epidemiology* 26: 561–570.
Bateson P and Martin P (2000) *Design for a Life: How Behavior Develops*. London: Vintage.
Buss DM (1994) *The Evolution of Desire: Strategies of Human Mating*. New York: Basic Books.
Dawkins R (1976) *The Selfish Gene*. Oxford: Oxford University Press.
Fodor J A (1983) *The Modularity of Mind*. Cambridge, MA: MIT Press.
Hamilton W D (1996) *Narrow Roads of Gene Land*, vol. 1. Oxford: W H Freeman.
Heyes C and Huber L (2000) *The Evolution of Cognition*. Cambridge, MA: MIT Press.
Hinde R A (1984) Why do the sexes behave differently in close relationships? *Journal of Social and Personal Relationships* 1: 471–501.
Krebs J R and Davies N B (1981) *Introduction to Behavioral Ecology*. Oxford: Oxford University Press.
Krebs J R and Dawkins R (1984) Animal signals: mind-reading and manipulation. In: Krebs J R and Davies N B (eds.) *Behavioral Ecology*, 2nd edn, pp. 380–402. Oxford: Blackwell.
Laland K N and Brown G R (2002) *Sense and Nonsense: Evolutionary Perspectives on Human Behavior*. Oxford: Oxford University Press.
Pfaff D W (1997) Hormones, genes, and behavior. *Proceedings of the National Academy of Sciences of the United States of America* 94: 14213–14216.
Pinker S (1994) *The Language Instinct*. London: Penguin Books.
Segerstråle U (2000) *Defenders of the Truth: The Sociobiology Debate*. Oxford: Oxford University Press.
Tooby J and Cosmides L (1992) The psychological foundations of culture. In: Barkow J H, Cosmides L and Tooby J (eds.) *The Adapted Mind*, pp. 19–136. New York: Oxford University Press.
Trivers R L (1985) *Social Evolution*. Menlo Park, CA: Benjamin/Cummings.
Wilson E O (1975) *Sociobiology: The New Synthesis*. Cambridge, MA: Harvard University Press.
Wilson E O (1976) Author's reply to multiple review of Sociobiology. *Animal Behavior* 24: 716–718.

31
Creationism

Michael Ruse

Christianity, the theistic religion growing out of Judaism, which accepts Jesus of Nazareth as the Messiah, the son of God, who was crucified for Man's sins, takes the Holy Bible – the Old Testament and the New Testament – as authoritative. Christians believe that the Bible is the infallible word of God, yet the Christian tradition has always insisted that the Bible needs interpretation and that one should never simply accept its words in an unreflective, literal sense. Even in AD 400, Saint Augustine, the greatest of the Church Fathers, stressed that the early books of the Old Testament particularly are not written in the language of educated men and may well stand in need of allegorical understanding. In the sixteenth century, the great Protestant reformers likewise acknowledged that the words of scripture need careful study to discern their true import. Highly significant here was John Calvin, who developed the notion of 'accommodation', saying that God modifies his language for his audience, and it is for us to make out the true rather than simply the surface meaning (McMullin, 1985; Ruse, 2001).

In the light of this attitude, it is no surprise that when, in 1859, Charles Darwin published his great evolutionary work, *On the Origin of Species*, the main objections from Christian believers had little or nothing to do with such issues as the 6 days of creation or the Flood of Noah – almost everyone knew that these were metaphorical at best. Rather, the concern was with such issues as the status of humankind, given evolution. Could one still, for instance, think that we humans were made in the image of God, in the light of the fact that we are apparently simply modified monkeys? And even these issues rapidly became less troublesome and threatening as believers of many denominations – Catholic and Protestant – agreed that the really fundamental matters for the faithful, such as the immortality of the soul, have little or nothing to do with empirical science (Moore, 1979).

There was one part of the Western world, however, which stood apart from the mainstream, where religion took an idiosyncratic and simplistic (if understandable) turn. The American South, beaten in the Civil War and distrustful of and hostile toward the ideas and fashions of the northern states, especially those threatening the old comfortable ways of living and thinking, increasingly turned inward. It moved away from the new science of the age and from those elements of religious thought that accepted and indeed welcomed such science and all for which it was thought to stand. Thus there was an increased reliance on the Bible taken literally and, since evolution was both identified with the thinking of the northern states and seen as antithetical to true religion, Darwin's ideas soon took on the mantle of heresy, false to true science and dangerous to true religion. Teachers started to lose their jobs for teaching that the earth is ancient and that Adam may not have been the first man.

After the First World War, with the enactment of Prohibition and thus emboldened by their success in the fight against alcohol, these antievolution, religious enthusiasts – who became known as 'fundamentalists' – pushed for the passage of laws forbidding the teaching of natural origins. One of the states where the biblical literalists were successful was Tennessee. This led, in 1925, to the notorious Scopes trial, where a young teacher (John Thomas Scopes) was brought to task for teaching evolution. Prosecuted by three-times presidential candidate William Jennings Bryan and defended by noted agnostic and deadly advocate Clarence Darrow, Scopes was eventually found guilty, although the very light fine of $100 was overturned on appeal. But the true drama of the story was the defense of the Bible given by Bryan and its literal interpretation, made a mockery by Darrow. The evolutionists lost the case but they won the battle as – thanks particularly to the savage reporting of *Baltimore Sun* reporter H. L. Mencken – the whole of America laughed at the outmoded arguments of the backward-looking critics of modern science. Bryan boasted that he took his stand on the Rock of Ages rather than the age of rocks, and that about summed things up (Ginger, 1958).

Evolution triumphed in Tennessee, and the 1930s and 1940s were to see magnificent forward moves in our understanding of the evolutionary process. In 1937, the Russian-born, American-residing geneticist Theodosius Dobzhansky published his *Genetics and the Origin of Species*, and a whole new era of research was opened up. Yet, paradoxically, in the years following Scopes' trial, in the classrooms of the USA, evolution withered and died. Concerned about sales, publishers pressured their authors, so that increasingly textbooks suppressed reference to evolution, and Darwin's place as a seminal figure in Western thought was denied or ignored. Moreover, this was a state of affairs that lasted right through to the 1950s, when the Russian success with Sputnik spurred a push to revitalize American science education and finally a whole new generation of evolution-friendly biology textbooks began to appear.

However, with the textbooks came a revitalized opposition to evolution, epitomized by a work written by a hydraulic engineer (Henry Morris) and a theologian (John Whitcomb). The bible of the antievolutionist movement, *Genesis Flood*, appeared in 1963. Here was the definitive case for 'Creationism', an absolutely literalistic reading of Genesis: 6 days of creation, 6000 years ago, with miraculous appearance of plants and animals, humankind last; and a worldwide flood at some point later, a deluge that destroyed all but a few living beings. The politics of the situation had changed from those in Tennessee in 1925. The Supreme Court of the US had ruled that evolution may be taught in state-supported schools and that, to the contrary, religion may not be taught in such schools. Much therefore was made of the supposedly scientific case for Creationism or for the movement under the favored alternative and persuasively defined name of 'creation science'. Gaps in the fossil record were given prominent treatment, as were the supposedly inadequate ways in which conventional scientists date the age of the universe.

But, for all its pretensions as modern science, in major respects *Genesis Flood* has its roots in the past as much as the present, for it draws heavily on the ideas of a Seventh-Day Adventist, the Canadian George McCready Price (1870–1963). For the Adventists, a literal 6 days of creation (with the same length of day as experienced by us) is a crucial part of their theology, given the high place they give to the final day of rest. Price (who had no geological training) had therefore long been advocating a very literalist reading of the Bible, and he had given much time and effort to the forcing of the empirical facts to fit the picture. As with many dispensationalist sects (believing in epochs that ended with major upheavals), Armageddon figures highly in Adventist theology, and one finds a corresponding (and complementing) fascination with the earlier time of worldwide destruction, Noah's Flood. For Price, as for later creationists, including Whitcomb and Morris (1961), the sequential fossil record is confirmation of the Flood, being not a record of life through long eons of time but an artifact of the rising waters – the dinosaurs are found below the mammals because, being large and cumbersome, they were caught further down the mountainsides than the more nimble, warm-blooded denizens of the earth, who scrambled up to the mountain peaks before they were trapped by the ever-increasing waves.

Led by Henry Morris and his assistant Duane T. Gish – they became masters of the debate, taking on evolutionists and delighting massive audiences of Christian sympathizers – the new Creationism gained an increasingly high profile. Eventually, sympathetic legislators introduced bills to mandate the teaching of creation science along with evolution in the classes of the American Union. Arkansas in 1981 was the place and date of the ultimate showdown, where the American Civil Liberties Union – the organization dedicated to the upholding of the US Constitution – brought suit against a law that insisted on the teaching of literal Creationism alongside evolution. And again the nation's eyes were fixed on a southern courtroom, although this time the judge found decisively against Creationism, ruling its aspirations to becoming part of biology curricula to be unconstitutional (Gilkey, 1985).

Even as we move into the twenty-first century and a new millennium, Creationism has not vanished. Indeed, the last decade of the twentieth century has seen Creationism flourishing as never before. A new generation of enthusiasts – evangelical Christians as before, but better educated and more sophisticated in argument – has attacked evolution and argued that origins, including human origins, stand outside the normal course of nature. Initially modern Creationists – led by Phillip Johnson (1991), a lawyer, Michael Behe (1996), a biochemist, and William Dembski (1998), a philosopher and mathematician – contented themselves with attacking evolution. They brought forward spruced-up versions of supposedly empirical arguments that appeared in *Genesis Flood*, in Price's writings before this, and indeed in the writings of antievolutionists right back to the days of Darwin himself: If evolution be true, why does no one ever see it in action? How did life start? Could the inorganic really turn into flourishing life, without the intervention of a divine artificer? Why are there so many gaps in the fossil record, and where are the necessary intermediates? (Archaeopteryx, the ideal 'missing link' between the reptiles and birds, apparently counts for little) how could chance make the sophisticated functioning of the hand and the eye? And so forth.

But then, perhaps spurred by the objections of the evolutionists, increasingly the new Creationists mooted, formulated, and promoted an alternative form of origin – an alternative form that meshes smoothly and comfortably with a very literalistic reading of the early chapters of Genesis. In the eyes of today's antievolutionists, the adaptive complexity of the living world is simply too subtle to be the product of blind law. In its intricate functioning nature, it apparently exhibits 'irreducible complexity'. And if blind law is inadequate for the job, what alternative is there? None other, argue the Creationists, than some form of 'intelligent design'. There had to be an intervention – or interventions – in the normal course of nature, that brought on the true marks of the living. There has to be more than the inexorable grinding of soulless laws of regularity. Hence, normal science is not enough: the world shows the results of the direct intervention of a conscious intelligence.

Expectedly, the intelligent-design theorists have been criticized by conventional scientists, no less severely than were the earlier generations of Creationists (Miller, 1999). The notion of irreducible complexity is argued to be conceptually confused and empirically unwarranted – the examples used by recent Creationists prove readily decomposable and explicable in orthodox evolutionary terms. When we cannot see how something occurs, it does not follow that nature is likewise limited. The evolution of the eye seems impossible, until you put in sequence the line of existing eyes, from little more than vague awareness of light and dark right through to the sophisticated eyes of humans and hawks. Moreover, the very idea of design itself, much out of step with the whole of modern science, is not needed and raises as many problems as it supposedly solves. Is the design something which happened in the distant past, held in check until needed? How could this holding operation have operated in any organisms, let alone those more primitive than modern forms? Is the design rather something that was inserted in the recent past, or even in the present, in which case why do we not have direct evidence of its occurrence?

Truly, however, as with early Creationists, although the surface arguments are scientific (or more accurately, pseudoscientific) – promoted as much to evade the US Constitution's ban on the teaching of religion in state schools as for any genuine scientific motives – what drives the antievolutionists of today is as much a matter of morality and culture as of anything purely empirical. Evolution is seen as representative of the sexual, political, and moral laxity of the present day, and the hope is for a return to a more conservative, biblically based ethic. Although it is a parody of history and of the rather mild, liberal, commonplace moral beliefs of most of today's scientists, repeatedly evolution is accused of promoting violence, pornography and family destruction. Therefore, given the ongoing social and political triumphs of the religious right in the United States, it is likely that, for all of the scientists' refutations, we shall see the continued existence – perhaps even flourishing – of the Creationist movement for many years to come (Ruse, 2000).

References

Behe M (1996) *Darwin's Black Box: The Biochemical Challenge to Evolution*. New York: Free Press.

Dembski W (1998) *The Design Inference: Eliminating Chance through Small Probabilities*. Cambridge: Cambridge University Press.

Gilkey L B (1985) *Creationism on Trial: Evolution and God at Little Rock*. Minneapolis, MN: Winston Press.

Ginger R (1958) *Six Days or Forever? Tennessee v. John Thomas Scopes*. Boston, MA: Beacon Press.

Johnson P E (1991) *Darwin on Trial*. Washington DC: Regnery Gateway Publishers.

McMullin E (ed.) (1985) *Evolution and Creation*. Notre Dame, IN: University of Notre Dame Press.

Miller K (1999) *Finding Darwin's God*. New York: Harper and Row.

Moore J (1979) *The Post-Darwinian Controversies: A Study of the Protestant Struggle to Come to Terms with Darwin in Great Britain and America, 1870–1900*. Cambridge: Cambridge University Press.

Ruse M (2000) *The Evolution Wars*. Santa Barbara, CA: ABC-CLIO.

Ruse M (2001) *Can a Darwinian be a Christian? The Relationship between Science and Religion*. Cambridge: Cambridge University Press.

Whitcomb J C and Morris H M (1961) *The Genesis Flood: The Biblical Record and its Scientific Implications*. Philadelphia, PA: Presbyterian and Reformed Publishing Company.

32
Genetics, Reductionism and Autopoiesis

Steven P. R. Rose

Consider a 'simple' observation: a frog jumping into a lake. How should biologists explain it? A physiologist sees a linear chain of events: some processes in the frog's brain result in impulses passing down a motor nerve to the leg muscles, which contract causing the jump. Developmental biologists, ethologists and evolutionists offer wider causes. How the frog jumps depends on how nerve and muscle have been wired up during development; why the frog jumps may be because it has seen a snake in a nearby tree and is attempting to avoid it; this avoidance reaction has been evolutionarily selected as an aid to survival. Molecular biologists offer a very different explanation: the frog jumps because its muscles contract; these contractions are the consequence of the molecular construction of the muscle, that is, the actin and myosin sliding filaments of which it is composed. Actin, myosin and the cellular organization in which they are embedded are the products of specific structural and regulatory genes.

There is nothing 'wrong' with any of these explanations (Rose, 1998). They are all valid ways of looking at the phenomenon, and which one a biologist will use will depend on the purposes for which the question is being asked. However, there is a powerful tradition in science that urges us to believe that the molecular, genetic explanation is somehow more fundamental, more 'scientific', than the others. As it has been put: physiological or developmental accounts are merely descriptions, whereas molecular accounts are explanations. This tradition, or philosophical position, is known as reductionism. It seems so obvious to many natural scientists that we take it almost for granted, even though for many, if not most biological questions, reductionism is of little help. The origins of reductionist thinking lie in the birth of Western science, or rather physics, in the mid-seventeenth century. Physics revealed 'laws of nature', a world composed of elementary particles without color or form, interacting via fundamental forces. For physicists, everything we observe around us in our sensory world, from the solidity of the ground to the jumping of the frog, is a mere illusion derived from the interactions of such particles.

It is the task of natural sciences to 'reduce' all such observations, the subjects of the sciences of chemistry, biology, psychology, even the social sciences, to the laws of physics. Indeed, practitioners of such 'soft sciences' as biology and psychology are sometimes described as suffering from 'physics envy' in their inability to force the phenomena they study into a molecular straightjacket.

Nowhere has this reductionist enthusiasm been more obvious than in the scientific claims made concerning the implications of sequencing the human genome (Malik, 2000). The metaphors – the 'code of codes', the 'human blueprint', the 'book of life', the 'human instruction manual' – spell it out. Not merely are there supposed to be genes 'for' every aspect of bodily construction, but also for such more nebulous aspects of human social life as intelligence, sexual orientation, susceptibility to drug abuse, aggression, religiosity, voting behavior and tendency to mid-life divorce (Ciba Foundation, 1996). When the leader of the public sequencing project, Francis Collins, spoke in a press interview in February 2001 of how in the future people would be able to carry their genetic code with them on a digital video disk (DVD), enabling a competent doctor or geneticist to foretell their future fate, he played precisely to this view of the power of the genes. It was, oddly, left to Craig Venter, of Celera Genomics, the driving force behind the rival, private, sequencing effort, to claim in another interview (McKie, 2001) that the finding that the human genome contained only 30 000 genes should dispel such genetic determinist visions. Just why this should be truer for 30 000 genes than it was when the betting was on the genome containing 100 000 genes is, however, obscure.

The origins of such genetic determinist thinking lie deep within the history of genetics itself. For Mendel, the appearance of green or wrinkled peas (the phenotypes, as they were later known) was the consequence of the presence in the pea plant of 'hidden determinants'. Later it became customary for geneticists to speak of 'genes for' particular phenotypes, from eye color to Huntington disease. In the 1930s, Beadle and Tatums' famous 'one gene, one enzyme' formulation was merely a way of rewriting the simplistic 'one gene, one phenotype' concept in biochemical terms. Of course, every geneticist knows that 'gene for' talk is at best a shorthand. There is not really even a 'gene for' such a simple trait as eye color. Even ignoring the fact that eye color can only be expressed in an eye constructed by the concerted action of tens, if not thousands, of genes during development, there is a difference in the metabolic pathways that lead to blue compared with brown eyes, as in the latter there is a specific enzyme missing; a gene 'for' blue eyes means 'one or more genes in the absence of which the metabolic pathway that leads to pigmented eyes terminates at the blue-eye stage'. This formulation reintroduces what has been lost from the simple genetic shorthand: the concept of development.

One of the problems is that the word 'gene' has taken on many different meanings. For molecular geneticists a structural gene is a strand of deoxyribonucleic acid (DNA), perhaps interspersed with introns, perhaps alternatively spliced, which, under appropriate regulatory conditions, can be transcribed and edited by the enzymic machinery of the cell to result in the synthesis of a particular protein. In this sense, a gene is neither self-replicating nor some sort of 'master molecule', it is an information store on which the cell can draw during development and in its normal function. By contrast, for evolutionary biologists, a gene is an abstract accounting unit, a formalism that determines phenotypic properties that can be subject to selection; indeed evolution is these days defined not in terms of phenotypic change but as a change of gene frequencies in a population (Singh et al., 2001).

It is this population, rather than molecular, view of genes that enables writers such as Richard Dawkins (1982) to speak of genes for 'bad teeth' or 'altruism', and that has led to so much confusion, and for psychometricians and behavior geneticists to contemplate genes for complex social behaviors. But once we recognize that there is no one-for-one relationship between a gene in the molecular biologist's sense and any complex phenotype, it becomes clear that such gene talk is guaranteed to confuse. First, the expression of any gene is environment dependent; that is, genes show a norm of reaction to the environment. The classic example is the mutation that gives rise to the inability to metabolize phenylalanine, and the irreversible mental retardation associated with the resultant phenylketonuria: change the environment by removing phenylalanine from the diet, and the infant develops reasonably normally. Second, no gene acts in isolation from the rest of the genome; in this sense the environment for any gene includes all other genes (Lewontin, 1993). Thus, even in the case of the dominant autosomal mutation that results in Huntington disease, the severity of the disease varies because the gene product, the protein huntingtin, acts in interaction with many other receptor and related proteins, all themselves of course gene products (Cattaneo et al., 2001).

Third, and perhaps most importantly, most complex phenotypic traits are the consequence of developmental interactions between genes, genomes and the multiple levels of internal and external environment. Clearly it is impossible to specify the hundred billion nerve cells or their hundred trillion synaptic connections genetically; one is dealing here with combinatorial interactions. That is why attempts to partition out contributions of genes and environment, nature and nurture, misunderstand such relationships. It is more helpful to understand developmental processes as ones of autopoiesis, in which organisms, and perhaps especially humans, actively construct themselves out of the raw materials of their genes and environment (Oyama, 1985).

In their working practices, all biologists surely understand this. So it becomes interesting to ask why there is so much enthusiasm for such metaphors as those of the 'selfish gene', and especially for trying to explain human behavior and social organization in genetic terms. Partly, the enthusiasm is a consequence of the dramatic expansion of genetic knowledge over the past decades, and with it the, as yet unproven, claims of genetic manipulation to cure disease and enhance life. It perhaps flows partly from a recognition that the world we live in is beset by problems (of social injustice, sexual inequality, racial tensions, ethnic wars), which we seem unable to resolve by the efforts of politicians or social scientists. Maybe, therefore, the world has to be unequal and unjust, because it is written in our genes. In this sense, genetic determinism would suggest, we may not live in the best of all imaginable worlds, but we live in the best of all possible ones; to change it for the better will go against our genetic human nature (Rose and Rose, 2000). But to believe this is profoundly to misunderstand how genes function and their role in autopoiesis. It is precisely in our nature, enabled by our genes, to create different and better societies and qualities of life. I am not sure that this is what Venter meant, but it is how we would do well to interpret him.

References

Cattaneo E, Rigamonti D, Goffredo D, et al. (2001) Loss of normal huntingtin function: new developments in Huntington's disease research. *Trends in Neuroscience* 24: 182–188.

Ciba Foundation (1996) *Genetics of Criminal and Antisocial Behavior*. Chichester, UK: John Wiley & Sons.

Dawkins R (1982) *The Extended Phenotype*. London: W H Freeman & Co.

Lewontin R C (1993) *The Doctrine of DNA: Biology as Ideology*. Harmondsworth, UK: Penguin.

Malik K (2000) *Man, Beast and Zombie: What Science Can and Cannot Tell Us about Human Nature*. London: Weidenfeld and Nicolson.

McKie R (2001) Revealed: the secret of human behaviour. The *Observer*, London, February 11.

Oyama S (1985) *The Ontogeny of Information*. Cambridge: Cambridge University Press.

Rose H and Rose S (eds.) (2000) *Alas, Poor Darwin: Arguments Against Evolutionary Psychology*. London: Jonathan Cape.

Rose S (1998) *Lifelines: Biology, Freedom, Determinism*. Harmondsworth, UK: Penguin.

Singh R S, Krimbas C B, Paul D B and Beatty J (2001) *Thinking about Evolution*, vol. II: *Historical, Philosophical and Political Perspectives*. Cambridge: Cambridge University Press.

33

Personal Identity: Genetics and Determinism

Ruth Chadwick

Introduction

The question of what constitutes personal identity is not a new one, but it has been given a new twist in the light of developments in human genome analysis. Under one interpretation, the analysis of the genome will reveal everything there is to know about human beings. There is a distinction to be made between personal identity of the individual and collective identity, whether at the level of the species or population group or subgroup. There is another distinction to be drawn, however, between personal identity in a deep sense and in a superficial sense. Sometimes when we say that X 'is not the same person', we simply mean that we notice some superficial change in the way X appears or behaves, while holding that in some deep sense the 'essence' of X still goes on. However, this might be precisely what is in dispute, with regard, for example, to persons who are in a persistent vegetative state: Should we say that X still exists, in a deep sense, though in a radically different state of being, or that X no longer exists, that the 'essence' of X is gone? There is a more radical view, which is that there is no such thing as the 'essence' of a person, no 'core' that makes one 'me', and that personal identity is a matter of degree or has to be constructed.

Physical account

Attempts to establish wherein lies the essence of the person may be broadly divided into the physical and the psychological. A physical account of the person depends on spatiotemporal continuity of the physical aspect of or some part of the person. In the pre-genome era, the most popular version of the physical account was the view that a person's identity consists in his/her brain. A problem with such an account concerns the possibility of division and of how to account for that; for example, what would be the correct description of the situation in which a person's brain were divided into two and each half transplanted into a different body? Such thought experiments have generated philosophical puzzles (Parfit, 1984). A new version of physical account, however, has been the interpretation of personal identity as consisting in the genes. The apparent advantage of genes as the guarantors of personal identity in some deep sense is that, while all the cells of a human body die and are replaced over time, the genes persist throughout the life of a human being in every cell of the body, even in the case of persistent vegetative state. The disadvantage, however, is that insofar as this is a physical account, it is subject to the problems to which all physical accounts of personal identity are subject, namely the problem of how to account for division. Whereas the philosophical puzzles generated by the possibility of dividing a brain and transplanting the two halves into different bodies are thought experiments, we know that the deoxyribonucleic acid (DNA) in one fertilized egg can and does split into two when identical twins are formed, and this has led to difficulties about establishing when a person's life story begins (Williams, 1990).

Psychological account

On the psychological account of the person, a person's identity consists in their mental life and this guarantees their continuity over time – through memory, for example. (It is the importance of a person's mental life that has led to the popularity of the brain as a physical criterion.) It might appear that if a person's identity lies in their psychological attributes, then genes have less of a role to play than on a physical account. What would be important, however, would be the extent to which genes or certain genes play a role in the development of these attributes. Here the issue of genetic determinism arises, the question being to what degree our genes determine the kind of character and psychological characteristics that we develop. Arguably the genetic complement in the embryo is a necessary condition for the development of the person and their psychological states, but it does not follow that it fully determines that development or that genetic reductionism is correct. A belief in freedom and responsibility of the individual would be compatible with a view of the individual's genome as setting the parameters within which the person develops (Miller, 1998).

A soul equivalent

Perhaps, however, the attraction of DNA as a candidate for the 'essence' of the person is as something over and above the set of physical or psychological attributes that constitutes the person as we encounter him or her. It has been suggested that DNA has taken on a soul-like role as the location of the essence of the self (Nelkin and Lindee, 1995). It has also taken on the role of guarantor of immortality. We find a degree of immortality in passing on our genes to our descendants. Reproductive cloning offers a vision of passing on all of them, instead of only half, and thus may appear more attractive to those in search of continued existence. Unlike a soul, however, DNA has a physical existence. What makes DNA attractive as a soul-like candidate is its double aspect, as both physical matter and as 'information'. Whereas we know that all the cells of the body are replaced over a period of years, in each new replacement cell the nucleus will contain the same set of genes. It is this that makes cloning by somatic cell nuclear transfer a possibility and which has led to the metaphor of the 'blueprint'.

Identity as a construction

There is an another view that there is no such thing as the core or essence of the person which guarantees that X is the same person throughout all the changes and experiences of life. That there is such an entity as the soul cannot be established, and genes are subject to mutation and change. It is not clear, therefore, that the question about where, if anywhere, the essence of the person is located can be answered.

So why, if at all, is personal identity important? What matters is arguably how we think of ourselves; how we construct our identity. This happens both at the individual level and the collective level. Our genes may play a very important role in our sense of who we are. This phenomenon can be seen in the individual case, for example, in disputes about genetic parentage, and at the collective level in establishing ideas of community which have some genetic basis – this might be an ethnic group or a disease group, for example.

Identity over time

Another way in which personal identity might matter practically relates specifically to identity over time, as when we ask whether the person in the dock is the same person as the person who committed atrocities 50 years ago, with a view to ascertaining whether he or she should be punished. DNA fingerprinting has an increasingly prominent role to play in this process, but using this technique as a means of identification does not of itself establish that the essence of the person consists in the DNA, just as it does not follow from the use of fingerprints as a means of identification that 'I' am my fingerprints.

There might, however, be other questions about identity over time. If there is some relation between genes and personal identity, then the question arises as to how much change in the genome would be possible, for example, by gene therapy, without bringing about an identity change (Elliott, 1993). There are different possibilities: (1) any change in the genome brings about an identity change, (2) a change in a certain proportion brings about an identity change and (3) a change in a certain key part brings about an identity change. If, on the other hand, personal identity does not consist in (some aspect of) a person's genome, then it would in principle be possible to bring about a change in the genome without that constituting a change in the identity of the person.

To a certain extent these questions await further empirical evidence about the genetic basis of certain human traits; but the issues concerning how humans regard themselves, and the philosophical questions concerning free will and determinism in human life will remain, because the question is whether, even if every event has a cause, human free will and responsibility are still possible.

References

Elliott R (1993) Identity and the ethics of gene therapy. *Bioethics* 7: 27–40.

Miller III H (1998) DNA blueprints, personhood, and genetic privacy. *Health Matrix* 8: 179–221.

Nelkin D and Lindee SM (1995) *The DNA Mystique: The Gene as Cultural Icon*. New York: W H Freeman.

Parfit D (1984) *Reasons and Persons*. Oxford: Oxford University Press.

Williams B (1990) Who might I have been? In: Chadwick D, Bock G, Whelan J, et al. (eds.) *Human Genetic Information: Science, Law and Ethics*, pp. 167–179. Chichester, UK: John Wiley.

34
Racism, Ethnicity, Biology and Society

Hannah Bradby

Pitfalls in using the terms race and ethnicity

Race and ethnicity are often used as synonyms to refer to difference between groups of people. Race tends to be used for difference that can be read from appearance, and is assumed to concern phenotype or physical difference with a biological basis. Ethnicity is used to refer to those aspects of shared group identity that are cultural, such as language, religion, marriage patterns, dress and food. The assignment of race categories by observers contrasts with the greater emphasis on the self-assignment of ethnic categories. It is sometimes assumed that ethnic groups are simply subdivisions of the broader racial groupings with the (problematic) implication that biological differentiation can never be overwritten by cultural demarcation.

In genetic terms it is widely agreed that there is no such thing as race: humanity cannot be divided into distinct groups on the basis of the genetic variation that corresponds to the typologies defined by race scientists. Humans are a fairly homogenous species in terms of genetic variation (Jones, 1981), with about 85% of identified human genetic variation accounted for by differences between individuals and only 7% due to differences between the so-called races (Lewontin, 1993). Although these differences can be useful in tracing human evolutionary history, two individuals who share a skin color, hair and nose type that would place them in a single racialized category may be no more similar in genetic terms than two individuals from different racialized groups. This seemingly simple point, which is largely undisputed in scientific discussion, underlies misunderstandings of the nature of differences between ethnic groups in many discourses.

The consensus over the nonexistence of a genetic basis for categorizing humanity into discontinuous races has led commentators from both sociological and biological backgrounds to suggest that the term 'race' has no place in scientific vocabulary. To utilize the term race is to perpetuate and confirm a notion that was conceived as part of a racist and sexist scientific project and which has been utilized in research that has denigrated and exterminated groups defined as inferior. Presumed inferiority in the hierarchy of races has underpinned attempts to exterminate complete groups (such as the Jews and the Roma in the Holocaust) and to prevent or hinder the reproduction of others. Justified by and justifying the extreme denigration of racialized groups has been cruel and abusive research that often resulted in deaths. For example, scientific research under the Nazis experimented on human beings who died as an inevitable side effect of investigations. Although not necessarily the primary aim of the experiments, the deaths of denigrated group members were a result of the racist eugenic ethic that informed the research, and this ethic was in turn reinforced by the experimental findings.

Eugenics and racism were embedded in scientific thinking until after the Second World War, when the scale of the Holocaust was acknowledged publicly and the consequences of a racist categorization of humanity and a strong eugenic science became apparent. Eugenics as a science was largely abandoned, but racist practice in science was not. For example, in 1932 in Macon County, Alabama, the infamous Tuskegee syphilis experiment was started in which public health service physicians withheld treatment from infected men and observed the course of the disease. All of the infected men were Black, and by 1972 over 100 had died despite the widespread use of penicillin by 1946. The deaths associated with medical research in the Tuskegee experiment suggest that despite the public documentation of the horror of the Holocaust, science is not immune from the prejudices that justify discrimination elsewhere in society. Racism constructs certain categories of people as inferior and, in extreme cases, as unworthy of preservation. The consequences of these attitudes in terms of the neglect and abuse of those deemed unworthy have been documented in science as in other spheres of life.

Although race is no longer current as a scientific analytic category, and it is rare to hear public support for the existence of a biological hierarchy of races, the persistence of race categorization is remarkable. There is an argument to suggest that the continuing existence of racism means that race has a social reality that cannot be ignored in social science. The problem with retaining the vocabulary of race for describing and analyzing social variation, even if it is conceived as social in origin, is that notions of essential difference and biologically based hierarchy have proved impossible to jettison.

Research that uses a racialized category, albeit for the best motives, contributes, albeit unintentionally, to the idea that the differences between groups are immutable, essential and relevant to other aspects of human diversity. It is difficult to avoid this implication when discussing difference between races, because the concept of race was developed as a hierarchical model with types of northern European (the progenitors of race and eugenic 'sciences') at the pinnacle. The influence that the sciences of race and eugenics had on Western thinking is considerable and still being teased out.

One example of the penetration of eugenics into current scientific thinking is the use of the term Caucasian, usually as a synonym for White in scientific papers. Caucasian was a term coined by Blumenbach when he was engaged in a project of measuring skull volumes to demonstrate racial hierarchy. People from the Caucasus were deemed to have the biggest brain volume and therefore represented the purest form of 'the most beautiful race of men' (quoted in Jones, 1994). It seems unlikely that researchers who currently use the term intend to mean that the so-called Caucasians have bigger brains and so are superior to comparative groups. The lack of anyone's positive identification with the term 'White' and the term's apparent scientific respectability shown by its widespread use, seem to override the disadvantages of using a category that has little geographical, historical or social merit as a descriptor of populations of northern European descent. It is an institutional inability or unwillingness to think through the implications of using such terminology, rather than individualized racism that is the problem.

Simply abandoning the vocabulary of race would go some way towards addressing the problem of the extensive permeation of racialized models in Western thought. By deliberately avoiding the term race, researchers and commentators are forced to define with greater precision the dimensions of human variation under consideration. For example, one research project might investigate the effect on mental health of being subject to racism owing to skin color and hair type among young men of African descent. Another project might look at the effect of race on mental health. The second project is much more likely to conflate biological and social effects through a lack of linguistic and conceptual clarity. Disciplined use of language makes it more difficult to imply (but not show) that 'race', 'racial difference', 'racial stock' or 'inherited difference' is responsible for an observed difference in some measured outcome.

However, avoiding the term race does not, unfortunately, avoid racialized thinking altogether, because 'ethnic' is so often used as a euphemism for race. The term ethnic (without reference to majority or minority) is used to mean alien and/or exotic as in 'ethnic food' or 'ethnic clothes'. The term can take on a more pejorative tone such as in the description of people as 'ethnics', meaning 'not like us' (but leaving the membership of the group of 'us' unspecified). Widespread agreement that, in theory, ethnicity refers to a shared cultural or religious identity has not prevented the term from picking up baggage from racialized thinking. The slippage between the vocabulary of race and that of ethnicity has impeded the development of means of conceptualizing human diversity without recourse to essentialism. Use of the unwieldy hybrid 'race–ethnicity' perhaps indicates the reluctance to jettison a dimension of essential difference related to biological inheritance when referring to cultural difference.

Ethnicity can be defined as the real or probable, or in some cases mythical, common origins of a people with visions of a shared destiny that are manifested in terms of the ideal or actual language, religion, work, diet or family patterns of those people (Bradby, 1995). In this definition, ethnicity is a characteristic of groups and individuals that is contingent, labile, hybridizing and complex. As with any complex theoretical concept that is subject to empirical research, however, a measurable proxy must be identified that will inevitably encapsulate rather less of the contingent, negotiated complexity involved. The particular difficulty with operationalizing ethnicity as a research variable is the existence of the commonsense notion of race that is often assumed to be easily measurable by any observer.

The conceptualization of ethnicity has not been decoupled from racialized thinking, as illustrated by the comparison of ethnic groups with Caucasians in contemporary research. The persistence of racialized thinking, sometimes under the guise of the vocabulary of ethnicity, underlies some of the well-documented abuses of research into human diversity of a cultural nature and health.

Attention to terminology is thus more than just a matter of avoiding the offence of individual sensibilities. Particularly in the field of genetics, precision and clarity in defining the dimension of human difference that is under consideration are crucial. The danger of research and practice informed by models of human diversity that are based on racialized categorizations, whether expressed in terms of racial or ethnic groups, is that they invite essentialist interpretation. The expression of racialized ideas in the language of ethnicity can impede the identification of problematic underlying assumptions.

Racialized thinking in research and practice

Sickle cell and anti-African racism in the United States

When a screening program for sickle cell anemia was introduced in 1972 in the United States, the genetic basis of the disease was well established, including an understanding of the environmental conditions that lead the recessive gene responsible for the condition in the homozygous state to increase in frequency in malarial areas. The screening program was initially welcomed by Americans of African descent as a means of addressing a long-neglected health problem, but it soon became clear that its effects on people's life chances were going to be largely negative. Sufferers (homozygotes) and carriers (heterozygotes) of the sickle cell condition were identified and treated in a way that has been described as both stigmatizing and punitive (Duster, 1990). For example, being a carrier of the trait resulted in termination of employment, increased insurance premiums and delay in adoption proceedings.

Although irrational and morally unacceptable, discrimination against those with the disease is predictable. Discrimination against people with many different illnesses, including hemoglobinopathies, in employment and insurance is justified in financial terms owing to the burden of supporting an individual who may have a greater than average need of healthcare. Regardless of whether this justification is acceptable for those who are homozygous for the sickle cell gene, it does not apply to heterozygotes because there are no symptoms or health penalties associated with carrying a single copy of the gene. Indeed, it confers some health benefits in those areas where malaria is endemic. In seeking to explain the punitive and stigmatizing aspects of this screening program, the anti-African racism that underpinned slavery and a highly segregated society cannot be ignored. When the screening program was instigated, the segregation of people of African origin in public life was very recent history. The systematic exclusion of people from, for example, particular institutions of learning on the basis of their skin color and hair type rested on an assumption of inferiority in various respects: intellectual, moral and cultural. Evidence of a faulty gene, even if it was only a single copy of a recessive gene, in the context of a long history of presumed inferiority of 'African racial stock', reinforced and perpetuated existing racist discrimination.

Rickets and racism in Britain

In the nineteenth century, rickets was prevalent among British inner city dwellers living and working in squalid conditions. The observation that many of the sufferers were Irish (labor migrants to the developing industrial centers) led a contemporary commentator to propose that the propensity to rickets and to red hair were linked and could be attributed to the poor quality of the Irish racial stock (Cooper, 1984).

By the late twentieth century rickets had been identified as a disease that results from vitamin D deficiency. Labor migrants from south Asia were, like the Irish nearly two centuries earlier, attracted by employment opportunities in British cities. Having been all but completely eradicated among the general population with the introduction of fortified foods, rickets reappeared among people of south Asian origin. The shock of a nineteenth-century disease of poverty appearing in an industrialized nation was arguably contained by it becoming known as 'Asian rickets'. The transfer of the descriptor 'Asian' from the sufferers to the disease implied that its origins lay with the people themselves. In this case, race was not straightforwardly seen as causing the disease, but rather cultural practices such as diet and clothing were held to deprive Asians of adequate vitamin D (Donovan, 1984). However, the importance of skin color as an indicator of inferiority is seen, despite the mainly cultural causes identified for rickets, in the way that the greater pigmentation of Asian compared with northern European skin was blamed for an inability to synthesize vitamin D in the British climate.

The preponderance of cultural over biological features identified as causing rickets means that it might be included in what has been referred to as the 'new racism'. However, there is very little novelty in the anti-Asian bias of the response to rickets, even if it is culture rather than genetic material that was ultimately held accountable. The identification of cultural practices as being responsible for the disease implied that Asian rickets was essentially different from the non-Asian variety and that south Asians themselves were essentially different from those Britons who had previously suffered rickets.

Can thinking in racialized categories be avoided?

Ethnicity is a complex, contingent and labile characteristic, unlike race, which was conceptualized as essential and immutable. The challenge of considering the implications of ethnicity for health and operationalizing such a complex concept in research has often been sidestepped by falling back on the cruder, familiar, racialized categories, although these might be referred to using the vocabulary of ethnicity.

The necessity of a concise classification for use in the census means that broad classifications are unlikely to disappear in the near future and there is merit in retaining comparability with census data. However, the dangers of reproducing and reinforcing racialized distinctions in the British population continue to be a concern. The *British Medical Journal* has been active in promoting a discussion of these and other issues around appropriate terminology for categorizing ethnicity in research and practice. The best general advice in operationalizing ethnicity is to avoid any assumption that observed difference is due to biological difference and to relate the rationale for any ethnic categorizations to the research question under investigation with great care and precision.

To move away from nineteenth-century racialized models has not proved easy but, given the well-documented dangers of conceiving of human diversity in fixed terms, it is conceptual work that is essential.

References

Bradby H (1995) Ethnicity: not a black and white issue, a research note. *Sociology of Health and Illness* 17(3): 405–417.

Cooper R (1984) A note on the biologic concept of race and its implications in epidemiological research. *American Heart Journal* 108: 715–723.

Donovan J L (1984) Ethnicity and health: a research review. *Social Science and Medicine* 19: 663–670.

Duster T (1990) *Backdoor to Eugenics*. London: Routledge.

Jones J S (1981) How different are human races? *Nature* 293: 188–190.
Jones S (1994) *The Language of the Genes*. London: Flamingo.
Lewontin R C (1993) *The Doctrine of the DNA*. London: Penguin.

Web links

British Medical Journal. http://bmj.bmjjournals.com/
Tuskegee.http://www.healthsystem.virginia.edu/internet/library/historical/medical_history/bad_blood/

Part 6

Reproduction, Cloning and the Future

Introduction

Angus Clarke and Flo Ticehurst

This final section addresses some of the most contentious issues around human genetics. Although many genetics researchers would see a clear distinction between genetics and experimental embryology, the media and the public often elide the two and treat them as intimately connected. There clearly are links between genetics and embryology, but while the issues that appear most regularly in the news relate largely to cloning, projections into the future see the two areas as effectively integrated in the Brave New World of genetically modified embryos.

Morgan and Ford (chapter 35) outline the ethical concerns that, it is often argued, need to be addressed by regulation in relation to *in vitro* fertilization (IVF). They highlight scientific and professional responsibility as another area that should be addressed because modern science and technology have greatly enhanced the power of human beings and their dominion over the world.

The authors describe different approaches to statutory regulation, illustrated with examples. They discuss the way in which the recommendations of the Warnock Committee were developed into the 'radical laissez-faire approach' of the UK's Human Fertilization and Embryology Act of 1990, establishing the Human Fertilization and Embryology Authority to oversee the new reproductive technologies. They proceed to criticize several aspects of the current system of regulation in the UK, including the competing but incompatible considerations and interests between which the HFEA has to adjudicate. They indicate that the lack of national or international consensus on many of the points of principle makes the role of the HFEA very difficult indeed. They argue that legislation may need to be modified and to set a looser framework than currently applies.

Harris (chapter 36) discusses the right or entitlement to reproductive liberty and asserts that there is no consensus as to the nature and scope of this right. He examines other accounts of reproductive liberty and discusses the freedom to pass on one's genes as an important principle. Those who seek to deny the moral claims of others, he argues, must show good and sufficient cause – and he then considers whether it can ever be proper to deny access to artificial reproductive technologies.

Harris then turns to consider pre-implantation genetic diagnosis to select for non-medical traits, arguing that this is not morally problematic *in itself* although some might argue that this could be against the public interest or lead to psychological harms for the resulting child(ren). He asserts that selection *for* particular qualities/traits is not to imply that those individuals without these are of any less worth; although this argument may be rational it may ignore the reality of some emotional responses.

Dickenson (chapter 37) presents two feminist approaches to the new reproductive technologies, which argue for the broader issues to be considered, and not just the consequences for the individuals immediately involved in a particular case. Feminist analysis has been of great importance in bringing into prominence within biomedical ethics the justice-related issues of power, oppression and exploitation. In particular, she argues that it is women whose bodies are invaded and manipulated in genetic testing and in IVF-related procedures.

Dickenson then discusses the differences between feminist and conventional views of the ethical issues arising in these contexts. She points out that it is women who tend to take responsibility within the family for decisions about reproduction and related matters, and manage the sharing of genetic information with relatives. Conventional abstract rules need to be considered within the social context of the person undergoing genetic testing.

Davis (chapter 38) assesses the social, ethical and legal issues relating to fetal sex selection. She argues that these can only be evaluated within the specific cultural and historical context. The motives underlying sex selection are very varied, in response to economic contingencies, social status, cultural practices or a desire for 'family balancing' (itself a thoroughly loaded term) and laws against sex selection simply cannot be enforced.

Feminist arguments against misogynistic sex selection uphold women's right to choose and have sympathy for those forced to select sex in order to secure marriage and family life. More conventional ethical arguments against sex selection have been closely tied with the abortion debate. However, more commentators are coming to regard foetal sex selection as acceptable for non-medical reasons if it is not done in such a way as to devalue women but rather for 'family balancing' (but we do advise caution in using this phrase that sounds so reasonable and all-too-smooth).

Petersen (chapter 39) discusses the media coverage of the creation of Dolly the sheep and how this has drawn public attention to the promises of genetic manipulation and cloning. He also considers projected fears of what this will bring.

The cloning of Dolly was a scientific breakthrough that challenged the belief that a differentiated cell could not be reprogrammed. Polly received much less attention but was not only cloned but also genetically altered so that she would express human clotting factor IX protein in her milk – which achievement was hoped would be one step along the road to a treatment for hemophilia B. It is easy to be distracted from the fact that cloning was performed as a way to achieve a 'Pharming' breakthrough – so Polly was of much more scientific and commercial significance than Dolly.

Political responses to the cloning of Dolly came quickly and moves were made by a number of countries to outlaw human cloning and withdraw funding from research on cloning. Petersen examines the public unease about cloning and the media accounts that obscured the meaning of 'cloning' in this context. Since then, claims from so-called maverick scientists that they have cloned, or are on the way to cloning, humans have served to fuel the media reports with controversy – perhaps heightening public concerns as well. The arguments in this area have led to a focus on the differences between reproductive and therapeutic cloning, and distinguishing between these two applications of the same technology is seen as a way to emphasize the benefits offered by developing cloning techniques for treatment of disease.

Lindsay and Barsam (chapter 40) ask whether cloning experiments and the like are harming animals. They assert that animal welfare is clearly compromised by transgenesis and animal cloning techniques. Furthermore, the animals used in these experiments 'are being used as if they were objects of no intrinsic worth, artifacts, and that human relations with them should be circumscribed by economic considerations alone.' This is a challenge to philosophy – animals can certainly suffer, so an ethical framework such as utilitarianism could be readily extended to incorporate their interests – and is also a challenge to many theological views as well.

Farrelly (chapter 41) describes how new genetic and biological knowledge challenges the notion of distributive justice (the divisions of benefits and burdens in society) and how this raises the question within political philosophy as to which goods the state should ensure are distributed fairly. Currently, he says, the genes we have are a matter of luck – no-one can be held responsible for the unfair distribution of genetic advantage and disadvantage, and there's nothing we can do collectively (as a society) about this. If our knowledge about how genes work increases to the point where we can intervene, however, then this will no longer be the case. In Farelly's view, decisions regarding the regulation of biotechnology will determine who receives the greatest share of benefits.

This raises numerous questions, including: which genetic advantages/disadvantages should be included within distributive justice? to whom should genetic enhancement be made available? and, what weight should we put on the value of freedom in this 'post-genomic revolutionary world'?

Farelly then discusses procreative liberty, which he points out suffered gross violation in the name of eugenics. He suggests that the new genetics raises questions about the weight we must place on this as our ability to intervene successfully in the 'natural lottery of life' increases.

Kolleck (chapter 42) considers problems that may arise when we assess future consequences of enhancement using germ-line alterations. 'Enhancement' can be interpreted variously in different social contexts, but cannot be defined in absolute terms. She asks how the broad implementation of germ-line alterations may interact with society and possibly change it; we can only make reasonable guesses by assuming that future society won't be very different from our own now.

Developments in reprogenetics expands our control over future generations. If germ-line modification is proved safe (at least for the individual directly manipulated) – then other therapeutic uses may come to be considered and this may blur the line between treatment and enhancement. Issues of justice may well then arise – of access to the technology, and of fairness over access to enhancement. Kolleck addresses the question of how behaviour and personality may be modified for political ends or simply to fit in with social values.

One crucial point made by Kolleck is that our knowledge of the consequences of interventions may initially be imperfect, and we may underestimate the complexity of gene-environment interactions and even unpredictability of results. These factors may, in the long term, limit the application of technologies. What damage may be wrought by our hubris before then, however?

35
In Vitro Fertilization: Regulation

Derek Morgan and Mary Ford

Introduction

Societies are in the throes of a 'reproductive revolution'. Assisted conception technologies and procedures present radical challenges to our understandings of the family. New family forms have become available, necessitating a vigorous re-examination of the legal, biological and social constructs of parenthood. As a result, bioethics, law and society each face new challenges, including the examination of whether and how to regulate the reproductive revolution (Lee and Morgan, 2001).

Reasons to regulate

Knoppers and Le Bris (1991) have identified five main ethical concerns in regulation:

- a respect for human dignity
- the need to safeguard human genetic material
- quality control
- the inviolability of the human person
- the inalienability of the human body.

Whether all of these are truly classifiable as 'ethical concerns' is perhaps debatable, and the list is not exhaustive. Another, indisputably ethical, reason can be found in Hans Jonas's argument that the enormously enhanced power which modern science and technology have helped to bring to human beings and their dominion of the world brings with it a change in responsibility. This responsibility is a 'correlate of power and must be commensurate with the latter's scope and that of its exercise' (Jonas, 1984).

Chalmers has proposed that one of the most salient arguments for regulation is the fact that science and technology frequently call for the expenditure of public monies, which in itself would justify the introduction of a regulatory framework. The prevention of harm, particularly in connection with community safety and public health, the welfare of children born of assisted conception technology and the disclosure of information in a democratic society are adduced as further reasons for stipulating regulatory oversight. Finally, because assisted conception involves the medical profession in the creation of children within modern family units, the development and deployment of family law is engaged (Chalmers, 1994).

Different approaches to regulation

Unsurprisingly, many jurisdictions have concluded that some form of regulatory control (usually through specially framed and implemented legislation) is desirable, although the nature of that regulation differs markedly. States take different approaches to the question of where the responsibility for regulation ought to lie. A state may rely upon:

- a private ordering approach, based upon individual control, responsibility and power
- professional self-regulation and control, through the medical profession, local research and institutional review committees
- community control, through national ethics committees and the courts
- legislative and regulatory control
- a combination or blending of one or more of these approaches.

Statutory regulation

Several different legislative models can be identified. These include:

- the 'radical laissez-faire approach' embodied in the UK Human Fertilization and Embryology Act 1990
- the 'cautious regulatory system' created by the Danish legislation of 1987–1997
- the 'prohibitive licensing system' of Austria
- the 'liberal constitutional approach', envisaged by the Canadian Law Reform Commission's proposals on 'medically assisted procreation' (Morgan and Nielsen, 1993).

The model an individual state chooses to adopt will depend upon permutations and nuances in tradition, religion, culture, economics and wealth.

Statutory approaches to regulation have possible advantages, including the potential to form consensus and create other social benefits like certainty and stability. In democratic states, such approaches are also likely to offer political advantages such as public accountability and legitimacy. The benefits derived from the adoption of any particular statutory system of regulation will depend upon, among other things, the specific type of statutory model adopted (see above), and the political climate, culture and traditions of the state in which it is enacted. Disadvantages of statutory regulation might, paradoxically, include either vagueness or precision. An example of the former is the Australian Gene Technology Act 2000 §1923B (definition of cloning 'a whole human being'), while of the latter, the UK Human Fertilization and Embryology Act 1990 §1 definition of 'embryo' and whether it is applicable to cell nucleus substitution where 'fertilization' is never 'complete' because fertilization does not take place.

Additionally, there may be conflicts of interest, difficulties in moving from previous forms of regulation to statutory regulation, problems associated with a lack of moral consensus, such as perceptions of moral hazards or slippery slopes, and claims that the law in question lacks moral authority, and adverse reactions to paternalism and/or compulsion (e.g. compulsory counseling). Again, much will depend on the type of statutory system enacted, and its ability to reflect the prevailing political climate.

Approach adopted in the United Kingdom

In the United Kingdom, the Warnock Committee was established in response to the rapid advance of biomedical technology and its growing use in private clinics. The Committee was charged with considering 'recent and potential developments in medicine and science related to human fertilization and embryology', and the report it published in 1984 proposed a regulatory structure which has formed the basis for the UK approach to regulation ever since. The Warnock recommendations were substantially enacted in the Human Fertilization and Embryology Act 1990, the culmination of nearly two decades of scientific research and ethical and philosophical debate. Section 5 of the Act provided for the creation of a regulatory body, the Human Fertilization and Embryology Authority (HFEA), whose 21 members are chosen for their knowledge and experience in fields including science, medicine, law and religion.

The principal statutory functions of the HFEA include the licensing and monitoring of clinics that carry out certain artificial reproductive procedures for research and treatment purposes, and the regulation of the storage of gametes together with the policing of activities relating to gametes and embryos. It also has several other important roles under the law, including the formal registration of information concerning assisted conception. The Authority operates under a Code of Practice which guides the conduct of licensed activities, and also acts as a source of information and advice about assisted conception for the general public, patients and donors. It is also empowered to advise the Secretary of State for Health on a number of matters relating to embryology; however, the Authority has so far taken a relatively restricted view of its advisory functions.

The 'radical laissez faire' model embodied in the 1990 Act illustrates benefits and drawbacks of the model as it operates within the UK context. Endorsement of the UK approach comes from Margot Brazier, who argues that it ensures public accountability, promotes high standards of medical practice (thereby reassuring those seeking treatment), promotes consensus between regulators, clinicians and scientists, which together work to the benefit of patients, and generally promotes British reproductive medicine (Brazier, 1999). However, the UK approach has been criticized by those who claim that the regulatory body (HFEA) is unduly influenced by scientists working in the field. It has been charged that the composition of the HFEA excludes those who are sceptical of or who have reservations about the wisdom of the 'reproductive revolution' itself. The requirement in the Act to prioritize the welfare of the child appears, it is argued, to be undermined by the HFEA's workload and its ambitious attempt to 'safeguard all relevant interests'. In practice this means mediating between the potentially competing interests of patients, children, clinicians, scientists, future generations and the wider public. It is also said that the HFEA has become 'an arbiter' of fundamental ethical issues on which there is no widespread agreement (Smith and Sutton, 1995). Each of these criticisms must be weighed and judged seriously.

International consensus and harmonization

The variety of approaches to statutory regulation raises the question of whether some sort of international harmonization or convergence is possible or even desirable. Although it is possible to identify areas of general consensus among the different approaches to assisted reproduction, such approaches differ across continents and even within groups of states (such as Europe and the Council of Europe), and it seems unlikely that a significant degree of harmonization will be achieved.

The Council of Europe's Convention on Human Rights and Biomedicine

The Council of Europe's Convention on Human Rights and Biomedicine was adopted by the Committee of Ministers on 19 November 1996 and opened for signature on 4 April 1997. The Convention sets out only the most important principles; additional safeguards and more detailed questions will be spelled out in protocols, of which the first, on the prohibition of cloning, was opened on 12 January 1998. The Convention makes provisions with regard to:

- the priority of the interests and welfare of the human being over those of society (Article 2)
- equitable access to healthcare (Article 3)
- observation of professional obligations and standards in carrying out health interventions, including research (Article 4)
- free and informed consent by a person before an intervention in the health field (Article 5)
- respect for a person's private life in relation to health information (Article 10)
- a prohibition on discrimination on the grounds of genetic heritage (Article 11)
- protection against human genome modification unless for therapeutic purposes (Article 13)
- a prohibition on sex selection, unless for preventing the transmission of a serious sex-linked disease (Article 14)
- only limited research on the human embryo, ensuring where the law permits research 'adequate protection of the embryo' (Article 18)
- a prohibition on the creation of an embryo only for research purposes (Article 19)
- a prohibition on the use of the human body and its parts for financial gain (Article 21).

Some member states (e.g. Germany) have as yet refused to sign the Convention, holding that its provisions are insufficient and some (e.g. the United Kingdom) that it is too stringent and incompatible with their domestic law (in both cases, incidentally, on the question of the status of the human embryo).

Article 29 of the Convention provides that the European Court of Human Rights in Strasbourg may give Advisory Opinions on the interpretation of the Convention, thus beginning the process of tying the Convention to European human rights jurisprudence. This provision presents the most significant opportunity yet to work toward supranational consensus on issues of assisted reproduction and of biomedicine more generally. While there are substantial areas of international consensus on the principles and practices of assisted reproduction, there are also many areas of seemingly unbridgeable difference. Admittedly, the deep ideological differences between different parts of the world will almost certainly preclude the achievement of complete consensus on these issues, but there is still an important practical reason continually to pursue greater harmony between states; as Brazier (1999) notes, when local rules differ

> those wealthy enough to participate in reproduction markets can readily evade their domestic constraints. If I can order sperm on the internet, or hire

> a surrogate mother from Bolivia, are British regulators wasting their time? The international ramifications of the reproductive business may prove to be a more stringent test of the strength of British law than all the different ethical dilemmas that have gone before.

As such, harmonization is potentially a powerful weapon against the growth of 'procreative tourism'. It is still important for individual states to regulate in the absence of the desired consensus, however, since the adoption by one state of its own legislation may mean a greater possibility of influencing emerging supranational agreements. Although it may be true that wise government does not always legislate at the first opportunity, it is perhaps also true that, while it is important to seek supranational consensus, it may sometimes be desirable to act without it, since such action might itself precipitate the desired consensus.

Local differences

Some of what is distinctive or notable among the vast number of state regulations that exist in relation to assisted reproduction may now be observed.

In Greece, for example, parental rights are tied very explicitly to the need for parental responsibility. Moreover, the permissibility of assisted reproduction procedures depends, in part, upon 'prevailing moral conditions'. Both of these features evince the importance of community, and show how communitarian values like social responsibility and shared norms are central to the ethical thinking which informs and justifies the authority of the legal system in Greece.

The German and Austrian approaches are notably restrictive, and Italy has recently considered a similarly restrictive system of regulation. Scandinavian countries such as Denmark, Sweden, Norway and Finland express their scepticism about the general wisdom of reproductive technologies in varying cautious regulatory models rather than in outright prohibitions or liberal acceptance. The level of control in the Middle and Far East varies; Muslim countries mainly ban gamete donation, while Sinic countries mainly allow it. At the liberal end of the spectrum, Hong Kong is considering a model based substantially upon the United Kingdom's 1990 Act.

Despite any potential benefits of global harmonization, the current philosophy in North America would frustrate any such ambitions. Neither Canada nor the United States of America has any national or federal system of regulation in respect of assisted reproduction, with these matters, and healthcare matters generally, reserved (as also in countries such as Australia) to the provinces and the states respectively. Although several draft Uniform Laws have been proposed in the United States by the American Law Institute, these deal predominantly with the family law implications of assisted reproduction technology, like paternity, rather than with the regulation of assisted reproduction itself. The only federal Acts in force deal with 'quality control' matters, requiring the publication of annual reports by clinics, rather than with substantive matters of principle or practice.

There are several reasons for this lack of federal intervention, particularly in the United States. First, assisted reproduction raises ethically controversial issues, and the federal government, mindful of the abortion controversy and its ability to dominate political debate and divide national opinion, may be reluctant to present another potentially divisive issue for public scrutiny. The rate of development of biomedical technologies and the entrepreneurial character of medicine in the United States are additional factors restraining intervention. Finally, the United States Supreme Court has, over the course of its abortion decisions, moved from an individualistic position in Roe to an augmentation of the role of state interest in Casey, indicating that it increasingly favors locating power over such issues with state rather than federal authorities.

The future

How should the regulation of assisted reproduction be developed in order to accommodate new technologies, international differences and oscillations in public opinion, not to mention an absence of ethical consensus? Given the unpredictability of science and the temporality of social mores, perhaps the best that regulators can hope to do is to identify some sound, if basic, principles. For example, attempts at harmonization will probably always be inappropriate, save in the most extreme international circumstances. It is probably also true that wherever the potential for ethical controversy exists, it will be actualized. This means that criticism of regulation on the basis that it imposes one particular ethical view in the absence of popular or academic consensus is probably unavoidable, and as such should not represent a barrier to legislative action.

There is a temptation to believe that recognizing one view at the expense of others is simply the enforcement of moral majoritarianism. Accordingly, legislation is sometimes asked to portray or reflect a weakened rather than an expansive ethical or moral conception. Perhaps ethically based legislation should seek the fulcrum which will ensure a proper balance between a minimum level of humanity and moral standards and freedom for individual men and women to live according to the (possibly stricter) dictates of their own consciences.

References

Brazier M (1999) Regulating the reproductive business? *Medical Law Review* 7: 167–193.

Chalmers D (1994) Government's role in human sexuality and reproduction. In: *Reproductive Medicine: Beyond 2000*, pp. 26–27 (symposium at Monash University, Melbourne, 1994).

Jonas H (1984) *The Imperative of Responsibility*. Chicago, IL: University of Chicago Press.

Knoppers B M and Le Bris S (1991) Recent advances in medically assisted conception: legal, ethical and social issues. *American Journal of Law and Medicine* 17: 329–381.

Lee R G and Morgan D M (2001) *Human Fertilization and Embryology: Regulating the Reproductive Revolution*. London: Blackstone Press.

Morgan D and Nielsen L (1993) Prisoners of progress or hostages to fortune? *Journal of Law, Medicine and Ethics* 21(1): 30–42.

Smith S and Sutton A (1995) *The Human Fertilization and Embryology Authority: A Critique of its First Reports (1992–94)*. London: Comment on Reproductive Ethics (CORE).

36
Reproductive Choice

John Harris

What is medical technology for?

Our question concerns the ethics of using assisted reproductive technologies (ART) to influence nonmedical traits; but this immediately raises another question: when and why is it appropriate to use ART to influence medical traits or for therapeutic purposes? If we know why this might be appropriate then we might have some idea about nontherapeutic uses.

Medical technology, that is the use of technology for medical and therapeutic purposes, is morally justified by the good that it will do, by the importance of that good, and by the fact that it is desired by those whom it benefits or (where patients cannot request or consent to its use) is in their best interests. The urgency of the use of medical technology is proportional to the magnitude of the good that it will do and the use of the technology will be ethical if it will do the good that the patients want or when patients cannot request it, its use will be ethical if it does good and its use is in the patient's best interests. Medical goods are often important because they protect life, ameliorate pain or suffering, restore mobility and so on, all very important goods. However, note that there is an ambiguity in what we mean when we say that the use of technology is justified by the good that it does. There is ambiguity between the question of whether the deployment of public resources for the achievement of the particular good is justified and the question of whether individuals are justified in accessing or others are justified in offering the technology. The use of public resources may be justified if the technology does a good that ought to be done, although pressure on resources will always influence what is actually funded. However, the individuals will be justified in accessing the technology and others justified in providing it if it does no harm or no significant harm, even if there are no moral imperatives for doing that harmless or marginally harmful thing.

The use of medical technology in plastic surgery is a good example here. If the surgery is necessary to prevent or mitigate suffering, then it will be justified both to do it and to spend public resources on it. However, if it is purely a matter of personal preference ('I'd like a smaller nose – but I am not made miserable by my existing one'), then again the deployment of the technology, while not required by morality and while not in the public interest, is not wrongful in any way. However, if the individual wants the slogan 'Death to all Americans' indelibly burnt into her forehead, then although only mildly harmful to herself it is immoral for independent reasons, for example, that it constitutes incitement to racial hatred, is against the public interest and arguably harmful or is a danger to others.

Note again, that questions of scarce resources aside, the provision of services that require technology is not really a moral issue unless some harm can be shown to result from their provision. There is nothing special about 'medical' technology save that it is assumed to partake of the general justifications and 'prestige' of the practice of medicine more generally. Without those general justifications it requires its own. And here the question of with whom the burden of proof lies assumes importance: who has to justify the use of technology and to whom must it be justified?

The presumption in favor of liberty

In the United Kingdom and the United States and in many other democracies, there is a presumption in favor of liberty. This means that the burden of justifying their actions falls on those who would deny liberty and not on those who would exercise it. This idea is constitutive of liberalism and hence of all liberal democracies. Indeed I believe that it is part of all societies that treat people as equals, and hence of all democracies properly so-called; but to argue for that proposition is beyond the scope of this article. If this is right, the presumption must be in favor of the liberty to access ART unless good and sufficient reasons can be shown against so doing. However, suppose this presumption is not accepted. Can anything else be said about the liberty to access ART which might give credibility to a presumption of liberty?

Reproductive liberty

When people express their choices about procreation, even in unusual or idiosyncratic ways, they are claiming an ancient, if only recently firmly established, right of a very fundamental sort. This right is found in all the principal conventions on human rights. It has been expressed as the right to marry and found a family or as the right to privacy and to respect for family life (European

Convention on Human Rights, 1953; International Covenant of Civil and Political Rights, 1976; Universal Declaration of Human Rights, 1978, Article 16).

Some see the right or entitlement to reproductive liberty as derived from the right to reproduce *per se*, others as derivative of other important rights or freedoms. (I defended this liberty, albeit in a somewhat convoluted form, in Harris (1992); for a more explicit and more elegant defence see Robertson (1994).)

There is no consensus as to the nature and scope of this right; however, it is clear that it must apply to more than conventional sexual reproduction and that it includes a range of values and liberties that normal sexual reproduction embodies or subserves. For example, John Robertson outlining his understanding of this right suggests:

> The moral right to reproduce is respected because of the centrality of reproduction to personal identity, meaning and dignity. This importance makes the liberty to procreate an important moral right, both for an ethic of individual autonomy and for the ethics of community or family that view the purpose of marriage and sexual union as the reproduction and rearing of offspring. Because of this importance the right to reproduce is widely recognised as a *prima facie* moral right that cannot be limited except for very good reason.
>
> (Robertson, 1994)

Ronald Dworkin has defined reproductive liberty or procreative autonomy as 'a right to control their own role in procreation unless the state has a compelling reason for denying them that control'.

> The right of procreative autonomy has an important place ... in Western political culture more generally. The most important feature of that culture is a belief in individual human dignity: that people have the moral right – and the moral responsibility – to confront the most fundamental questions about the meaning and value of their own lives for themselves, answering to their own consciences and convictions ... The principle of procreative autonomy, in a broad sense, is embedded in any genuinely democratic culture.
>
> (Dworkin, 1993)

Dworkin's and Robertson's accounts both center on respect for autonomy and for the values that underlie the importance attached to procreation. These values see procreation and founding a family as involving the freedom to choose one's own lifestyle and express, through actions as well as through words, the deeply held beliefs and the morality that families share and seek to pass on to future generations.

The freedom to pass on one's genes is widely perceived to be an important value; it is natural to see this freedom as a plausible dimension of reproductive liberty, not least because so many people and agencies have been attracted by the idea of the special nature of genes and have linked the procreative imperative to the genetic imperative. Whether or not this suggestion is ultimately persuasive, it is surely not possible to dismiss the choices about reproduction and access to the relevant technologies that constitute the point of claiming reproductive liberty as a simple and idle exercise of preference. Reproductive choices, whether or not they prove to be protected by a right to procreative liberty or autonomy, have without doubt a claim to be taken seriously as moral claims. As such they may not simply be dismissed wherever and whenever a voting majority can be assembled against them. Those who seek to deny the moral claims of others (as opposed, possibly, to the exercise of their idle preferences) must show good and sufficient cause.

Is there such 'good and sufficient cause' to deny access to artificial reproductive technologies?

Without listing them all, the main technologies currently available are, first, technologies that generate an embryo: *in vitro* fertilization (IVF), intracytoplasmic sperm injection (ICSI) and cell nuclear transfer (reproductive cloning). Then there are technologies that allow embryo selection such as IVF, preimplantation genetic diagnosis (PIGD) and various methods of testing *in utero*. There are genetic tests available prior to conception: carrier testing and the like. And finally in the further future there may be genetic manipulation available to select for nonmedical traits.

It is difficult to deal briefly with all the possible uses of such technologies to influence nonmedical traits, but perhaps a general observation can indicate a possible approach. If we can identify states of affairs (nonmedical traits) that would be morally problematic of themselves, we might be secure in judgments as to which of such states of affairs or traits it would be morally problematic deliberately to produce. The answer seems to be only those traits that would be harmful to the individual produced or harmful to others. Thus it would not be a morally problematic event if a boy rather than a girl were produced (or vice versa), and it would not be morally problematic if a child with a particular skin color, hair color, eye color or a range of useful abilities – sporting prowess, musical talent, intelligence, etc. – were to be born or created. It could not be said that children with any of these features would be born in a harmed condition or at any disadvantage whatsoever, neither would it be plausible to claim that they would be in any way harmful or dangerous to others. If it is not wrong to wish for a bonny, bouncing, brown-eyed, intelligent baby girl with athletic potential and musical ability, in virtue of what might it be wrong to use technology to play fairy godmother to oneself and grant the wish that was father to the thought?

There is of course room for claims that the choice of such features would not be in the public interest, but this raises the question of just how something harmless or even positively beneficial in itself might become contrary to the public interest when deliberately chosen. It is sometimes claimed that those who would produce children with particular nonmedical traits would have burdensome expectations of those children – expectations that would in fact harm the children (Harris, 2000; Holm, 1998). Notice here that such expectations are not peculiar to the use of technology to select for particular traits, but are a feature of all burdensome parental expectations. If we object to such expectations we must take care to demonstrate their unreasonableness and protect children from them. Among such expectations would have to be included those concerning religious observance, career path, values and occupation.

What objections could there then be (which do not refer to the expectations that parents might have that children would use these traits in particular ways) to the creation of nonmedical traits? There are other sorts of objections that might be made, but such objections are of necessity more 'remote' and more speculative than objections that concern the welfare of the child to be born or the welfare of others. 'Remote' in the sense that they must appeal to abstract notions like 'the public interest' or 'public acceptability', not harm to the individuals concerned; and speculative in that they may refer to possible harmful effects of patterns of choice of nonmedical traits on society or baseless speculation about the harmful effects on the individual that knowledge that they were selected to have such traits may have. I say 'baseless speculation' not because it might not be proved right in the event, but baseless in the sense that there is no prospective reason to expect harmful rather than beneficial effects to be the ones that actually occur.

Some people have been attracted to the idea of talking as if the disabled are simply differently abled and not in any way harmed. Deafness is often taken as a test case here (Journal of Medical Ethics, 2001). Insofar as it is plausible to believe that deafness is simply a different way of experiencing the world, but by no means a harm or disadvantage, then of course the deaf are not suffering from any disability. However is it plausible to believe any such thing? Would the following statement be plausible: 'I have just accidentally deafened your child, it was quite painless and no harm was done so you needn't be concerned or upset'? Or suppose a hospital were to say to a pregnant mother 'Unless we give you a drug your fetus will become deaf, since the drug costs £5 and there is no harm in being deaf we see no reason to fund this treatment'. But there is harm in being deaf and we can state what it is.

Imagine a child whose deafness could have been successfully treated, saying the following to the parents who denied the treatment: 'I could have enjoyed

Mozart and Beethoven and dance music and the sound of the wind in the trees and the waves on the shore, I could have heard the beauty of the spoken word and in my turn spoken fluently but for your deliberate denial.' In response it might be suggested that 'One may acknowledge the joy that (these things) bring others without insisting that the inability to perceive them is a harm or a deficit. After all, many persons are "deaf" to the pleasures of classical music (or jazz, or reggae, or rap, etc.) and yet none assume their limits of comprehension reflect a deficit or harm.' But to be 'deaf' to the pleasures of classical musical is to be deaf in inverted commas, not really deaf. Musical taste can be educated, but not so hearing for the profoundly deaf.

Deafness is a harmed condition properly so-called but equally to say so does not imply that the deaf do not have lives that are thoroughly worth living and are the moral equals of anyone. This may be seen if we consider the issue not of disabilities but of enhancements. Suppose some embryos had a genetic condition that conferred complete immunity to many major diseases – HIV/AIDS, cancer and heart disease for example – coupled with increased longevity. We would, it seems to me, have moral reasons to prefer to implant such embryos given the opportunity of choice. However, such a decision would not imply that normal embryos had lives that were not worth living or were of poor or problematic quality. If I would prefer to confer these advantages on any future children that I may have, I am not implying that people like me, constituted as they are, have lives that are not worth living or that are of poor quality nor that they are in any the sense lesser or inferior as persons.

Preimplantation genetic diagnosis

Finally we should note the use of PIGD in selection. In addressing the ethics of PIGD, a comparison is sometimes made with abortion. The purpose of such a comparison is to suggest that PIGD must be justified using criteria comparably stringent with those required to justify abortion and that in particular a woman's motives for requesting PIGD must be minutely scrutinized. However, this comparison is fallacious. The fallacy involved in the comparison is that a decision to abort must, in most jurisdictions, be justified and comply with the law, whereas a decision not to implant embryos requires no legal or moral justification whatsoever. The decision not to implant *in vitro* embryos is within the unfettered discretion of any woman.

A woman cannot be forced to implant any embryos and she can also usually determine the fate of those embryos, requiring them to be destroyed or donating them for research or to another woman for IVF. How might it be justifiable to deny a woman information on her *in vitro* embryo, which would enable her to make a rational and informed decision as to what to do with it? Suppose she were denied tests that she sought, and implanted an embryo that turned into a damaged child. However slight the damage, the mother (and according to some, also the child) would have moral and perhaps also legal grounds for complaint that the damage was due to the denial of access to information.

References

Brownlie I (ed.) (1971) *Basic Documents on Human Rights*, pp. 211–232, 338–363. Oxford: Clarendon Press.

Dworkin R (1993) *Life's Dominion*, pp. 166–167. London: HarperCollins.

Harris J (1992) *Wonderwoman & Superman: The Ethics of Human Biotechnology*, chapters 2 and 3. Oxford: Oxford University Press.

Harris J (2000) The welfare of the child. *Health Care Analysis* 8(1).

Holm S (1998) A life in shadows. *Cambridge Quarterly of Healthcare Ethics* 7: 160–162.

Journal of Medical Ethics (2001) Symposium: equality and disability. *Journal of Medical Ethics* 27(6): 370–392.

Robertson J A (1994) *Children of Choice*. Princeton, NJ: Princeton University Press.

United Nations (1978) Universal Declaration of Human Rights, Article 16. New York, NY: United Nations.

37 Feminist Perspectives on Human Genetics and Reproductive Technologies

Donna L. Dickenson

Feminist ethical perspectives: two strands

Feminist ethical perspectives have brought two separate but equally important insights to bear on human genetics and the new reproductive technologies (NRTs). The first, which is the one that is more properly termed 'feminist', is concerned with power in the doctor–patient relationship and, more broadly, with the ways in which NRTs – while appearing to afford women greater reproductive freedom and more control over their bodies – have the potential to exploit or oppress women. The second, which is widely known as 'feminist' but is more properly termed 'feminine', is connected to the 'ethics of care' (e.g. Larrabee, 1993). It concentrates primarily on relationship, which is often neglected in analyses of decisions about genetic testing that focus only on the choices of the affected individual. Although many feminist commentators combine both approaches, the second approach is distrusted by some feminists of the first approach because it seems to confirm the sexist, feminine stereotype that women are more concerned with relationship and are more naturally caring than men.

In the area of genetics and NRTs, both approaches coalesce in urging us not to be concerned solely with the individual clinician or patient, but with the wider social and political picture. It has been written of genetic testing (d'Agincourt-Canning, 2001, pp. 236–237) that

> Feminist ethics ... urges us to examine the implications of this technology within existing and potential patterns of oppression. It asks us to examine the specific circumstances of people involved in testing, to take into account their actual experiences and concerns and to assess the significance of disclosure of genetic information on personal and social relationships.

Feminist analysis: power and exploitation

Conventional bioethics frequently lacks a political dimension. Although the common principlist approach (Beauchamp and Childress, 2001) includes justice as its fourth element, the usual emphasis is on the first principle, autonomy. It has taken a feminist analysis – for example, through the work of the large subgroup of the International Association of Bioethics on Feminist Approaches to Bioethics – to bring justice-related issues of power, oppression and exploitation to the forefront of biomedical ethics.

In relation to genetics and NRTs, a key insight of feminism has been that neither of these is gender-neutral. 'The goal thus pursued is to identify gender-specific differences that might be overlooked or underestimated if a gender-neutral stance were maintained' (Mahowald, 2000, p. 6). Because women bear a disproportionate share of the risks and burdens associated with NRTs, it is right that their situation should be considered specifically. For example, if there were to be a general movement toward preimplantation genetic diagnosis (PGD) on the basis of minimizing hereditary genetic disorders, women would be asked routinely to undergo superovulation and egg extraction for the purpose of *in vitro* fertilization (IVF), which is normally a prerequisite for PGD. Both these procedures carry risks for women, in addition to being invasive in a way that sperm donation is not invasive for men. It is women who undergo amniocentesis, α-fetoprotein testing and other prenatal tests for genetic disorders such as Down syndrome, even when IVF is not required.

Similarly, in relation to the new stem cell technologies and therapeutic cloning, it is not often recognized that the enucleated eggs required depend on the reproductive labor of women, and that poor women or women from undeveloped countries may become exploited as sources of enucleated eggs for 'harvesting' (Dickenson, 2001). Likewise, the debate on human cloning has been preoccupied with individual identity on the one hand and the status of the embryo on the other hand. It is rarely pointed out that contract or surrogate mothers would be needed to produce human clones and that it took 267 attempts by 'surrogate mother' ewes to produce Dolly. Perhaps one does not need to be a feminist to make these obvious scientific points, but the nonfeminist literature has been slow to notice them.

Feminists have also delineated ways in which women are made to bear primary ethical and legal responsibility for the health of their fetuses and babies. For example, in the United States, women in some states have been prosecuted in criminal courts and imprisoned for causing harm to the fetus through, for example, drug addiction. Although evidence suggests that defective sperm are equally implicated in many genetic disorders, the prevailing discourse is in terms of female responsibility (Daniels, 2002). Here again, feminism draws attention to a power imbalance that means that NRTs and genetics are not gender-neutral.

Earlier second-wave feminism was often associated with the right to abortion. Although this right is not to be taken for granted, particularly in the light of decisions made by the United States Supreme Court, feminist thought is in fact far more wide-ranging. Indeed, if the right to abort becomes a duty to abort – as might be the assumption if genetic abnormalities in the fetus are identified – one could argue that women's rights to choice are served badly. Although the status of the embryo or fetus has been assumed to be the main issue both in stem cell technologies and in much of the abortion literature, an important contribution of feminist thought has been to direct attention away from the false assumption that this is the only ethical issue. In an influential article, Thomson (1971), with her famous violinist example, defended abortion while conceding, for argument's sake, that the fetus could be likened to a person. McDonagh (1996) similarly conceded fetal personality *in arguendo* but defended abortion by using an argument based on lack of consent. In relation to stem cell technologies, I have argued likewise that ethical debate does not turn only on the moral status of the embryo or fetus but also on issues of justice and exploitation, and of women's property in their reproductive labor (Dickenson, 2001, 2002).

Relationship in genetics

The core notion in the concept of a 'different voice' developed by the psychologist Carol Gilligan (1982) is that, in addition to the predominant ethic of rights, which is often but not necessarily associated with men, there is an equally valid ethic of responsibility, caring and relationship, which is evident in the outlook of many women but not confined to women. In relation to genetics, the key point is that decisions to undergo testing should not be conceived solely as matters of individual choice; instead, although it is the individual who must give legal consent, the decision should be made in a wider context, for example, within the family.

Feminist commentators have developed survey evidence of the ways in which the obligation to disclose genetic information determined through testing for hereditary disorders is gendered (d'Agincourt-Canning, 2001). Women in this study typically took primary responsibility in disclosing genetic information, perceiving it to be a moral duty for the benefit of other family members. More generally, it may be asked whether the wider availability of genetic tests will burden women disproportionately with moral quandaries, because typically it is women who are expected to assume responsibility not only for their own health but also for that of their families (Richards, 1996).

Another sense in which a feminist outlook differs from the conventional view that genetic information is private lies in the argument that our genome and gametes, with their symbolic and literal links to both our ancestors and our descendants, are not owned but are, in a sense, only lent (Dickenson, 1997, chap. 7). Here again is a focus on relationship, which is shared perhaps with other approaches such as narrative ethics (Lindemann Nelson and Lindemann Nelson, 1995).

Feminist ethics is also concerned that abstract rules, such as the 'duty to disclose', need to be considered in the social context of the person undergoing genetic testing. Although this may seem an obvious statement, in feminism it gains theoretical weight from a general preference for 'standpoint theory' (Mahowald, 2000) over the purportedly objective 'view from nowhere' (Nagel, 1986).

Conclusion

Feminist writers draw our attention to the ways in which NRTs and techniques of the new genetics may have different impacts on men and women. However useful that reminder may be, it is also necessary to go beyond the mere fact of difference to the ethical issue of whether differential treatment of men and women is unjust. Feminism is concerned deeply with gender justice.

Mahowald (2000, pp. 70–71) writes that:

> Gender justice thus refers to a situation in which men and women are equal to each other because they are equally valued but not the same. Because of unchangeable biological differences, women experience burdens and risks that men do not experience in reproduction and genetics, and those burdens and risks are not equal. Both informally and formally, however, measures can be introduced to reduce the inequitable impact of their differences.

According to a feminist analysis, genetics and NRTs pose a risk when they ignore or even worsen those differences in burdens between men and women that actually could be alleviated.

References

Beauchamp T and Childress J (2001) *Principles of Biomedical Ethics*, 5th edn. Oxford: Oxford University Press.

d'Agincourt-Canning L (2001) Experiences of genetic risk: disclosure and the gendering of responsibility. *Bioethics* 15: 231–247.

Daniels C (2002) Between fathers and fetuses: the social construction of male reproduction and the politics of fetal harm. In: Dickenson D L (ed.) *Ethical Issues in Maternal–Fetal Medicine*, pp. 113–130. Cambridge: Cambridge University Press.

Dickenson D L (1997) *Property, Women and Politics*. Cambridge: Polity Press.

Dickenson D L (2001) Property and women's alienation from their own reproductive labour. *Bioethics* 15: 205–217.

Dickenson D L (2002) Commodification of human tissue: implications for feminist and development ethics. *Developing World Bioethics* 2: 55–63.

Gilligan C (1982) *In a Different Voice: Psychological Theory and Women's Development.* Cambridge MA: Harvard University Press.

Larrabee M J (ed.) (1993) *An Ethic of Care.* New York: Routledge.

Lindemann Nelson H and Lindemann Nelson J (1995) *The Patient in the Family.* London: Routledge.

Mahowald M B (2000) *Genes, Women, Equality.* Oxford: Oxford University Press.

McDonagh E L (1996) *Breaking the Abortion Deadlock: From Choice to Consent.* Oxford: Oxford University Press.

Nagel T (1986) *The View from Nowhere.* Oxford: Oxford University Press.

Richards M (1996) Families, kinship and genetics. In: Marteau T and Richards M (eds.) *The Troubled Helix: Social and Psychological Implications of the New Human Genetics,* pp. 249–273. Cambridge: Cambridge University Press.

Thomson J J (1971) A defense of abortion. *Philosophy and Public Affairs* 1: 47–66.

38
Sex Selection

Dena S. Davis

Introduction

People have always speculated on what influences the conception and birth of boys or girls; in many cultures, people have attempted to control or influence the outcome of that process. Since the second half of the twentieth century, it has become possible to select the sex of one's offspring by selective abortion; at the beginning of the twenty-first century, techniques for sex selection that occur before implantation, or even before conception, are becoming available. How those techniques are used and the ethical, legal and social implications of that use can only be evaluated within cultural and historical contexts.

Techniques of sex selection

Premodern

Folk prescriptions abound for selecting the sex of one's offspring. These include: putting an axe under the bed before intercourse to conceive a boy; hanging the man's overalls on one bedpost or another; eating sweet or sour foods; and placing the bed to face in a particular direction. Hebrew and Greek sages agreed that girls came from the left testicle and boys from the right; French noblemen were advised to tie off or even amputate their left testicles in order to ensure male heirs. In many cultures, men blamed their lack of male heirs on their female partners. Even today, in many parts of the world, it is common for men to divorce or displace wives who do not produce sons.

Abortion

Until recently, the only reasonably sure way of selecting the sex of one's offspring was to ascertain the sex of a fetus *in utero*, by amniocentesis or ultrasound, and to abort the undesired fetus. This remains the most common technique and is commonly used in a number of developing countries.

Preconception sex selection

In the 1970s and 1980s, techniques such as douching and strategic timing of intercourse sought to exploit the differences between X- and Y-bearing sperm to influence the sex of the conceptus. Although a number of Western couples made use of these inexpensive, low-technology strategies, their success was never proved (Davis, 2001). In 1998, researchers in the USA announced a new and more successful technology that sorted the male partner's sperm in the laboratory and then used artificial insemination to impregnate the female partner with sperm carrying the desired trait. Although their technique is not completely effective, it improves the odds for conceiving a girl to more than 90% and is somewhat less successful at conceiving boys (Belkin, 1999).

Preimplantation genetic diagnosis as a concomitant of *in vitro* fertilization (IVF) is another way to ensure a fetus of a desired sex. In this technique, embryos created through IVF are tested for their genetic characteristics, including sex, and only those of the desired sex are implanted in the womb. This technique is so unwieldy and has such a high rate of pregnancy failure that it is unlikely to be used for sex selection alone, in the absence of some other problem.

Motives for sex selection

In some cultures, daughters are significant economic burdens, especially if they are destined to leave home and join their husband's family. Sons are therefore an important form of old-age security for their parents. Sons may also contribute to the parents' standing in the community and may be crucial for performing religious rituals that ensure a parent's well-being in the hereafter. Economist Amartya Sen has calculated that, due to feticide, infanticide and simple neglect, there are 100 million missing females around the world (Kristof, 1991). It is not only poor families who seek to avoid daughters. In contexts where people are achieving middle-class status by decreasing their family size, the total number of children desired falls more swiftly than the total number of desired sons, leaving even less 'room' for daughters (Das Gupta and Bhat, 1997).

In North America, there is strong evidence that people no longer prefer boys over girls. Numerous studies show that when people express interest in sex selection, it is to 'balance' their family by seeking to conceive a girl if they already have a boy, or vice versa (Davis, 2001).

Legal issues

Many countries in both the developed and developing worlds, including some states within India, have laws forbidding abortion for sex selection, or even forbidding the use of screening techniques such as ultrasound if the purpose is to determine the sex of the fetus (in the absence of a sex-related genetic disease). Most of these laws are ineffective, as it is impossible to police a couple's motivation for screening. Furthermore, when a person has a legitimate reason for screening, such as advanced maternal age, it is usually considered her legal right to know all the information that results from the screening, of which fetal sex is inevitably one part.

Ethical issues

Feminist philosophers find the problem of sex selection to be an enormous challenge. On the one hand, they have sympathy with the women in those countries for whom the birth of too many daughters and not enough sons may spell divorce or even death. They also are staunchly supportive of a woman's right to decide for herself why she should have an abortion. Among genetic counselors in the USA, a 1989 survey found that a majority would perform prenatal diagnosis for the purpose of sex selection or refer the couple to someone who would (Fletcher and Wertz, 1990). On the other hand, as most sex selection around the world is directed against girls and expresses a profound sexism and even misogyny, feminists deplore the practice (Powledge, 1981).

For many ethicists, simply the fact that sex selection is accomplished by abortion is enough to make it immoral. A common ethical position is to support abortion for 'serious' reasons such as fetal anomalies or the burden on the family of raising another child; aborting a fetus of the 'wrong' sex is often held up as the most egregious of 'trivial' reasons.

Another argument against sex selection is that it skews the normal gender ratio in a society. In India, for example, a 1991 census found 92.9 females for every 100 males (Balakrishnan, 1994, p. 269). Although one might think that 'market forces' would act to make women more desirable and improve their lives, what happens instead is that women are increasingly valued for their domestic and reproductive capacities alone and find their lives increasingly narrowed (Guttentag and Secord, 1983). On the other hand, if sex selection will happen anyway, through infanticide and neglect, perhaps abortion is the more humane alternative. Even in countries where feticide and infanticide account for the loss of many girls, the most significant decrease in girls occurs between the ages of 1 and 4 years, because boys are favored in receiving nutrition and medical care (Balakrishnan, 1994, p. 269).

Sex selection in North America at the beginning of the twenty-first century presents a very different set of ethical issues. First, with the advent of effective means for preconception sex selection, the argument will shift away from the focus on abortion. Prochoice feminists may then feel more free to criticize sex selection, but other commentators, who relied on abortion as their main argument against sex selection, will have to rethink their position. Second, most people in North America, if they were to use sex selection at all, would probably use it only to balance the numbers of children of each sex in their family (the exceptions appear mainly to be among people who have immigrated from countries where the pressure to produce boys is strong). There is even some evidence that people who avail themselves of preconception selection techniques are more likely to be trying to have girls (Belkin, 1999). Thus, it is no longer obvious that sex selection devalues women or will lead to societal imbalance. In the absence of these arguments, some commentators assert that sex selection is morally neutral (Wertz and Fletcher, 1989).

Other commentators continue to find that sex selection is 'sexist', but they must make this argument in more nuanced terms. Selection can be sexist if it is motivated by rigid notions of gendered behavior, so that parents who want an 'assertive' child will make sure to have a boy, while parents seeking a 'loving' child, or perhaps one who values music over athletics, will want a girl. Parents who believe that only boys can go on fishing trips with their fathers and who make sure that they have at least one boy for that reason will miss out on the discovery that girls can enjoy fishing as well and will perpetuate gender stereotypes by selecting children to meet those expectations (Bayles, 1984; Pogrebin, 1980). Even parents who want a girl so as to oppose gender stereotyping and raise the first woman president of the United States are still seeing her primarily in terms of gender. Some ethicists caution that parents who invest substantial resources in making sure that they have a girl or a boy may be insufficiently attuned to the child's unique characteristics and individual flourishing (Davis, 2001; Ryan 1992).

References

Balakrishnan R (1994) The social context of sex selection and the politics of abortion in India. In: Sen C and Snow R (eds.) *Power and Decision: The Social Control of Reproduction*, Harvard Series on Population and International Health, pp. 267–286. Cambridge, MA: Harvard University Press.

Bayles M (1984) *Reproductive Ethics*. Englewood Cliffs, NJ: Prentice-Hall.

Belkin L (1999) Getting the girl. *New York Times Magazine* 25 July.

Das Gupta M and Bhat M P N (1997) Fertility decline and increased manifestation of sex bias in India. *Population Studies* 51: 307–315.

Davis D (2001) *Genetic Dilemmas: Reproductive Technology, Parental Choices, and Children's Futures*. New York: Routledge.

Fletcher J C and Wertz D C (1990) Ethics, law, and medical genetics. *Emory Law Journal* 39: 747–809.

Guttentag M and Secord P F (1983) *Too Many Women? The Sex Ratio Question*. Beverly Hills, CA: Sage Publications.

Kristof N D (1991) Stark data on women: 100 million are missing. *New York Times* 5 November.

Pogrebin L (1980) *Growing Up Free: Raising Your Child in the 80s*. New York: McGraw-Hill.

Powledge T (1981) Unnatural selection: on choosing children's sex. In: Holmes H B, Hoskins B B, Gross M (eds.) *The Custom-made Child? Women-centered Perspectives*, pp. 93–100. Clifton, NJ: Humana Press.

Ryan M (1992) The argument for unlimited procreative liberty: a feminist critique. In: Campbell C (ed.) *What Price Parenthood? Ethics and Assisted Reproduction*, pp. 83–90. Aldershot, UK/Brookfield, VT: Dartmouth Publishing Company.

Wertz D C and Fletcher J C (1989) Fatal knowledge? Prenatal diagnosis and sex selection. *Hastings Center Report* 19: 21–27.

39
Dolly and Polly

Alan Petersen

Introduction

A series of genetic experiments involving sheep, undertaken at Roslin Institute, a government laboratory in Scotland (UK), received a great deal of media coverage in the late 1990s. These experiments triggered an extensive debate among scientists, ethicists, politicians and members of the lay public about the ethics and utility of genetic manipulation, particularly cloning. The announcement in *Nature* in February 1997 (Wilmut et al., 1997) that scientists had cloned a sheep, named Dolly (after the singer Dolly Parton), provided the catalyst for a wide-ranging discussion about cloning technology and its potential applications. The level of media interest in Dolly and subsequent public reaction, in particular concerns about the implications for human cloning, has been phenomenal. Much less media attention has been given to Polly, a sheep born a year after Dolly, which is both cloned and genetically transformed (Wilmut et al., 2000, pp. 2, 18, 20). Polly carried an introduced gene that codes the protein factor IX that is involved in blood clotting, deficiency of which causes a form of hemophilia (Wilmut et al., 2000, p. 2). The production of transgenic animals such as Polly is seen to have potentially significant therapeutic implications, particularly for the treatment of hemophilia. Dolly continues to attract media interest, despite subsequent doubts about the success of the experiment, after reports that she has shown signs of premature aging and that she has succumbed to arthritis, a condition uncommon in a sheep her age. News reports about developments in cloning technology and its applications, and about efforts of so-called maverick scientists to produce human clones, often refer to Dolly as evidence of the established 'fact' of cloning technology. She has become firmly embedded in public discourse about cloning and biotechnology in general, providing a focus for the expression of both fascination about the possibilities of science and fears about where it might be leading.

Scientific significance of Dolly and Polly

The scientific significance of Dolly, according to scientists, is that she was the first animal of any kind created from a fully differentiated cell using a process called nuclear transfer. This involves the insertion of the nucleus from an adult cell into an egg cell that has had its nucleus removed and which is then encouraged to grow by electrical stimulation (Tagliaferro and Bloom, 1999, p. 187). In the case of Dolly, researchers took a cell from the mammary gland of a 6-year-old Finn Dorset ewe, which had been pregnant for 3.5 months, and fused it with the egg cell of a Scottish blackface ewe. As the scientists Ian Wilmut and Keith Campbell, who created Dolly, explain, in their own account of the experiment, while Dolly was not the first mammal ever to be cloned, she was the first to be cloned from an adult body cell (Wilmut et al., 2000, p. 15). Dolly was the culmination of a long history of efforts by biologists to manipulate the process of cell differentiation and in particular to examine the genomic potential of older embryonic cells. Experiments in the late nineteenth century involving the physical separation of the cells of two-cell sea urchin embryos and of amphibians paved the way for experiments in the mid- and late-twentieth century in nuclear transplantation, involving insects, fish, mammals and eventually Dolly (McKinnell and Di Berardino, 1999). According to the editors of *Science*, the birth of Dolly was a major scientific breakthrough, in that it made obsolete 'preconceived limits' in the field of biotechnology and was seen to 'profoundly change the practice or interpretation of science or its implications for society' (Bloom, 1997, p. 2029). The apparent success of this experiment seemed to overturn one of the most firmly held beliefs of biology; namely that, once a cell had differentiated into its specialized form, it could not be 'reprogrammed', or induced to develop into any other specialized cell.

Although Polly has received much less media and public attention than Dolly, her birth was also of great scientific significance, if not of greater significance, by virtue of her being genetically transformed as well as cloned (Dolly-style). As noted, Polly was genetically transformed so that she would express human clotting factor IX protein in her milk. The production of transgenic animals such as sheep and pigs via the process of nuclear transfer is seen as more efficient than other methods in producing animals that may be bred as a source of human proteins used for medicines. Genetic identity achieved through nuclear transfer, scientists argue, contributes to the consistency of the medicinal product (Schnieke et al., 1997, pp. 2132–2133).

Political and public response to Dolly

Although Dolly's birth can be seen as the realization of an idea that had been discussed for more than half a century, her birth was treated as a great surprise (Turney, 1998, p. 214) and caused a worldwide furor. For example, both the director of the US National Human Genome Research Institute and the president of the US National Academy of Social Sciences were reported to have been 'totally caught off guard' by the announcement (Butler and Wadman, 1997, p. 9). The political response was swift. A number of jurisdictions moved quickly to outlaw human cloning or to withdraw funds for research on cloning, and politicians sought advice on the issue from government bioethics committees. In the USA, President Clinton instructed the National Bioethics Advisory Commission to report to him on the issue of human cloning within 90 days. In Europe, a number of national leaders also sought advice from their respective bioethics committees (Butler and Wadman, 1997, p. 8). In the UK, the Human Genetics Advisory Commission, which reported to ministers on issues arising from new developments in human genetics, and the Human Fertilization and Embryology Authority, undertook a consultation exercise on cloning in order to identify the ethical issues involved (see Web links section). Calls for an outright ban on cloning came from diverse quarters, including the Vatican – which argued that humans have a right to be 'born in a human way, and not in the laboratory' – and the US Biotechnology Industry Organization, which represents 700 companies (Butler and Wadman, 1997, p. 8).

Even before details of the research that led to Dolly were published in *Nature*, the implications of the research for the cloning of humans, as well as for medicine and agriculture, were drawn. (The story was broken by a science editor of the *Observer*, who obtained information from a source other than *Nature*, thus technically avoiding breaking the embargo that was in place until the expected date of publication of the research article, which was 27 February (Kolata, 1997, p. 30; Wilmut et al., 2000, pp. 243–244).) In the article, Wilmut himself never spelt out the human cloning implications of the research, nor did the accompanying editorial. The title of Wilmut's article certainly said nothing about clones; it was called 'Viable offspring derived from fetal and adult mammalian cells' (Kolata, 1997, p. 27). However, the ramifications of this experiment, not only for medicine, but also for the cloning of humans, was immediately grasped by many people and became the focus of a great deal of media attention and public discussion, with concerns expressed about whether science was in danger of 'going too far'. Dolly was the catalyst for wide-ranging debates about the ethical, legal and social implications of human cloning.

Dolly's reproduction from a fully differentiated cell rather than through the union of a sperm and egg seemed to unsettle some deeply held assumptions about 'nature', reproduction and 'individuality'. For many, her birth seemed to provide incontrovertible proof that one could 'turn back the biological clock' and that asexual reproduction was possible. Many people feared that if this technology were applied to human beings, people would 'lose their identity' and that the 'bonds of the family' would be threatened. Some writers have suggested that human cloning may permit the individual to achieve a kind of 'pseudoimmortality', by allowing their genetic likeness to live on in the form of a cloned offspring (Rifkin, 1998, p. 218). News media coverage has drawn liberally on popular images of cloning as duplication or cheap imitation of a complete original; for example, by reference to the potential for the production of a 'master race' and to the 'mass production of identical people' (Petersen and Bunton, 2002, p. 127). In public discussion, such popular cultural representations of cloning have tended to obscure the more specific, biological meaning of cloning (Silver, 1998, pp. 105–106). The suggestion that cloning technology may be used for human reproductive and therapeutic purposes raises questions about the meaning of life, the limits of the natural and the definition of normality.

Despite the proliferation of writings and discussion on cloning since Dolly's birth, ensuing debates have offered little insight into such basic questions as to whether cloning has the implications imagined by many people and indeed whether the claims about the success of the Dolly experiment have been adequately verified. Wilmut and his colleagues have been more circumspect than many others in their assessment of their research, admitting that the cell that produced Dolly might not have been fully differentiated (Wilmut et al., 2000, pp. 248–249). The effectiveness of the nuclear transfer procedure can also be questioned, given that only one of 277 embryos that began with mammary gland nuclei became a live lamb (Wilmut et al., 2000, p. 239). However, such qualifications or doubts have been overlooked or downplayed in media reports on the issue, which have tended to be preoccupied with the applications and implications of the technology. Periodic claims by so-called maverick scientists of their intentions to clone a human being, such as those of US physicist Dr Richard Seed and the Italian fertility specialist Dr Severino Antinori, have served only to fuel media interest and, arguably, to heighten public concerns about cloning.

Purported therapeutic implications

In view of widespread concerns about the implications of the Dolly experiment, many scientists have sought to emphasize the medical benefits of cloning research and to openly reject its applications for human reproduction. Scientists have made extensive use of the media to defend and explain their work. In news articles on cloning research that were published or broadcast after the birth of Dolly, scientists who were quoted or cited often sought to distinguish therapeutic cloning – implicitly seen as 'good', useful and legitimate – from reproductive cloning – seen as 'bad', dangerous or illegitimate (Petersen, 2001, p. 1265). Suggested therapeutic applications include the creation of human cells and tissues that can be used to repair damaged organs, and the growth of laboratory cultures of heart, skin and blood vessels on which to test new pharmaceuticals (Sexton, 2001, p. 163). Cloning is also seen to have great potential in xenotransplantation, the process of transplanting animal organs into humans. Dolly-style cloning could be used to make as many identical copies of a transgenic animal as needed within a single generation, thus creating a plentiful supply of a desired therapeutic agent (Cooper and Lanza, 2000, p. 104). In theory, a patient's own cells (e.g. skin or retinal tissue) could be cloned, providing them with grafts that would not be rejected, overcoming the immunological reaction to xenografting (Cooper and Lanza, 2000, p. 106). In the light of Dolly, Polly and other recent genetic developments, some scientists have begun to consider seriously the prospect of engineering the human germ line so as to eliminate the most debilitating diseases (Stock and Campbell, 2000), raising the spectre of eugenics.

Social and commercial context of the Dolly and Polly experiments

Amid the voluminous discussion about the Dolly experiment and its significance, there has been relatively little discussion about the context shaping cloning research and the commercial interests driving the work that led to Dolly and Polly. Scientists in this and other areas of research are increasingly dependent on private sources of funding as governments retreat from their commitment to basic research. What is clear from published accounts of Dolly and related experiments is the crucial role played by her creator's sponsor, PPL Therapeutics Limited, who see great potential in 'pharming', that is, genetically altering animals so that they produce therapeutically valuable materials, usually proteins. Pharming is seen as a way of producing drugs more cheaply than is possible by other means. PPL Therapeutics provided the cells from which Dolly was made and some of the financial support, and was the principal creator of Polly (Kolata, 1997, pp. 182–189, 197; Wilmut et al., 2000, pp. 233–235). As Wilmut et al. have pointed out, PPL has survived and flourished because of its sponsored work on genetically transformed sheep, and those responsible for the sponsorship 'were interested in cloning only as an adjunct to their work on genetic transformation' (Wilmut et al., 2000, p. 234).

The question of whether Dolly's premature death at age six and a half years old from a progressive lung disease (she was put down), in February 2003 (Meek, 2003), will lead to a wide-ranging assessment of the claims made about the technology that created her, remains to be seen. However, there is little doubt about the impact of media coverage of Dolly on public awareness of the issue of cloning and on debates about the implications and ethics of biotechnology applications, thus on these grounds alone assuring her a firm place in the history of reported scientific breakthroughs.

References

Bloom F E (1997) Breakthroughs 1997. *Science* 278: 2029 (editorial).
Butler D and Wadman M (1997) Calls for cloning ban sell science short. *Nature* 386(6620): 8–9.
Cooper D K C and Lanza R P (2000) *Xeno: The Promise of Transplanting Animal Organs into Humans*. Oxford: Oxford University Press.
Kolata G (1997) *Clone: The Road to Dolly and the Path Ahead*. London: Penguin.
McKinnell R G and Di Berardino M A (1999) The biology of cloning: history and rationale. *Bioscience* 49(11): 875–885.
Meek J (2003) Dolly the sheep is put to sleep, aged only six. *Guardian*, 15 February: 8.
Petersen A (2001) Biofantasies: genetics and medicine in the print news media. *Social Science and Medicine* 52: 1255–1268.
Petersen A and Bunton R (2002) *The New Genetics and the Public's Health*. London: Routledge.
Rifkin J (1998) *The Biotech Century: How Genetic Commerce Will Change the World*. London: Phoenix Press.
Schnieke A E, Kind A J, Ritchie W A, et al. (1997) Human factor IX transgenic sheep produced by transfer of nuclei from transfected fetal fibroblasts. *Science* 278: 2130–2133.
Sexton S (2001) If cloning is the answer, what was the question? Genetics and the politics of human health. In: Tokar B (ed.) *Redesigning Life? The Worldwide Challenge to Genetic Engineering*. London: Zed Books.
Silver L M (1998) *Remaking Eden: Cloning and Beyond in a Brave New World*. London: Weidenfeld and Nicolson.
Stock G and Campbell J (2000) *Engineering the Human Germline: An Exploration of the Science and Ethics of Altering the Genes We Pass to Our Children*. New York: Oxford University Press.
Tagliaferro L and Bloom M V (1999) *The Complete Idiot's Guide to Decoding Your Genes*. New York: Alpha Books.
Turney J (1998) *Frankenstein's Footsteps: Science, Genetics and Popular Culture*. New Haven, CT: Yale University Press.
Wilmut I, Campbell K and Tudge C (2000) *The Second Creation: The Age of Biological Control by the Scientists Who Cloned Dolly*. London: Headline Book Publishing.
Wilmut I, Schnieke A E, McWhir J, et al. (1997) Viable offspring derived from fetal and adult mammalian cells. *Nature* 385(6619): 810–813.

Web links

www.hfea.gov.uk/AboutHFEA/consultations

40

Cloning of Animals in Genetic Research: Ethical and Religious Issues

Andrew Linzey and Ara Barsam

Harm to animals

The claim is often made that genetic research, and specifically cloning, involves no risk of harm to animals. Director of the Roslin Institute, Grahame Bulfield, argues that animal welfare is about the physical or psychological state of the animal 'rather than the way it is bred'. It is not the technique that is crucial, but the 'resultant effect on the animal'. For example, 'if an animal is seriously lame...either because of breeding technique or transgenesis, then that is not acceptable' (Bulfield, 1997). He informs us that a study of the behavior of transgenic sheep at Roslin could find 'no differences between them and control animals in eight measures in three husbandry situations' (Bulfield, 1997).

Bulfield's article provides no details concerning the purported measures and so we are unable to assess this claim; but that cloning techniques do involve mild to severe risks, both actual and potential, is confirmed by his colleague at Roslin and creator of Dolly, Ian Wilmut. Cloning techniques involve, *inter alia*, death through malformed internal organs and, of the survivors, gross abnormalities. Indeed, Wilmut poignantly provides an example of the fate of one lamb born through transgenesis at his institute in Roslin 2 years after the technique used to clone Dolly had been successfully established: 'It [the lamb] could run about perfectly normally – but it hyperventilated all the time; it panted night and day. We tried to treat it, but in the end decided it was kinder to put it down' (Wilmut, 2001).

Some proponents have argued that to produce Dolly there needed to be experiments in order to successfully establish the cloning technique. But what Wilmut's disclosure shows is that, even 2 years subsequently, cloning experiments are being performed that result in palpable harm. Bulfield's claim that animal welfare is not compromised by transgenesis or breeding techniques is simply contradicted by the evidence. Because animal welfare is intrinsically compromised by cloning techniques, Bulfield's assertion that considerations of animal welfare can be distinguished from the techniques themselves – 'the way it is bred' – is untenable.

Selective breeding

It is often said that the new technologies (and specifically cloning) are no different in nature from long-established practices such as selective animal breeding. Proponents point out that humans first domesticated feral animals for their use between 10 000 (dogs) and 4000 (farm animals) years ago and that such changes, specifically in the past 200 years, have significantly changed the genetic make-up of farm animals. But, as Bulfield himself acknowledges, this 'classical' form of breeding is not without ethical and welfare problems. For example, obesity, lameness and poor fertility in modern strains of farm animals are the results of selective breeding. These techniques have been the subject of intensive ethical scrutiny and been found wanting. There is now a significant number of ethicists who oppose modern intensive conditions precisely because they involve the infliction of substantial harms for nonessential purposes (e.g. increased growth). To give one example: David DeGrazia offers a comprehensive critique of current animal usage from a utilitarian perspective and concludes that: 'it is clear that the institution of factory farming, which causes massive harm for trivial purposes, is ethically indefensible' (DeGrazia, 1996, p. 284).

It is therefore odd to argue that we should be allowed to intensify genetic modification when its recent history has been the subject of such widespread moral critique. It is nevertheless argued that transgenesis has the potential to eliminate such welfare problems, because 'it manipulates only a single gene whose function is often well understood' (Bulfield, 1997). However, these words disguise the fact that transgenesis causes actual harm, often of a severe kind, to farm animals. Although it may be that the function of one gene is sometimes well understood, it is not always the case. This means in practice that every proposed modification carries with it the real or hypothetical risk of inflicting direct harm. As in the case of the lamb, animals are repeatedly subjected to experimentation to discover the limits of their physiological adaptability even years after a particular technique has been apparently successful. In short, transgenesis is a further move in a morally dubious direction.

Economic benefits

Proponents of animal cloning often appeal in a straightforwardly utilitarian way to the notion of 'benefit' to the human species. Their understanding of what constitutes benefit is often remarkably narrow; in many instances it amounts to little more than economic gain. Bulfield again, writing from a British perspective, offers a frank admission in this regard: 'if we allow other countries access to technology which we ignore, we will soon lose our market position' (Bulfield, 1997). In other words, cloning is about extending national (in this case specifically British) economic self-interest. He further tells us that it is not only 'the level of production that is important but the efficiency . . . In addition, cloning from adult cells (as with Dolly), would permit the replication of (for example) a proven high yielding and productive dairy cow' (Bulfield, 1997, pp. 15, 16). Put more simply, cloning is good because it enables us to design animals for no other purpose than as bigger and better meat machines or laboratory tools.

Economic considerations deserve their due in moral debate, but the idea that profit unambiguously represents what is morally good is flawed. Indeed, the failure to see the distinction between moral and economic considerations lies at the heart of the debate. If animals were simply nonsentient things then it might be appropriate to classify them as economic commodities. But as they are not, to do so constitutes a category mistake. It implies that animals are mere artifacts, with no intrinsic worth, and that human relations with them should be circumscribed by economic considerations alone.

Not satisfied with simply exploiting animals (e.g. via selective breeding), we now presume to change their nature in order to do so more profitably. In the words of one scientist: 'we can design the whole carcass, if you like, from embryo to plate to meet a particular market niche' (Street, 1995).

Cloning and genetic research violates the intrinsic worth of animals

Since Dolly, scientists have cloned mice, cattle, goats and pigs. Wilmut points out that very few cloned embryos survive to birth and, of these, many die shortly thereafter. Survivors are often grotesquely large or have other defects. To this end, Wilmut states that 'attempting to clone a human would be extremely cruel for the woman and the children involved' (Wilmut, 2001). That statement begs some important questions: why should there be a total moral distinction between our treatment of children and animals? On what grounds can one justify the 'extremely cruel' treatment of sentient animals, while at the same time opposing this research in human sentients as a matter of principle? If the appeal to utility can justify cloning animals, what rational grounds might there be for not cloning humans when there would be as much, or arguably more, benefit?

While of course there are important differences between children and animals, there are also a range of considerations that apply equally to both. By common consent, and affirmed by Wilmut, we find the infliction of harm and suffering on children morally outrageous. The grounds for extending special protection to children rests, *inter alia*, on appeals to their vulnerability, defenselessness, their inability to give or withhold consent and their moral innocence. But these considerations apply not only to newly born infants but also equally to animals. It is difficult to see on what impartial (i.e. nonanthropocentric) grounds we could regard children as subjects of a special solicitude that does not similarly apply to sentient animals.

Cloning and genetic research represents an instrumentalist view of animals

Between 1970 and 2000, there has been a marked shift in our perception of the value of animal life. We have begun to move away from the idea that animals are simply resources, machines or commodities to the realization that as sentient beings they have their own worth, dignity and rights. In the light of this, the comment made by Bulfield in justification of cloning that 'we have to continue to improve our efficiency to compete on a world stage' misses the mark precisely because it presupposes a wholly instrumental view of the worth of animals. The notion that animal cloning represents a morally progressive view is misplaced. Animals are not bettered or improved by cloning – indeed quite the reverse is the case. What is most dominantly shown in the act of cloning is that animals can be cloned, that is, they are beings that can be manipulated, controlled and exploited. In short, that they are here for our use.

It is sometimes argued that if it is right to farm and to kill animals for food then it must also be right to genetically manipulate them for research and/or farming. But the two cases are logically and morally distinct. Even if it can be shown that we may make use of animals in specific limited circumstances occasioned by genuine human need, it does not follow that we have the right to subordinate their life as their absolute masters. Control of deoxyribonucleic acid (DNA) is control of the very fabric of life, and cloning represents a new tier of instrumentalization: the concretization of the old subordinationist view that animals belong to us or are here for our use.

Cloning and genetic research represents a spiritually impoverished view of animals

Proponents of cloning and genetic research often appeal to 'benefits', hypothetical or real. What informs the British Science and Technology Committee's defence of cloning is, *inter alia*, a sense of the 'scientific vistas', the likely profits of the pharmaceutical industry and an appeal to medical spin-offs. These considerations, worthy as some may be, are not the only ones. What is not addressed in this utilitarian discourse, for example, is the debit side involved in developing techniques that treat animals as machines, or the likely social and institutional effects of so doing. Are humans really benefited by a wholly utilitarian and instrumentalist understanding of other sentient creatures? At least it is a question worth addressing.

Even in terms of a wholly utilitarian calculus, the debit side is not adequately weighed. On closer examination, many of these benefits prove to be of a largely indirect, long-term, overstated kind. Indeed the difficulty in securing adequate justification is accepted by the Committee in this revealing line: 'It is notoriously difficult to predict the benefits which will arise from a particular piece of research' (Science and Technology Committee of the House of Commons, 1997, p. vi). What is lost here is the ethical realization that such unpredictability and uncertainty count against the risking of actual harm to animals. An appeal to some putative – and indirect – future benefit does not constitute a case of moral necessity. Indeed even in strictly utilitarian terms, Jeremy Bentham, himself a founder of modern utilitarianism, pioneered the extension of moral consideration to animals on the grounds of sentiency alone. In his famous words, the crucial question is not 'Can they reason? nor, Can they talk? but, Can they suffer?' (Bentham, 1948).

There is also a theological challenge that needs to be addressed. Although traditional religious ethics have allowed some use of animals for human ends (e.g. for survival and self-defense), they have never countenanced the view that animals are nothing more than means to human ends and that our use of animals is unlimited. Theists have generally held that humans have a high place in creation and that human welfare is important; but it does not follow that human betterment is the only good in creation, or that God is exclusively, or even mainly, concerned with the well being of *Homo sapiens*. The notion of God, and specifically God's will, must place limits on purely human calculations of what is right and good for our own species. Estimations of our own needs and wants cannot constitute the sole criterion on which we base our relations with the natural world.

A similar limitation is found in Indian religious traditions, most notably in Buddhism and Jainism, in which the central moral imperative is *ahimsa* (nonharming, nonviolence). The Buddhist and Jain traditions speak of a feeling of equanimity toward all sentient beings. Both take the view that all beings are equal in that all desire happiness and the avoidance of suffering. While many world religious traditions have a poor historical record in relation to animals, it nevertheless remains true that they offer positive resources for a more ethically sensitive view of animals. To look upon other creatures as simply resources here for us is a spiritually impoverished attitude. The better part of all these traditions is united in the view that human benefit does not consist in material advance alone but in our capacity to live in harmony with other sentientbrae creatures.

References

Bentham J (1948) An introduction to the principles of morals and legislation [1789]. In: Harrison W (ed.) *A Fragment on Government and an Introduction to the Principles of Morals and Legislation*, 1823 edn, pp. 411–412. Oxford: Blackwell.

Bulfield G (1997) Biotechnology and farm animals. *Bulletin of Medical Ethics* 131: 15–18.

DeGrazia D (1996) *Taking Animals Seriously: Mental Life and Moral Status*. Cambridge: Cambridge University Press.

Science and Technology Committee of the House of Commons (1997) *The Cloning of Animals from Adult Cells*, vol. 1. London: HMSO.

Street P (1995) Fast life in the food chain, 1992. In: D'Silva J and Stevenson P (eds.) *Modern Breeding Technologies and the Welfare of Farm Animals*, p. 17. Petersfield, UK: Compassion in World Farming.

Wilmut I (2001) Dolly's creator says no to human cloning. The *Guardian*, 29 March: 11.

41
Distributive Justice and Genetics

Colin Farrelly

Advances in genetic and biological knowledge bring us closer to a world where we will have the ability, or at least a much greater ability than we currently have, to manipulate our genetic make-up. With this new ability will come new questions concerning what the demands of distributive justice are. Distributive justice concerns the just division of benefits and burdens in society. The political, social and economic institutions of our society influence the distribution of many different kinds of benefits and burdens. These institutions confer on us, for example, certain rights and freedoms (e.g. the right to vote, freedom of expression) and bring about a certain distribution of goods such as income and wealth. Our institutions also ensure that we fulfill certain obligations such as paying our taxes and respecting the rights of others. The index of goods one believes the state should be fairly distributing is a much-debated topic in political philosophy, and the new genetics promises to raise a number of new concerns for these debates.

As of the year 2002, the genes we have are the result of the 'natural lottery' of life. No one has the ability to manipulate the genes we are born with, and thus the different advantages and disadvantages that our genes confer on us are the result of brute luck. Some people are born with genetic diseases or have a higher risk of developing certain diseases than other people. Some are born with genes that increase their chances of developing valued physical and behavioral traits. These people have advantages that others, for example those whose genetic profiles impede their ability to develop these valuable traits, do not have. No one is responsible for this unfair division of the advantages and disadvantages that our genes confer on us. There is nothing we could do, collectively as a society, about it. However as our knowledge of how genes work increases, and with it the prospect of being able to successfully intervene in the natural lottery of life, this will no longer be the case. The decisions we make regarding the regulation of biotechnology will determine who receives the greatest share of the benefits these technologies confer:

> Macro decisions determine (1) what kinds of health-care services will exist in a society, (2) who will get them and on what basis, (3) who will deliver them, (4) how the burdens of financing them will be distributed, and (5) how the power and control of these services will be distributed.
>
> (Daniels, 1985)

What will the demands of distributive justice be in the postgenetic revolutionary world? Will our genes be viewed as one of the social goods the state should fairly distribute? This position may seem more reasonable to assert as biotechnology progresses further toward human genetic manipulation. The prospect of intervening in the natural lottery of life raises fundamental questions on both intragenerational and intergenerational justice.

The new genetics raises a number of pressing and complex questions for theories of distributive justice. Firstly, we must decide which genetic advantages or disadvantages should be included within the domain of distributive justice. Does justice require that we fairly distribute genes that influence every conceivable advantage and disadvantage, ranging from the risk of disease to the likelihood of developing valued physical and behavioral traits? One extreme would be to argue that all (or most) of the advantages genes confer on us should be within the domain of distributive justice. Such a position would be consistent with recent egalitarian theories of justice which have become: ' . . . dominated by the view that the fundamental aim of equality is to compensate people for undeserved bad luck'– being born with poor native endowments, bad parents, and disagreeable personalities, suffering from accidents, and illness, and so forth.' (Anderson, 1999).

Those attracted to such a version of 'luck egalitarianism' might find the notion of 'genetic equality' appealing, as it applies the principle of equal opportunity wholesale to our natural endowments. However, such a proposal assumes that it would be possible to employ very sophisticated levels of genetic intervention as well as to have fantastic knowledge of how genes influence these various assets. Even if we had the knowledge and technology to take the idea of pursuing genetic equality seriously, there are other concerns: 'The fact of value pluralism and the fact that the value of traits is relevant to social conditions call for caution about any commitment to genetic equality' (Buchanan et al., 2000).

People have diverse conceptions of what is of value in life, and thus there could never be a consensus on what physical and behavioral characteristics are most valuable. Furthermore, the traits we value change over time. We do not have the foresight to know which traits will be viewed as an asset or deficit in the future.

If the idea of equalizing all potential genetic advantages is one extreme, the opposite extreme would be to limit the demands of justice to a narrow range of advantages such as being disease-free or having a minimum risk of disease. Such a proposal does not face the same problems that genetic equality faces, as everyone would agree that such an aim is valuable and this would not change over time. However as our powers to manipulate our genes increase, limiting the demands of justice to the issue of disease may seem too restrictive. If genetic enhancements permit us to increase our potential for living longer or being more intelligent, for example, we may view it as unjust if some people cannot enjoy these advantages simply because they cannot afford such enhancements? 'Justice may require regulating the conditions of access to genetic enhancements to prevent exacerbations of existing unjust inequalities' (Buchanan et al., 2000).

Whatever distributive principle one believes equality requires in the postgenetic revolutionary world, be it genetic equality, a genetic decent minimum, or some other principle, the pursuit of such a principle must be tempered by considerations of utility and freedom. These two concerns are particularly important concerns given the costs involved in making the benefits of genetic intervention a reality and the potentially intrusive nature of such interventions. While considerations of equality might dictate that we should pursue a principle of equal access to genetic interventions, the viability of such a policy will depend on budget constraints. Given the other commitments that egalitarians may also endorse (e.g. equal opportunities in education, a right to decent healthcare in general), a determination of the just allocation of public funds for such programs must take into consideration the utility generated by each of these various programs and the costs involved in making these utilities a reality. Genes are just one factor among many (e.g. environment) that influence our level of well-being, and thus a determination of how much public resources we should allocate to pursuing genetic manipulation must consider the budget constraints on other important programs.

Furthermore, the prospect of genetic intervention also raises fundamental questions regarding the weight society ought to place on the value of freedom. Concerns about the injustices of past eugenic movements, for example, must be taken seriously. One of the most influential accounts of distributive justice in the twentieth century was John Rawls' theory of 'justice as fairness'. In *A Theory of Justice*, Rawls criticized utilitarianism. As a public philosophy, utilitarianism is deficient, argues Rawls, because it treats questions of distributive justice as questions of efficient administration. Our desire to achieve a fair distribution of genes must be balanced against the interests we have in protecting certain fundamental rights and freedoms. Rawls claims that 'justice denies that the loss of freedom for some is made right by a greater good shared by others' (Rawls, 1971). One fundamental liberty is procreative liberty. Procreative liberty is 'freedom in activities and choices related to procreation' (Robertson, 1986). Eugenic policies such as involuntary sterilization are gross violations of procreative liberty and justice rules such coercive measures out. However the new genetics does raise a number of questions concerning how much weight we should place on procreative liberty:

> [A]s a society we must address when, if ever, we should adopt policies that encourage in a variety of ways, or require by law, that individuals obtain and make use of genetic information and interventions to prevent harm to their prospective children and other related individuals.
>
> (Buchanan et al., 2000)

As our ability to successfully intervene in the natural lottery of life increases, the tension between respecting reproductive freedom and fulfilling the demands of the duty to prevent harm across generations also increases. How we shall achieve a reasonable balance between these potentially conflicting demands is perhaps one of the greatest challenges we are likely to face in the not-so-distant future.

References

Anderson E (1999) What is the point of equality? *Ethics* 109: 287–337.

Buchanan A, Brock D, Daniels N and Wikler D (2000) *From Chance to Choice: Genetics and Justice*. Cambridge: Cambridge University Press.

Daniels N (1985) *Just Health Care*. Cambridge: Cambridge University Press.

Rawls J (1971) *A Theory of Justice*. Cambridge, MA: Harvard University Press.

Robertson J (1986) Embryos, families, and procreative liberty: the legal structure of the new reproduction. *Southern California Law Review* 59: 501–602

42
Reprogenetics: Visions of the Future

Regine Kollek

Background

The term *reprogenetics* characterizes the conceptual and practical convergence of two lines of medical and scientific developments: human genetics, spurred by the theoretical and technical progress of the Human Genome Project, and artificial reproductive technologies, through which human embryos can be created *in vitro* and therefore are accessible to further interventions.

Existing technologies which are applied to introduce inheritable genetic modifications in such embryos still have many flaws and therefore may lead to physical problems in the future individual. Embryos or fetuses with abnormal features or developments following germ-line alterations would probably be discarded or aborted; this practice might lead to the view that such interventions are irresponsible and unethical. Furthermore, it is quite possible that the complexities and dynamics of the human genome and of genotype–environment interactions will preclude anything but the manipulation of the most simple phenotypes in a controlled manner. For the sake of argument, however, it is assumed that germ-line engineering procedures in humans are no more risky than natural conception. Parents could then choose to enhance their embryos with genes that they themselves do not carry.

If this could be done safely, what kind of possible futures might be imagined for the carriers of modified genes, enhanced traits or capabilities, and for the society they live in?

Context

Several problems arise when we attempt to assess future consequences of germ-line alterations. The first is that enhancement can mean different things in different interpretative and social contexts (Juengst, 1998). Second, benefits or harm resulting from interventions aimed at enhancing features or traits cannot be defined in absolute terms. Their interpretation depends to a large extent on the social or cultural contexts shaping the norms according to which human features or abilities are valued. Although these norms are generally characterized by considerable stability, they are also prone to change over time. Therefore, features of the human body or mind which are highly valued today might be irrelevant or even have negative social connotations in future societies.

The third problem is that we can hardly foresee how the effects of the broad implementation of germ-line alterations will interact with society and possibly change it. Reasonable guesses about future developments can therefore only be made by presuming that future societies will not differ much from current societies in their main characteristics. Scenarios developed against the background of liberal Western societies would not be coherent or meaningful in the context of paternalistic, authoritarian cultures or even dictatorships. Based on these assumptions it seems reasonable to assume that the application of future reprogenetics will be shaped by a plurality of ethical standpoints, by a scarcity of resources, by commercialism and by globalization (Norgren, 1998).

Furthermore, modern industrialized societies are characterized by a high degree of individualism. Old belief systems and guiding norms are losing ground and being replaced by pluralism with respect to religion or secular normative views. Autonomy and freedom of choice are among the most highly regarded liberties. Lifestyles are no longer bound to conventional roles but multiply as societies become more and more heterogeneous by education, permeability of traditional social structures, migration, mobility and so on. These developments pose quite a challenge to the individual. The lack of orientation and social embeddedness resulting from these changes needs to be balanced. In this situation, private and personal relationships become more important. Since large families have become increasingly rare, relationships between different generations are emphasized as strongly as the perceived dependency and feeling of mutual responsibility between members of vertical family lines. These developments are promoted by increasing life expectancy in Western industrialized societies.

Reprogenetics now promises to provide the means to stabilize and strengthen such family lines. It expands control not only over the process of procreation but also to future generations, providing an element of security and orientation in an insecure and ambiguous world (Kollek, 2000). It not only allows couples to choose freely the time for procreation and the number of children, but also the health status, sex, and – at least the assumption here – behavioral traits and other phenotypes of their offspring, independent of their own genetic endowment. Reprogenetics therefore seems to match perfectly individualized lifestyles characterized by the need and the desire to plan and structure one's own life course rationally and in accordance with the requirements of modern society. The question is whether reprogenetics can keep its promise of security and control or whether old uncertainties will simply be replaced by new ones.

Health purposes: the slippery slope

It is likely that germ-line interventions will first be carried out in order to prevent severe inherited disorders. Imagine the extremely rare case of both partners being homozygous carriers of the same recessive mutation, leading inevitably to a pathological condition in every child. If such a couple insists on having a healthy child of its own, germ-line therapy may be regarded as the method of choice.

Once it has been proven that germ-line alterations for therapeutic purposes are safe – at least in the individual directly manipulated – other applications will soon be considered, some of which will inevitably cross the line between treatment and enhancement, a line which is difficult to draw (Parens, 1998). In the context of prevention, the introduction of genes that confer resistance against widespread infectious diseases such as AIDS or offer protection against cancer or enhance the activity of detoxifying genes could be defended. Germ-line immunization to protect against severe diseases could be offered to groups or communities at risk. However, since germ-line interventions still require one or more cycles of *in vitro* fertilization (IVF), including hormone treatment, health insurers may soon be forced to refuse coverage for such treatments because too many people apply for it.

Only individuals or families with considerable financial means would then be able to afford such treatment, which might dramatically widen the gap in health status and life expectation between the more and the less well off (Silver, 1997), resulting in increased social tensions. In countries with public retirement insurance, well-off people living significantly longer than average may even be accused of exploiting the social system and living at the expense of others. To ensure equality of opportunity, some people may then claim a right to become enhanced and force the state to grant equal access to such treatments to everyone.

Financial restrictions and/or cultural preferences may also lead to unequal gender distribution of benefits from germ-line interventions. If a couple can afford genetic enhancement of one child only, they may invest in a boy rather than in a girl. This may especially be true if birth rates are controlled, or in cultures preferring a boy for the first-born child.

Enhancing looks: preprogrammed disappointment

As a result of a lack of consensus on reproductive matters in current and future democratic societies, and thanks to the support of proponents of a radical reproductive autonomy view (Robertson, 1994) and the effects of market forces, inheritable genetic modifications may not only be offered for health-related purposes, but also for cosmetic reasons. Some people would consider altering their skin color or body shape and size. Since favorable physical attributes are regarded as one of the preconditions for social and economic success, a woman undergoing IVF treatment because of fertility problems may also be interested in treatments enhancing the physical attributes of her future child.

Given predictable outcomes, this may finally increase the number of good-looking men and women as compared to current standards. Will this make them happier or even more successful? There are grounds for doubt. Such developments could, for example, reinforce social prejudice by promoting stereotypes of race and gender and supporting questionable standards for normalcy. Offering people opportunities to choose the phenotype of a child may therefore result in psychosocial pathologies, including deeper class and racial divisions in society (Krimsky, 2000).

But even more benign scenarios may have ambiguous outcomes. For instance, features conferred by genetic enhancement follow the ideals of others and thus differ from cosmetic surgery, where people decide for themselves. Children quite often do not have the same ideas about beauty as their parents and may be very dissatisfied with parents' choices. The resulting frustration may be even more profound than it would be if one's physical appearance was solely the result of natural processes. In the latter case, no one could be held responsible whereas in the case of disliked physical enhancements, parents or grandparents can be blamed for their decisions.

Ideals of physical attractiveness may also change with time. It is not unlikely that fashion and the media will some day become bored with interchangeable beauties and replace them with models considered more attractive because they are characterized by natural heterogeneity in shape, size and proportions and endowed with interesting deviations from previous norms. In that case, cosmetic enhancement will prove to have been, at best, a waste of money. At worst, it will have resulted in a group of unsatisfied beauties who wish they had been born as naturals or who might even sue their parents for denying them a genetic outfit that is the product of chance. Living with the knowledge that one's own hereditary factors have been programmed therefore may restrict an individual's right to an open future and undermine the essentially symmetrical relations between free and equal human beings (Habermas, 2003).

Controlling behavior: society strikes back

Violence can be imagined as a growing problem in future societies. Since governments and influential social groups in general are averse to social changes, because they fear the loss of privileges, they may provide massive funding for research in neurogenetics and behavioral genetics. Despite substantial drawbacks – many promising results cannot be replicated in later research – geneticists finally may succeed in convincing politicians that the diverse activities labeled under the umbrella term 'violence' – from rape to terrorism – are all manifestations of an individual's genetically determined aggression, and not regular human reactions to social and economic oppression.

Trials will be initiated involving twinned embryos from couples undergoing IVF treatment, in order to test the hypothesis of neurogenetic determinism. Couples will volunteer for this experiment because they are eager to support science and to promote societal evolution. In each twin-pair, one individual will have his or her neurometabolic outfit genetically altered to reduce aggressive tendencies. After several years of monitoring, however, it may become evident that behavioral outcomes do not correlate with genetic interventions. For example, many of the nonmanipulated twins do not exhibit violent behavior, even under adverse circumstances, and many of those manipulated unexpectedly do. Interestingly, among those twins who carry germ-line alterations and exhibit less aggressive behavior, fewer individuals may prove to be successful in business or medicine, presumably because they lack the aggressive approach which is needed to survive in such competitive environments.

Children who realize that they have been intentionally manipulated by their parents to suit their interests and the interests of society may experience psychological distress. They may feel that their destiny has been arbitrarily preprogrammed. Some will deliberately choose to behave in contrast to the expected outcome of genetic modification.

Neuroscientists and behavioral geneticists will finally agree with psychologists and sociologists that the expression of violence and how it is valued depends, to a large extent, on the social context. The same violent act may be socially desirable or condemned; a soldier shooting a suspected terrorist may receive a medal or be charged with murder (Rose, 1998). Since results of germline interventions aimed at controlling personality traits are so much amenable to societal values, there is a growing consensus that socially complex interactive processes can neither be reduced to the properties of individual neurometabolic patterns or genes, nor manipulated on the level of genetics with controlled outcomes. After several decades of experimentation, the practice of genetically interfering with behavioral traits will be abandoned.

Modesty and wisdom: let future generations decide

One could also imagine a completely different scenario. In the coming 10 or 20 years, not only will knowledge about the human genome increase dramatically, but also our understanding of the complexities of genotype–phenotype relationships. We will learn how the expression of genes is modulated by other genes and environmental factors. Various neuroscientific disciplines will show how personal and social parameters retroact on the expression of genes and behaviors. Although the technology and skills needed to perform genetic modifications with high precision will have been developed, it will also have become evident that the resulting social outcomes will not be predictable. Thanks to this insight, researchers and specialists in molecular and reproductive medicine will become increasingly reluctant to interfere with the human germ-line, especially for the purpose of enhancement. A call for an international ban on germ-line interventions will enjoy massive support in many countries on the part of a broad spectrum of societal groups as well as scientists. Such an agreement will be signed and become internationally effective before the rights of individuals yet to be born have been violated.

References

Habermas J (2003) *The Future of Human Nature*. Cambridge: Polity Press.

Juengst E T (1998) What does enhancement mean? In: Parens E (ed.) *Enhancing Human Traits: Ethical and Social Implications*, pp. 29–47. Washington, DC: Georgetown University Press.

Kollek R (2000) Technicalization of human procreation and social living conditions. In: Haker H and Beyleveld D (eds.) *Ethics in Genetics in Human Procreation*, pp. 131–152. Aldershot, UK: Ashgate.

Krimsky S (2000) The psychosocial limits on human germ-line modifications. In: Stock G and Campbell J (eds.) *Engineering the Human Germ Line: An Exploration of the Science and Ethics of Altering the Genes We Pass to Our Children*, pp. 104–107. New York: Oxford University Press.

Norgren A (1998) Reprogenetics policy: three kinds of models. *Community Genetics* 1: 61–70.

Parens E (1998) Is better always good? The enhancement project. In: Parens E (ed.) *Enhancing Human Traits: Ethical and Social Implications*, pp. 1–28. Washington, DC: Georgetown University Press.

Robertson J A (1994) *Children of Choice: Freedom and the New Reproductive Technologies*. Princeton, NJ: Princeton University Press.

Rose S P R (1998) Neurogenetic determinism and the new euphenics. *British Medical Journal* 317: 1707–1708.

Silver L M (1997) *Remaking Eden: Cloning and Beyond in a Brave New World*. New York: Avon Books.

Web links

Frankel M S and Chapman A R (2000) Human inheritable genetic modifications: assessing scientific, ethical, religious, and policy issues. Prepared by the American Association for the Advancement of Science http://www.aaas.org/spp/sfrl/projects/germline/report.pdf

Nuffield Council of Bioethics (2002) Genetics and human behaviour: the ethical context. Published by the Nuffield Council of Bioethics, London http://www.nuffieldbioethics.org

Index

abortion *see* termination of pregnancy
adaptation, 232, 234, 242
adverse drug reactions, 8
adverse selection, 199
Alzeimer's disease, 5, 97, 102
  susceptibility to, 102
American Society of Human Genetics, 149
animal experimentation, 302–8
  Buddhist view, 306
  instrumentalist view, 305
  Jain view, 306
utilitarian perspective, 303
animal welfare, 302
anonymisation, 8
anonymity, 65
anthropology, 24, 50, 52
  classical, 50
  collection of blood, 50
  'salvage', 52
antievolutionists, 247–50
anti-Semitism, 136
anxiety, 14, 43
  disease risk and, 14
archived samples, genetic analysis of, 68
assisted reproductive technologies, 276, 279
Auschwitz, 138
autonomy, 17, 61, 64, 66, 91–2, 115, 128, 281, 287, 313
autopoiesis, 218, 253, 254
autosomal recessive, 11
autosomes, 18

ß-thalassemia *see* thalassemia
Baur, Erwin, 136
Bayesian density discriminant functions, 124
behavioural biology, 239
behavioural genetics, 169, 173, 179, 204, 210–16, 313, 315–16
  adoption studies, 204
  alcoholism, 135, 137
  family studies, 204, 207
  heritability coefficient, 207
  homosexuality, 135
  inheritance of, 169
  lay understanding of, 179
  twin studies, 173, 204
  violence, 315–16
  see also criminal behaviour/criminality
behaviour geneticists, 253
Belmont Report, 25, 64
best interest, 163
Bible, 247
Binet, Alfred, 203
bio-colonialism, 51, 68
biological destiny/determinism, 167, 172
  see also determinism; genetic determinism
biological difference, 54
bio-piracy, 40, 46
biotechnology
  Iceland, 58
  industry, 13, 23, 43, 46, 66
  regulation, 269, 308
  revolution, 47–8
  US Biotechnology Industry Organisation, 298
bipolar disorder, 213
blood, 49–54
  collection of blood, 50
  donation, 51
  economic value, 53
  medical value, 53
  significance indifferent cultures, 51–3, 189–90
Blumental, David, 41
British Medical Association (BMA), 194

Calvin, John, 246
Campbell, Keith, 297
cancer, 20
  breast cancer, 5, 43, 97, 178
  colon, 5

colorectal, 100
gene therapy, 71
genetic risk assessment, 97
therapeutic trials, 14
Catholic Center Party, Weimar Germany, 136
Cavalli-Sforza, Luca, 49
Celera Genomics, 27–32, 39, 252
choice, 35, 192–7, 288
based on genetic information, 192–7
freedom of, 281, 313
informed choice, 94, 115, 163, 196
reproductive, 115, 133
women's right to choose, 158
chorionic villus sampling, 118
Christianity, 246
Chromatin, 6, 18
chromosomes, 6, 18
chromosomal anomalies, 11, 19
sex chromosomes, 19
chronic heart disease (CHD), 5, 78, 102
family history of, 13–14
genetic testing for, 78
perceptions of family history of, 102–4
susceptibility to, 102
citizen's juries, 16
civil liberties, 16
civil society organisations (CSOs), 37
clinical genetics, 10–14, 77
Clinical Genetics Society (UK), 93, 94, 148
clinical trials, 25
gene therapy, 71–2
cloning, 267, 268, 296–301
of animals, 302–8
human cloning, 269, 298
media representation, 299
medical benefits of, 299
positional cloning, 56
reproductive, 257, 269, 282, 299
therapeutic, 269, 287, 299
Collins, Francis, 30, 31, 37, 252
commercial exploitation, 46
commercialization, 23, 24, 39, 41, 43, 313
in the clinical context, 42
relation to geneticization, 43
commodification, 33, 40, 46, 54
of biological information, 33, 54
of the body, 54
of the gene-pool, 46
of the human genome, 40
communitarian values, 276
community consultation, 114–20
complementary DNA, 28
complex disease, 5, 13, 79, 102, 207
complexity, 207, 217, 226
see also genetic complexity
confidentiality, 91–2, 114, 168, 196
medical, 192
conngenital hypothyroidism, 12
consent, 15, 92, 163, 288
capacity (to consent), 163
community, 25
to genetic testing, 92
group, 24, 51–2
parental, 124
presumed, 57, 59
to sterilization, 163
see also informed consent
corporate interests, 59
corporate liberalism, 33, 35, 37
Council of Europe's Convention on Human Rights and Biomedicine, 274–5
creation science, 248
Creationism, 218, 246–50
Genesis Flood, 248–9
Scope's trial, 218, 247
Crick, Francis, 4
criminal behaviour/criminality, 16, 141, 143, 169, 204, 210–16
genetic defence, 169, 212
Mobley, Stephen, 210
and responsibility, 213
risk assessment, 214–15
XYY karyotype/condition, 143, 212
cultural endogamy, 144
cystic fibrosis (CF), 5, 12,13, 73, 144, 206
clustering in populations, 144
gene therapy, 73

Darwin, Charles, 3, 140, 162, 217, 232, 247
data access and ownership, 30–1
Human Genome Project, 30–1
Icelandic database, 59

data protection , 25
  Icelandic database, 59–61
Dawkins, Richard, 218, 238, 246, 253
deafness, 283
decision making , 132, 145
  informed, 127, 196, 284
  risk and, 96
declaration of Helsinki, 25, 64
deCODE Genetics, 24, 56–63
degenerative diseases, 5
deindustrialization, 143
DeLisi, Charles, 28
dementia, 14
  family history of, 14
democratic debate, 16
  public engagement, 16, 17, 180
deoxyribonucleic acid (DNA), 4, 41, 65, 221, 253
  in cultural discourse, 171
  explanation of, 18
  patented, 41
  see also DNA sequences
designer babies, 36
determinism, 255–8
  neurogenetic, 316
  see also biological determinism; genetic determinism
developmental biologists, 251
developmental processes, 218, 238–45
  role of genes in, 242–4
diabetes mellitus type 1, 110
diabetes mellitus type 2, 5, 79, 102
  Barker's hypothesis, 109
  foetal insulin hypothesis, 109
  Freinkel's fuel mediated teratogenesis hypothesis, 109
  genetic etiology of, 54
  genetic susceptibility, 108–9
disability, 283
  medical model of, 158
  rights *see* disability rights
  social model of, 157
disability rights, 157–60
  abortion debate, 157–60
  antidiscrimination legislation, 157
  discrimination, 159
  movement, 11, 12, 132, 157–60, 163
  views of prenatal screening, 157
discrimination, 11, 15, 133, 159, 215, 263, 275
  criminal behaviour, 215
  see also disability rights; genetic discrimination
disease causation, 5
  lifestyle, 103
disease causing mutation, 7
disease susceptibility, 15
  see also genetic susceptibility
distributive justice, 269, 308–11
DNA *see* deoxyribonucleic acid
DNA sequences, 9
  explanation of, 18
Dobzhansky, Theodosius, 247
Dolly, the cloned sheep, 48, 223, 268, 287, 296, 304
dominant inheritance, 19
double helix, 18
Down syndrome, 82, 287
Drosophila, 6
  genome sequence, 31
Duchenne muscular dystrophy (DMD), 10, 12

economic considerations, 23, 304
  Icelandic economy, 58
economic forces, 46
  in eugenics, 138
  see also market pressures
economic injury, 65
education, 14, 15, 124
  discrimination, 15
  genetics, 80, 124, 180–1
  medical, 14
  nursing, 14
  emergence, 217, 226, 235
  of complexity, 227
enhancement, 160, 284, 312
  see also genetic enhancement
epilepsy, 137
equality of opportunity, 128
equity, 17
ethical principles, 16–17
ethics
  biomedical, 268, 287
  medical, 194
  narrative, 289
  religious, 306
Ethics Committee of the Human Genome Organisation, 148

ethnic communities, 123
  African American, 206
  Ashkenazi Jewish, 123, 206
  Cypriot, 114–21
  French Canadian, 123
  Hassidic, 128
  North American Indians, 51
  south Asian, 263
ethnicity, 219, 259–65, 262
ethologists, 251
eugenics, 35, 135–9, 140–6, 152, 162, 204, 219, 260, 300, 310
  meaning, in China, 152
  negative, 137–9, 141, 142
  positive, 131, 136, 141, 145
  and racism, 219
European Court of Human Rights, 275
euthanasia, 137–9
  coercive, 137–9
evolution, 3, 227–8, 232–7, 245–50, 316
  societal, 316
evolutionary biologists, 253
evolutionary psychology, 173, 218, 238–45
evolutionary theory, 218
  political implications, 244
evolutionists, 251
expressed sequence tags (ESTs), 28, 41
  patented, 41

familial cancer , 11, 13–14, 20, 91
  genetic risk assessment of, 97–8
family duties, 192–7
family history
  of cancer, 13–14
  disease risk and, 14
  lay understandings, 178
  role in risk estimates, 99
  role of women, 179
family life, respect for, 280
family tree, 59, 98, 185
  see also pedigree
fatalism, 15, 105, 167, 172
feminist perspectives, 286–90, 293–5
First World War, 136, 247
Fischer, Eugen, 136
foetal abnormality, 11, 12
foetal ultrasound scanning, 11
folk history, 53
forensic sciences, 144
founder effects, 56
fragile X syndrome, 162
free will, 214, 258
freedom, 269, 281, 310
  of choice, 281, 313
  individual, 17, 35, 167, 256
  reproductive freedom, 150, 155
  see also procreative liberty
Friedreich ataxia, 158

Galton, Francis, 3, 135, 140, 162
gamete donation, 195, 276
gender stereotypes, 294
gene(s), 6, 252
  different meanings of, 252
  explanation of, 17–18
gene expression, 6, 316
gene mapping, 56
  population, 24
gene therapy, 25, 42, 70–5, 160
  academic pioneers, 72
  germline, 70
  industry, 71
  Jesse Gelsinger case, 74
  monogenic disorders, 71
  public expectations, 73
  somatic gene therapy, 70
gene transcription, 6
gene transfer technology, 74
genealogical data, 25
  Icelandic, 56–63
gene–environment interactions, 9, 102, 106, 207, 211, 229, 235–6, 253, 270, 312
genetic activists, 48
genetic complexity, 207, 234
genetic conditions, causal attributions, 82
genetic counselling, 77, 81–95
  adult, 81
  for common diseases, 83
  definition, 81
  ethnocultural diversity, 84
  evaluation of, 87–8
  expectations of, 86
  psychological reactions to, 85–6
  reproductive, 82
genetic data (access to), 30–1, 59
  Icelandic database, 59
genetic data (ownership), 30–1
genetic databanks, 16, 24, 56–63

genetic determinism, 35, 172, 175, 212, 236, 244, 252, 254, 256
  and human rights, 175
genetic diagnosis, 10, 90
  progress in, 13
  technologies, 15
genetic discrimination, 12, 15, 35, 67
  and insurance, 199–200
  in psychiatric illness, 66
genetic enhancement, 269, 270, 310
genetic epidemiology, 153
genetic equality, 309
genetic essentialism, 35, 43, 167, 171, 174, 219
  and social policy, 74
genetic exploitation, 53
genetic heritage, 192, 195, 212
genetic homogeneity, 61
genetic identity, 167, 172
  see also identity
genetic imperative, 282
genetic information, 15, 69, 84, 168, 192–7, 288
  assimilation of, 84
  as a commodity, 69
  duty to disclose, 168, 288, 289
  institutional use, 16
  for policing, 16
  and privacy, 192–7
  public attitudes, 196
Genetic Interest Group (UK), 94, 158
genetic knowledge (nature of), 23
genetic manipulation, 268, 282, 309
genetic material (ownership of), 39, 40, 66, 69
genetic modification, 303, 312, 315
  for cosmetic reasons, 315
genetic mutation, 18
genetic research
  commercial interests, 15
  commercialization of, 23
  comparative genetic studies, 15
  comparative genomics, 10
  conduct of, 15
  conflicts of interest, 42
  duties to participants, 66
  environment, 41
  opposition to, 133, 158
  as a social activity, 21
  strategies, 5–10
genetic resources, 23
genetic rights, 172
genetic risk, 96–101, 198
genetic risk assessment, 13, 14, 78, 97, 214
  in cancer 97
  of criminal behaviour, 214–15
  and pedigree data, 98
genetic risk estimates (prediction models), 99
genetic stereotypes, 173
genetic susceptibility, 13, 24, 102–7, 135
  behavioural change, 106
  to common disease, 24, 78
  to complex disease, 229
  diabetes, 108
  and eugenics, 135
  fatalism, 105
  lay perceptions of, 78
  screening for, 13
genetic variation, 4, 5, 24, 65
Genetical Society (UK), 148
geneticization, 43, 183
  commercialization, 43
  of cultural models of kinship, 183
genetics differences (between populations), 15
genocide, 138
genohype, 112
genome, 6, 221
  human, sequencing of, 21–2, 27–32
genome-wide analysis, 7
germline immunization, 314
germline modification, 270, 314
  therapeutic purposes, 314
Gillick competency, 93
globalization, 313

haemoglobin disorders, 12, 79, 263
  prenatal and carrier screening, 114–20
  chorionic villus sampling, 118
  WHO Working Group on Hemoglobin Disorders, 118
Haemophilus influenzae, 29
Hamilton, Bill, 238
HapMap project, 24
Hardy–Weinburg law of genetic equilibrium, 96
harms and benefits, 196

healthcare delivery/services, 13, 16, 17, 309
equitable access to, 275
quality of services, 128
specialist health care services, 13, 14
hemoglobinopathies *see* haemoglobin disorders
Herrnstein, Richard, 204
Hitler, Adolf, 136, 138
HIV/AIDS, 15
Holocaust, the, 138, 204, 260
House of Commons Science and Technology Committee, 194
human behavioural ecology, 240
human dignity, 17, 23, 40, 61, 271, 281
human diversity, 260
human embryo, status of, 275, 287
human ethology, 240
Human Fertilization and Embryology Act, 267, 273
Human Fertilization and Embryology Authority (HFEA), 195, 267, 273, 298
Human Genetics Society of Australia, 148
Human Genome Diversity project (HGDP), 23–4, 47, 49–55, 205–6
Human Genome Project (HGP), 22, 23, 27–32, 65, 205, 221, 227
Human Genome Sciences Inc. (HGSI), 29
public consortium, 30
US legislation, 28
US National Human Genome Research Institute, 298
human population history, 49
human potential, 244
human rights, 16, 150, 167, 280
in China, 150
Universal Declaration on the Human Genome and Human Rights, 40
Huntington('s) disease (HD), 10, 42, 84, 91, 97, 137, 193, 200, 212, 252, 253
hygienics, 140
hypercholesterolemia, 13, 14,
genetic susceptibility testing, 79, 104–6
hypertension, 5, 13, 110

Iceland, 24, 54, 56–63
identity, 186, 214, 219, 255–8, 287, 298
as a construction, 257
ideas about, 53
and kinship, 186
racial, 190
personal, 255
see also genetic identity
in vitro fertilization (IVF), 184, 282, 292
feminist perspectives, 287
legislation, 267
regulation of, 271–8
statutory regulation, 272–3
see also assisted reproductive technologies
indigenous knowledge, 46, 51
individualized medicine, 8
individualism, 313
Industrial Revolution, 140
inequality, 215, 254
inequity, 61
information technology, 9, 57
informational harm, 65, 68
informed consent, 25, 53, 57, 59, 118, 275
competency/capacity, 163
in genetic research, 64–9
Nuremburg Code, 25, 64
as an obstruction to progress, 62
see also consent
inheritance
patterns of, 17–19
social and cultural understandings, 183–91
innovation, 17, 40
institutional review boards (IRBs), 65
insurance, 15, 16, 17, 169, 198–202, 262
companies, 24, 57, 159
and Huntington's Disease, 200
intellectual impairments/ disabilities, 133
discrimination, 161
sexuality and procreation, 161–70
stigmatization, 162

intellectual property rights, 4, 15, 46–8, 66
intelligence, 15, 162, 169, 203–9
  IQ tests, 203
  predictions about, 15
  and race, 203–9
intelligent design, 218, 249
intelligence quotient (IQ) tests, 203
irreducible complexity, 249
isolated populations, genetically, 56, 62, 125

Jenson, Arthur, 204
justice, 17, 128, 270, 287, 288, 289
  distributive, 269, 308–11
  gender, 289
  theories of, 309

Kaiser Wilhelm Institute for Anthropology, Human Heredity and Eugenics, 136, 138
Kaiser Wilhelm Institute for Brain Research, 138
Kallman Syndrome, 243
Kinsella, Kevin, 57
kinship, 140, 168, 178, 183–91
knowledge commons, 36

Law for the Prevention of Offspring with Hereditary Diseases (Germany), 162
lay understandings, 16, 102, 167
  of heredity, 177
Lenz, Fritz, 136
liberalism, 280
lifestyle, 14, 103, 110

malaria, 114
Malthusian prediction, 140
market pressures, 42–3
Maternal and Infant Healthcare Law of the People's Republic on China, 132, 147
maturity-onset diabetes of the young (MODY), 109
media, 268, 296
media representations
  of cloning, 299
  of genetics, 167, 171, 208, 211, 299
medicalization, 85, 215
Mendel, Gregor, 3, 4, 252
Mendelian inheritance, 4, 5, 9, 203, 234
  family history of, 14
  lay understanding of, 177–81
  see also single-gene disorders
Mengele, Joseph, 139
messenger RNA (mRNA), 6, 28
metabolic pathways, 252
metaphor in genetics, 217, 219, 221–31, 252
  gene as soul, 256
  genome as blueprint, 257
  genome as computer program, 222–4
microarray technologies, 7, 13
mitochondrial diseases, 20
mitochondrial genome, 20
molecular biologists, 251
Morel, Benedict, 162

National Human Genome Research Institute, 30
Native American Graves Protection and repatriation Act of 1990 (NAGPRA), 51
natural selection, 4, 217, 232–7
  descent with modification, 232–6
nature of life (beliefs about), 48
Nazi experimentation, 64, 260
Nazi movement, 135–9, 141, 142, 162
Neel, James, 108, 148
neurogenetics, 316
neutral drift, 236
NIH Recombinant DNA Advisory Committee, 71
non-paternity, 66
normality, 160, 299
Nuremburg Medical Trial, 138

ovarian cancer, 97–8, 178

patenting, patents, 22, 28, 31, 40, 46–8
  application process, 40
  of cell lines, 51
  Chakrabarty v. Diamond case, 36
  definitions, 40, 47
  of DNA, 41
  of expressed sequence tags (ESTs), 41
  genetic activists, 48
  Lacks, Henrietta, 51

laws, 36
Moore, John, 51
single-nucleotide polymorphisms (SNPs), 41
United States Patent Office, 47
patterns of human migration, 15, 23
pedigree, 98, 185
see also family tree
Penrose, L. S., 96
personality (predictions about), 15
pharmaceutical industry, 15, 24, 66, 72
pharmacogenetics, 8
pharmacogenomics, 8
phenotype–genotype relationship, 218, 234–5, 316
phenotypic data, 56–7, 59
phenylketonuria (PKU), 12
policy (shaping of), 16
political considerations, 16
political pressures/forces, 33–8, 46
popular culture (genetics in), 171
population (genetic differences), 15, 205
population genetics, 153
positional cloning, 56
poverty, 140
and diabetes mellitus type 2, 112
pre-implantation genetic diagnosis, 268, 282, 284–5, 287, 292
prenatal diagnosis, 114, 125, 293
primary health care, 13, 14
privacy, 60, 128, 168, 192–7, 280
definition, 192
and genetic information, 192–7
Icelandic database, 60
UK Human Rights Act, 193
procreation, 183
theories of, 183–8
procreative autonomy, 281, 282
procreative imperative, 282
procreative liberty, 270, 282, 311
see also under freedom
procreative tourism, 276
production of knowledge, 21
protection of vulnerable populations, 64–5
protein synthesis, 6
psychological consequences, 15, 65, 81–9
impaired self esteem, 15
public debates about genetics, 16–17
public engagement, 16, 17, 180
public interest, 193
public understanding of science, 168, 177, 180–1

race, 54, 143, 203–9, 219, 259–65
rates of incarceration (US), 143
racial hygiene, 135, 204
racism, 169, 219
racist views, 205
rational therapies, 14
recessive diseases, 12, 18, 19, 114, 118, 123
reductionism, 213, 217, 226, 236, 251
biological, 35
reproduction, 299
reproductive autonomy, 315
reproductive counselling, 125
reproductive health, 13
reproductive liberty, 267, 286
see also under freedom
reproductive technologies (new), 268
feminist perspectives, 286–90, 293–5
see also assisted reproductive technologies; reprogenetics
reprogenetics, 270, 312
responsibility
biological, for health, 16
female, 287
individual, 167, 213, 256, 258, 271, 287, 288
social responsibility, 16, 208, 276
retroviruses, 71
ribonucleic acid (RNA), 6, 18
rickets, 263
risk perception, 78
Rüdin, Ernst, 135

Sanger Centre, 30
Scallmayer, Wilhelm, 135
schizophrenia, 5, 54, 135, 137, 150, 153, 204
screening (genetic), 13, 122, 144
adult, 12
carrier, 12, 13, 114–20, 122–9,
implications of, 114–21; *see also* testing, genetic
cascade, 122, 126
for cystic fibrosis, 13
community-based screening programmes, 114–29

screening (genetic) – *continued*
consequences of, 13
definition, 144
eugenic implication, 144
neonatal, 12, 144
prenatal, 11, 12, 13, 114–20, 144
population, 11, 12, 13
pro-screening lobby, 13
Second World War, 142, 260
selective breeding, 142, 145, 303
selfish gene, 235–6, 238, 253
self-organization, 217, 225
sex selection, 268, 275, 291
family balancing, 268, 292
European regulation of, 275
sex-linked recessive disorders, 11, 19
sexually transmitted diseases, 135
sickle cell disease, 12, 114, 115, 144, 228–9, 261–2
causation, 228–9
clustering in populations, 144
prenatal diagnosis, 114
single-gene disorders, 9, 97
see also Mendelian inheritance
social behaviour, 238
cooperation, 238–9
social environment, 242
social epidemiology, 111
social goods, 309
social inequities, 35
social injustice, 244, 254
social policy, 167
Society of Pharmaceutical Medicine, 61
sociobiology, 173, 218, 238–45
political implications, 244
somatic cell nuclear transfer, 257, 282, 297
stem cell technologies, 287, 288
sterilization, 131, 136, 137, 141
compulsory, 137, 150, 162, 204
enforced, 131
involuntary, 141, 142, 311
stigmatization, 11, 15, 84, 86, 133, 159, 215, 262
of criminal behaviour, 215
of disability, 159
previous experience of, 84
strokes, family history of, 14
Sulston, John, 30
Tay–Sachs disease, 12, 37, 117, 206
carrier screening, 122–9
prenatal diagnosis, 125
technological fix, 36
technology, 33
genetic technologies, marketing of, 42
genomic technologies, development of, 28
as progress, 34
social relations model, 34
therapeutic, 15
use/abuse model, 34
termination of pregnancy, 11, 12, 268, 277, 284, 288, 293
religious opposition, 36
testing, genetic, 14, 42–3, 54, 119, 122, 198, 200, 286
adult, 91
antenatal, 94; routinization of, 158–9
carrier, 77, 90, 92, 282
of children, 77, 90–5
commercial pressures, 42–3
to determine tribal membership, 54
and insurance, 198, 200
predictive, 11, 77, 90–2, 104–5
premarital certificate, 119
prenatal, 11, 147, 163, 287
United States Secretary's Advisory Committee on Genetic Testing, 42
ß-thalassemia, 12, 114, 115, 235
carrier screening, 118, 122–9
clustering in populations, 144
Cyprus incidence, 116
foetal diagnosis, 126
gene therapy, 71
prenatal diagnosis, 117–18
and termination of pregnancy, 117
UK incidence, 116
The Institute for Genomic Research (TIGR), 29
thrifty gene hypothesis, 79, 108–13
see also diabetes mellitus type 2
thrifty genotype, 244
transgenesis, 269, 302, 303
transgenic animals, 296, 297, 302
transgenic research, 71
treatment (access to), 15
Trivers, Robert, 238, 240

tropical diseases, 15
tuberculosis, 135, 144
Tuskegee syphilis experiment, 260

UK Advisory Committee on Genetic Testing, 93
UK Biobank, 8
UK Human Genetics Advisory Commission, 298
UK Human Rights Act, 193
UK National Health Service, 117
UK Society for the Protection of the Unborn Child, 158
uncertainty, 97, 306
United Nations Educational, Scientific and Cultural Organization (UNESCO), 40
Universal Declaration on the Human Genome and Human Rights, 40
US Code of Federal Regulations, 64–6
US National Academy of Social Sciences, 298
US National Bioethics Advisory Commission, 298
US National Institutes for Health (NIH), 28–9, 39, 51, 71, 73

Venter, J. Craig, 27–32, 252, 254
views of minority groups, 120

Warnock Committee, 267, 273
Watson, James, 4, 29, 36
Weimar welfare state, 136
Wellcome Trust, role in HGP, 30, 31
Wilmut, Ian, 297, 302, 304
Wilson, E. O., 238, 240, 242
World Health Organization (WHO), 114, 118

xenotransplantation, 300
X-linked disorders, 193, 196